Corrosion-Resistant Linings and Coatings

CORROSION TECHNOLOGY

Editor
Philip A. Schweitzer, P.E.
Consultant
Fallston, Maryland

ADDITIONAL VOLUMES IN PREPARATION

Corrosion Mechanisms in Theory and Practice: Second Edition, Revised and Expanded, edited by Philippe Marcus

Corrosion-Resistant Linings and Coatings

Philip A. Schweitzer

Consultant
Fallston, Maryland

CRC Press
Taylor & Francis Group
Boca Raton London New York

CRC Press is an imprint of the
Taylor & Francis Group, an **informa** business

RC Press
Taylor & Francis Group
6000 Broken Sound Parkway NW, Suite 300
Boca Raton, FL 33487-2742

First issued in paperback 2019

© 2001 by Taylor & Francis Group, LLC
CRC Press is an imprint of Taylor & Francis Group, an Informa business

No claim to original U.S. Government works

ISBN-13: 978-0-8247-0554-1 (hbk)
ISBN-13: 978-0-367-39710-4 (pbk)

Visit the Taylor & Francis Web site at
http://www.taylorandfrancis.com

and the CRC Press Web site at
http://www.crcpress.com

Preface

Although many corrosion-resistant metals and alloys are available, they do not always represent the most practical or economical means to combat corrosion. In many situations it is better to select a less resistant material and provide some type of coating or lining to protect it from corrosion.

Linings are generally used to protect the interior of vessels, which may be of carbon steel or synthetic construction. They may be made of sheet, glass, or cement, or may be applied in liquid form.

Coatings are normally used to protect the exterior surfaces of vessels or structures exposed to corrosive atmospheres. Some coatings may also be used as linings. Included in this category are paints, organic coatings, metallic coatings, and monolithic coatings for concrete.

This book provides a thorough introduction to the various types of linings and coatings that can be used to provide corrosion resistance. Physical, mechanical, and corrosion resistance properties—as well as areas of application—are provided for each of the materials. The effectiveness of any lining or coating is dependent on proper surface preparation and installation or application. Accordingly, these details are included for each lining and coating material. Because concrete is widely used as a construction material and is subject to corrosion, the text includes information on coatings that provide protection from atmospheric and other, more aggressive corrosive atmospheres.

For all professionals involved in design and maintenance processes, this book should serve as a valuable aid in the selection of corrosion-resistant linings and coatings.

Philip A. Schweitzer

Contents

Corrosion-Resistant Linings and Coatings

1

Introduction to Linings

Linings are barrier applications used to protect carbon steel, concrete, or other substrates from corrosion by harsh chemicals. In addition to the protection afforded, linings maintain product purity and absorb impact to maintain particle size; those having a low coefficient of friction prevent particles from hanging up and provide an easily cleanable surface. Economics plays a part in determining whether it is more advantageous to line a carbon steel vessel or to select an appropriate corrosion resistant alloy as the base metal of construction. When high temperatures and concentrations of chemicals rule out a lining system, or when downtime for repair of a lining is unacceptable, the high alloy is automatically selected.

At times external corrosion as well as internal corrosion is a problem. Under such circumstances a carbon steel vessel may have an external coating applied. Alternatively, a fiber reinforced polyester (or similar polymer) vessel may be used and lined, with the vessel shell resisting the external corrosion.

When a lining is to be used for corrosion protection, it is necessary to review the corrosion rate of the immersion environment on the substrate. Assuming that the substrate is carbon steel with a corrosion rate of less than 10 mils per year (mpy) at the operating temperature, pressure, and concentration of corrodent, then a thin film lining of less than 20 mils may be used. For general corrosion this corrosion rate is not considered severe. However, if a pinhole should be present through the lining, a concentration of the corrosion current

1

TABLE 1.1 Fluoropolymer Lining Systems

Lining system	Thickness, in. (mm)	Maximum size	Design limits	Installation	Repair considerations
Sheet linings—adhesive bonded					
Fabric-backed PVDF PTFE FEP ECTFE ETFE PFA	0.06–0.09 (1.5–2.3) 0.08–0.12 (2.0–3.1) 0.06–0.09 (1.5–2.3) 0.06 (1.5) 0.06–0.09 (1.5–2.3) 0.09 (2.3)	No limit.	Pressure allowed. Full vacuum only at ambient temperature. Smallest nozzle size 2 in. (51 mm). Maximum temperature limited by adhesive, typically 275°F (135°C).	Neoprene or epoxy adhesive. Sheets welded with cap strips. Heads are thermoformed or welded.	Repair is possible but testing is recommended.
Loose sheet lining					
FEP, PFA	0.06–0.187 (1.5–4.75)	Determined by body flange.	Pressure allowed. No vacuum. Gasketting required between liner and flange.	Liner with nozzles hand or machine welded, then slipped inside housing.	Difficult.

Sprayed dispersions

Material	Thickness	Dimensions	Notes	Repair
FEP	0.04 (1.0)			
PFA	0.01–0.04 (0.25–1.0)			
PVDF	0.025–0.03 (0.06–0.76)			
PFA with mesh and carbon	0.08 (2.0)			
PVDF with mesh and carbon	0.04–0.09 (1.0–2.30)	8 ft (2.4 m) dia. 40 ft (12.2 m) length	Pressure allowed. Vacuum rating undetermined.	Primer and multiple coats applied with combination 2 psig equipment. Each coat is baked.	Hot patching is possible, but testing is recommended.

Electrostatic spray powder

Material	Thickness	Dimensions	Notes	Repair
ETFE	Up to 0.09 (2.3)			
FEP	0.01–(0.28)			
PFA	0.01 (0.28)			
ECTFE	0.06–0.07 (1.5–1.8)			
PVDF	0.025 (0.64)	8 ft (2.4 m) dia. 40 ft (12.2 m) length	Pressure allowed. Vacuum rating undetermined.	Primer and multiple coats applied with electrostatic spraying equipment. Each coat is baked.	Hot patching is possible, but testing is recommended.

density occurs as a result of the large ratio of cathode to anode area. The pitting corrosion rate will rapidly increase above the 20 mpy rate and through-wall penetration can occur in months.

When the substrate exhibits a corrosion rate in excess of 20 mpy, a thick film lining exceeding 20 mils in thickness is used. These thicknesses are less susceptible to pinholes.

THIN LININGS

Thin linings are used for overall corrosion protection as well as for combating localized corrosion such as pitting and stress cracking of the substrate. Thin fluoropolymer linings are used to protect product purity and provide nonstick surfaces for easy cleaning.

Among materials available for thin linings are those based on epoxy and phenolic resins that are 0.15–0.30 mm (0.006–0.12 in.) thick. They are either chemically cured or heat baked. Baked phenolic linings are used to protect railroad tank cars transporting sulfonic acid. Tanks used to store caustic soda (sodium hydroxide) have a polyamide cured epoxy lining.

Thin linings of sprayed and baked FEP, PFA, and ETFE are also widely used. They are applied to primed surfaces as sprayed waterborne suspensions or electrostatically charged powders sprayed on a hot surface. Each coat is baked before the next is applied. Other fluoropolymers can also be applied as thin linings. These linings can be susceptible to delamination in applications where temperatures cycle frequently between ambient and steam. Table 1.1 presents details about these linings. Fluoropolymer thin lining systems can also be applied as thick linings or as sheet linings.

THICK LININGS

When the corrosion rate of the substrate exceeds 10 mpy, thick linings exceeding 25 mils (0.025 in.) are recommended. One such lining is vinyl ester reinforced with glass cloth or woven roving. Linings greater than 125 mils (0.125 in.) thick can be sprayed or trowelled. Maximum service temperature is 170°F (77°C). These linings can be applied in the field and are used in service with acids and some organics.

Another thick lining material for service with many acids and bases is plasticized PVC. This has a maximum operating temperature of 150°F (66°C).

Spray and baked electrostatic powder coating of fluoropolymers, as described under thin linings, can also be applied as thick linings. One such lining is PVDF and glass, or carbon-fiber fabric.

SHEET LININGS

Solid sheet linings can also be applied either as an adhesive-bonded lining or a loose-fitting lining. Adhesive-bonded sheet linings on steel substrates consist of fabric backed sheets bonded to steel vessel walls with neoprene or epoxy based adhesive. Linings may be applied in the field or in the shop. Joints are heat welded using a rod of the same polymer as the sheets.

Some adhesive bonded sheet lined vessels have been rated to withstand full vacuum at ambient temperature. Vacuum ratings may vary with vessel size.

Dual Laminates

Dual laminate structures are composed of fiber reinforced plastic (FRP) built with fluoropolymer sheeting that serves as a lining. This lining is fabricated by machine and hand welding of fiber backed fluoropolymer sheets, like those described above for adhesive bonded systems. The fabric aids in bonding the sheeting to the FRP.

Loose Linings

Loose linings are predominately composed of fluoropolymer sheets which are welded into lining shapes, folded, and slipped into the housing. The lining is flared over body and nozzle flanges to hold it in place. Weep holes are provided in the substrate to permit release of permeants. This is an ASME code requirement. Vacuum is permitted in vessels having diameters up to 12 in. (300 mm). Larger diameter vessels cannot be used under vacuum conditions.

PERMEATION

All materials are somewhat permeable to chemical molecules, but plastic materials tend to be an order of magnitude greater in their permeability rate than metals. Gases, vapors, or liquids will permeate polymers.

Permeation is a molecular migration either through microvoids in the polymer (if the polymer is more or less porous) or between polymer molecules. In neither case is there any attack on the polymer. This action is strictly a physical phenomenon. In lined equipment permeation can result in

1. Failure of the substrate from corrosion attack.
2. Bond failure and blistering, resulting from accumulation of fluids at the bond interface when the substrate is less permeable than the liner, or from corrosion/reaction products if the substrate is attacked by the permeant
3. Loss of contents through the substrate and liner as a result of the eventual failure of the substrate. In unbonded linings it is important that

the space between the liner and support member be vented to the atmosphere, not only to allow minute quantities of permeant vapor to escape, but also to prevent expansion of entrapped air from collapsing the liner.

Permeation is a function of two variables, one relating to diffusion between molecular chains and the other to the solubility of the permeant in the polymer. The driving force of diffusion is the partial pressure gradient for gases, and the concentration gradient for liquids. Solubility is a function of the affinity of the permeant for the polymer.

There is no relation between permeation and the passage of materials through cracks and voids, even though in both cases migrating chemicals travel through the polymer from one side to the other.

The user has some control over permeation, which is affected by

1. Temperature and pressure
2. Permeant concentration
3. Thickness of the polymer

Increasing the temperature will increase the permeation rate since the solubility of the permeant in the polymer will increase; and as the temperature rises the polymer chain movement is stimulated, permitting more permeants to diffuse among the chains more easily. For many gases the permeation rate increases linearly with the partial pressure gradient, and the same effect is experienced with the concentration gradients of liquids. If the permeant is highly soluble in the polymer, the permeability increase many be nonlinear. The thickness will generally decrease permeation by the square of the thickness. For general corrosion resistance, thicknesses of 0.010–0.020 in. are usually satisfactory.

The density of the polymer, as well as the thickness, will have an effect on the permeation rate. The greater the density of the polymer, the fewer voids through which permeation can take place. A comparison of the density of sheets produced from different polymers does not provide any indication of the relative permeation rates. However, a comparison of the density of sheets produced from the same polymer will provide an indication of the relative permeation rates. The denser the sheet, the lower the permeation rate.

Thickness of lining is a factor affecting permeation. For general corrosion resistance, thicknesses of 0.010–0.20 in. are usually satisfactory, depending upon the combination of elastomeric material and the specific corrodent. When mechanical factors such as thinning due to cold flow, mechanical abuse, and permeation rates are a consideration, thicker linings may be required.

Increasing the lining thickness will normally decrease permeation by the square of the thickness. Although this would appear to be the approach to follow to control permeation, there are disadvantages. First, as the thickness increases, the thermal stresses on the bonding increase, which can result in bond failure. Temperature changes and large differences in coefficients of thermal expansion are the most common causes of bond failure. Thickness and modulus of elasticity of the elastomer are two of the factors that would influence these stresses. Second, as the thickness of the lining increases, installation becomes more difficult, with a resulting increase in labor costs.

The rate of permeation is also affected by temperature and temperature gradient in the lining. Lowering these will reduce the rate of permeation. Lined vessels that are used under ambient conditions, such as storage tanks, provide the best service.

Other factors affecting permeation consisting of chemical and physiochemical properties are:

1. Ease of condensation of the permeant: Chemicals that condense readily will permeate at higher rates.
2. The higher the intermolecular chain forces (e.g., Van der Waals hydrogen bonding) of the polymer, the lower the permeation rate.
3. The higher the level of crystalinity in the polymer, the lower the permeation rate.
4. The greater the degree of cross-linking in the polymer, the lower the permeation rate.
5. Chemical similarity between the polymer and permeant: When the polymer and permeant both have similar functional groups, the permeation rate will increase.
6. The smaller the molecule of the permeant, the greater the permeation rate.

The magnitude of any of the effects will be a function of the combination of the polymer and the permeant in actual service.

Vapor permeation of PTFE, FEP, and PFA are shown in Tables 1.2 to 1.4.

ABSORPTION

Polymers have the potential to absorb varying amounts of corrodents they come into contact with, particularly organic liquids. This can result in swelling, cracking, and penetration to the substrate. Swelling can cause softening of the polymer, introduce high stresses, and cause failure of the bond. If the polymer has a high absorption rate, permeation will probably take place. An approximation of the expected permeation and/or absorption of the polymer can be based on the absorption of water. These data are usually available. Table 1.5 provides the water absorption rates of the common polymers.

TABLE 1.2 Vapor Permeation of PTFE[a]

Gases	Permeation (g/100 in.^{2}/24 h/mil) at 73°F (23°C)	86°F (30°C)
Carbon dioxide		0.66
Helium		0.22
Hydrogen chloride; anh.		<0.01
Nitrogen		0.11
Acetophenone	0.56	
Benzene	0.36	0.80
Carbon tetrachloride	0.06	
Ethyl alcohol	0.13	
Hydrochloric acid 20%	<0.01	
Piperdine	0.07	
Sodium hydroxide 50%	5×10^{-5}	
Sulfuric acid 98%	1.8×10^{-5}	

[a] Based on PTFE having a specific gravity >2.2.

The failure due to absorption can best be understood by considering the "steam cycle" test described in the ASTM standards for lined pipe. A section of lined pipe is subjected to thermal and pressure fluctuations. This is repeated for 100 cycles. The steam creates a temperature and pressure gradient through the liner, causing absorption of a small quantity of steam, which condenses to water within the inner wall. Upon pressure release, or on reintroduction of steam, the entrapped water can expand to vapor, causing an original micropore. The repeated pressure and thermal cycling enlarges the micropores, ultimately producing visible water filled blisters within the liner.

In an actual process, the polymer may absorb process fluids, and repeated temperature or pressure cycling can cause blisters. Eventually corrodent may find its way to the substrate.

Related effects can occur when process chemicals are absorbed, which may later react, decompose, or solidify within the structure of the polymer. Prolonged retention of the chemicals may lead to their decomposition within the polymer. Although it is unusual, it is possible for absorbed monomers to polymerize.

Several steps can be taken to reduce absorption. Thermal insulation of the substrate will reduce the temperature gradient across the vessel, thereby preventing condensation and subsequent expansion of the absorbed fluids. This also reduces the rate and magnitude of temperature changes, keeping blisters to a minimum. The use of operating procedures or devices to limit the rates of process pressure reductions or temperature increases will provide additional protection.

TABLE 1.3 Vapor Permeation of FEP

	Permeation (g/100 in.^{2}/24 h/mil) at		
	73°F (23°C)	95°F (35°C)	122°F (50°C)
Gases			
Nitrogen	0.18		
Oxygen	0.39		
Vapors			
Acetic acid		0.42	
Acetone	0.13	0.95	3.29
Acetophenone	0.47		
Benzene	0.15	0.64	
N-Butyl ether	0.08		
Carbon tetrachloride	0.11	0.31	
Decane	0.72		1.03
Ethyl acetate	0.06	0.77	2.9
Ethyl alcohol	0.11	0.69	
Hexane		0.57	
Hydrochloric acid 20%	<0.01		
Methanol			5.61
Sodium hydroxide 50%	4×10^{-5}		
Sulfuric acid 98%	8×10^{-6}		
Toluene	0.37		2.93
Water	0.09	0.45	0.89

ENVIRONMENTAL STRESS CRACKING

Stress cracks develop when a tough polymer is stressed for an extended period of time under loads that are small relative to the polymer's yield point. Cracking will occur with little elongation of the material. The higher the molecular weight of the polymer, the less likelihood of environmental stress cracking, other things being equal. Molecular weight is a function of length of individual chains that

TABLE 1.4 Permeation of Various Gases in PFA at 77°F (25°C)

Gas	Permeation (cm^3 mil thickness/100 in.^{2} 24 h atm)
Carbon dioxide	2260
Nitrogen	291
Oxygen	881

TABLE 1.5   Water Absorption Rates of
Polymers

Polymer	Water absorption 24 h at 73°F (23°C) (%)
PVC	0.05
CPVC	0.03
PP (Homo)	0.02
PP (Co)	0.03
PE (EHMW)	<0.01
E-CTFE	<0.1
PVDF	<0.04
Saran	nil
PFA	<0.03
ETFE	0.029
PTFE	<0.01
FEP	<0.01

make up the polymer. Longer-chain polymers tend to crystallize less than polymers of lower molecular weight or shorter chains and also have greater load-bearing capacity.

Crystallinity is an important factor affecting stress corrosion cracking. The less the crystallization that takes place, the less the likelihood of stress cracking. Unfortunately the lower the crystallinity, the greater the likelihood of permeation.

Resistance to stress cracking can be reduced by the absorption of substances that chemically resemble the polymer and will plasticize it. In addition, the mechanical strength will be reduced. Halogenated chemicals, particularly those consisting of small molecules containing fluorine or chlorine, are especially likely to be similar to the fluoropolymers and should be tested for their effect.

The presence of contaminants in a fluid may act an an accelerator. For example, polypropylene can safely handle sulfuric or hydrochloric acids, but iron or copper contaminants in concentrated sulfuric or hydrochloric acids can result in the stress cracking of polypropylene.

LINING SELECTION

At this stage an economic evaluation must be undertaken. There are several options which can be considered; an economic evaluation should be performed based on life-cycle costs that takes into consideration the replacement intervals and downtime associated with linings, compared to the higher costs of equipment fabricated from more expensive alloys.

Before the economic evaluation can be performed, potential lining materials must be considered. In order to determine which lining material may be suitable for the application, several broad categories must be considered, specifically materials being handled, operating conditions, and conditions external to the vessel.

The following questions must be answered about the materials being handled.

1. What are the primary chemicals being handled and at what concentrations?
2. Are there any secondary chemicals, and if so at what concentrations?
3. Are there any trace inpurities or chemicals?
4. Are there any solids present, and if so what is their particle size and concentration?
5. If a vessel, will there be agitation, and to what degree? If a pipeline, what are the flow rates, minimum and maximum?
6. What are the fluid purity requirements?

The answers will narrow the selection to those lining materials that are compatible. Screening of the candidates can be accomplished by reviewing industry publications, manufacturers' data sheets, and publications listing chemical resistance data, e.g., Ref. 1. Table 1.6 will serve as a general quide for liner selection.

This next set of questions will narrow the selection still further by eliminating those materials that do not have the required physical/mechanical properties.

1. What is the normal operating temperature and temperature range?
2. What peak temperatures may be reached during shutdown, startup, process upset, etc.?
3. Will any mixing areas exist where exothermic or heat of mixing temperatures can develop?
4. What is the normal operating pressure?
5. What vacuum conditions and range are possible during startup, operation, shutdown, or upset conditions?
6. Will there be temperature cycling?
7. What cleaning methods will be used?

Finally, consideration should be given to the conditions external to the vessel or pipe.

1. What are the ambient temperature conditions?
2. What is the maximum surface temperature during operation?
3. What are the insulation requirements?
4. What is the nature of the external environment? This can dictate finish requirements and/or affect the selection of the shell material.

TABLE 1.6 Guide for Liner Selection

Acids less than 120°F/49°C	
Dilute	Concentrated
Fluoropolymer	PVC
Elastomers	Epoxy novolac
Borosilicate glass	Silicon epoxy
Silicon epoxy	Fluoropolymer
Epoxy novolac	Borosilicate glass
Furan	Baked phenolic
Polyester	Vinyl ester
Vinyl ester novolac	PP
Epoxy phenolic	
Plasticized PVC	
Epoxy amine adduct	
Baked phenolic	
Polyurethane	
Vinyl ester	
Vinyl	
PVC, CPVC, PE, PP	

Acids greater than 120°F/49°C	
Baked phenolic	Fluoropolymer
Fluoropolymer	Borosilicate glass
Epoxy novolac	Elastomers
Vinyl ester	Acid brick
Silicon epoxy	HDPE
Borosilicate glass	
Acid brick	

Solvents less than 120°F/49°C	
Aromatic	Chlorinated
Baked phenolic	Baked phenolic
Silicon epoxy	Silicon epoxy
Fluoropolymer	Fluoropolymer
Furan	Vinyl ester novolac
Vinyl ester novolac	Borosilicate glass
Epoxy amine adduct	
Borosilicate glass	
Vinyl ester	

TABLE 1.6 Continued

Solvents greater than 120°F/49°C	
Aromatic	Chlorinated
Fluoropolymer	Fluoropolymer
Borosilicate glass	Borosilicate glass
Baked phenolic	Baked phenolic

Bases less than 120°F/49°C	
Dilute	Concentrated
Fluoropolymer	Fluoropolymer
Epoxy novolac	Epoxy novolac
Epoxy polyamide	Silicon epoxy
Epoxy amine adduct	Epoxy phenolic
Epoxy phenolic	Vinyl ester
Coal tar epoxy	Elastomers
Vinyl ester	Epoxy amine
Elastomers	Furan
Vinyl ester novolac	Vinyl ester novolac
Borosilicate glass	Borosilicate glass
Chlorinated rubber	Epoxy amine adduct
Baked phenolic	Plasticized PVC
Polyester	PP
Polyurethane	PE
Furan	PVC
PP, PE, PVC	

Bases greater than 120°F/49°C	
Fluoropolymer	Fluoropolymer
UHMWPE	UHMWPE
Elastomers	PVC
PP, PVC	Epoxy novolac
Epoxy novolac	

Elastomers	Thermoplastic polymers	
Natural rubber	HDPE	High density polyethylene
Nitrile rubber	CPVC	Chlorinated polyvinyl chloride
Butyl rubber	PE	Polyethylene
Hypalon	PP	Polypropylene
EPDM rubber	PVC	Polyvinyl chloride
SBR rubber	UHMWPE	Ultra high molecular weight
Neoprene		polyethylene

5. What are the external heating requirements?
6. Is grounding necessary?

In addition to the above information it is also necessary to be aware of the vessel design details that can affect lining performance or installation. Dip tubes may be located too close to the lining for injection of concentrated chemicals. The location of steam coils may raise temperatures beyond the resistance of the lining in local areas. Agitators may subject the lining to high erosion or abrasion. The curing of high-temperature baked linings may be inhibited by heat sinks on the shell wall such as support saddles.

The size of the vessel must also be considered in the lining selection. If the vessel is too large, it may not fit in a particular vendor's oven for curing of the lining. Also, nozzle diameters 4 in. and less are too small to spray-apply a liquid lining in them.

If the vessel is to be lined in the field, there are other factors that must be taken into account, including scheduling of the installation, physical location of the vessel, availability of qualified applicators, the outdoor climate at the time of installation, purging and ventilation requirements, and the need for temporary insulation. These factors can further narrow the list of potential linings.

Pigments and reinforcing agents are quite frequently added to the lining. They contribute to the mechanical properties and impermeability of the lining. Water can diffuse through a polymer film, but not through inorganic lamellar pigments. Wet-ground mica, aluminum flake, glass flake, and other lamellar pigments force water and electrolytes to travel through a tortuous maze of platelet layers. This compares to a more direct route if spherical pigments are used.

Each manufacturer formulates his lining to produce what he considers the best properties for a specific application. For example, a formulator may sacrifice mechanical strength in order to achieve lower shrinkage and thermal expansion by adding a particular filler. Therefore, a product with lower mechanical properties may have better essential properties for a lining, e.g., thermal shock resistance, adhesion, and impact resistance.

LINING SYSTEMS

The majority of lining materials are organic in nature and composed of thermoplastic, or thermoset, polymers and elastomers. Inorganic lining materials include glass, brick, and mortars.

Organic Linings

Thermoplastic polymeric linings are available either as liquid applied or in sheet form. Liquid applied linings of vinyls or chlorinated rubber are finding fewer applications because of the hazards involved during application. For example, the vinyl resin requires 4–5 lb of ketone aromatic solvent per gallon to dissolve the resin sufficiently for spray viscosity. The polymer relies on solvent evaporation for curing. This makes it necessary to purge and ventilate vessels of volatiles to maintain an oxygen level above 19.5% and keep flammability measurements no greater than 10% of the lower explosive limits. These hazards of solvent vapor ignition and air pollution are responsible for the decline in use.

The fluoropolymer materials are the most important for the chemical process industry. There are two categories of fluoropolymers: the fully fluorinated and the partially fluorinated.

Fully fluorinated materials include PTFE (polytetrafluoroethylene), FEP (fluorinated ethylene-propylene copolymer) and PFA (perfluoroalkoxy). These materials have the higher temperature ratings (above 300°F/149°C). Of these the most important is PTFE. It has exceptional resistance to chemical attack, being inert to all reagents except molten alkali metals and fluorine. It has an upper temperature limit of 450°F/232°C.

Partially fluorinated materials include ECTFE (ethylene trifluoroethylene), ETFE (ethylene tetrafluoroethylene), and PVDF (polyvinylidene fluoride). The partially fluorinated materials have higher mechanical properties but lower temperature ratings (<300°F/149°C), and chemical resistance.

Fluoropolymer linings are installed using adhesive bonded fabric-backed sheets, sprayed dispersions, rotolining, electrostatic sprayed powders and by mechanically attaching loose sheet linings.

The fluoropolymers have greater chemical resistance, higher temperature resistance, and lower surface coefficients of friction than the thermosets. This is the result of the high density of the tightly packed small fluorine atoms on long polymer chains, the absence of chain branching, the high percentage of crystallinity, and the high bond dissociation energy of the carbon-fluorine bonds.

Thermoset resin polymeric linings find applications with moderate exposure to chemicals and temperatures. A maximum temperature of 130°F/54°C is typical. The epoxy structure is the base polymer resin for many of the thermoset polymeric resin systems. These include vinyl esters, novolacs, and silicon epoxies which are based on the repeating unit of the product of epichlorhydrin and a phenol homologue. Differences in polymer backbone structure, functionality, and curing agents results in a high cross-link density, high-energy bond formations, and atoms with sizes that prevent the entrance of solvating agents. Curing agents are responsible for cross-links, which form a three-dimensional network to increase properties. Aliphatic amines, cycloaliphatic amines, aromatic amines, amido amines, and polyamides are commonly used curing agents. Polyamides provide the best acid resistance, and aliphatic amines provide the best solvent resistance. Aromatic amines also produce the highest mechanical properties, and heat distortion temperatures and polyamides produce the lowest.

Phenolics, which are based on the polymerization product of phenol and formaldehyde, are another important class of thermoset linings. These polymers were the first to be synthesized and used commercially. Phenolic linings are resistant to all ordinary solvents, including ketones and chlorinated hydrocarbons. Application is by spraying in multiple coats to obtain a dry film thickness of 1–2 mils per coat, and then baked at 350–450°F/177–232°C.

Epoxy and phenolic resins can be blended to provide linings with a cross-linked copolymer that shows the characteristics of both polymers. Epoxy-phenolic linings are air cured with good chemical and temperature resistance and are more easily repaired than phenolics. However, they do not have the comparable solvent resistance.

Both natural and synthetic rubbers are used as elastomeric linings. The most commonly used synthetic elastomers are NBR (acrylonitrile-butadiene), Hypalon (chlorosulfonated polyethylene), EPDM (ethylene-propylene-diene monomer), EPT (ethylene-propylene-diene terpolymer), SBR (styrene-butadiene), and neoprene (polychloroprene). A maximum use temperature of 175°F/80°C is typical.

Trace organic solvents can cause swelling and penetration of the lining to attack the underlying adhesive.

Inorganic Linings

Inorganic linings are basically ceramic compounds. The most commonly used material is glass which is used to line process vessels. These glass linings are capable of handling all acids at elevated temperatures except hydrofluoric and phosphoric.

Another inorganic system is mortars for bonding acid and other types of brick. This composite system is used to line chimneys, wet scrubbers, and acid process tanks.

Cement is also used as a lining material for piping and vessels.

LINING SPECIFICATION

There are different considerations for specifying a sheet lining or a liquid applied lining. Generalities will be discussed here while more explicit details will be covered in Chapter 2: Sheet Linings and Chapter 5: Liquid Applied Linings. For sheet linings it is essential that the vessel design and fabrication techniques followed are suitable to accept a sheet lining. For example, the vessel should be of butt welded construction and all weld spatter removed and the surface properly prepared to accept the lining.

For liquid applied linings it is important to remember that for a successful installation it is necessary for a chemical reaction to occur in the material on the substrate. In order for the reaction to take place, it is necessary that correct reactants are used, that they are uniformly dispersed, used in correct proportions, that there is sufficient heat activation, that there are no inhibitors present, and the correct environment is supplied for the reaction to go to completion.

Lining specifications must include all of the details necessary to ensure the proper installation of the lining. More specific details will be found in later chapters.

Whoever prepares the lining specifications should be present during the installation and set specific times for inspection, for example:

1. When the shell is prepared
2. After the application of each coat
3. After each coat has cured
4. At final inspection

REFERENCE

1. Schweitzer, Philip A. *Corrosion Resistance Tables*, 4th ed., vols. 1–3, Marcel Dekker, New York, 1995.

2

Sheet Linings

For many years vessels have been successfully lined with various rubber formulations, both natural and synthetic. Many such vessels have given over 20 years of reliable service. These compounds are still in use today but with the development of newer thermoplastic and elastomeric materials the range of application has been expanded. The newer materials, particularly the fluorocarbons, have greatly increased the ability of linings to protect substrate from corrosive chemicals. However, in order for these lining materials to provide the maximum protection, it is important that the vessel shell be properly designed and prepared and the lining properly installed. Unless these precautions are taken, premature failure is likely to occur.

CAUSES OF LINING FAILURE

Linings properly selected, installed in a properly designed and prepared shell, and maintained will provide long years of satisfactory service. Despite this there are occasionally reports of lining failures that can be traced to one or more of the following causes, all of which could have been prevented.

Liner Selection

For a lining to perform satisfactorily it must have the required mechanical, physical, and corrosion resistant properties to withstand the application. Physical and mechanical properties can be found in tables. There are many sources for corrosion resistance data available, such as Ref. 1 or manufacturers' data sheets. When data is not available, corrosion resistance testing should be conducted to guarantee the resistance of the material. Such testing should duplicate all conditions that the liner will undergo when installed.

Inadequate Surface Preparation

Proper surface preparation is critical. If not done properly, poor bonding or mechanical damage to the liner may result. It is important that surface preparation specifications be followed. Prior to lining installation, the surface should be inspected to guarantee that the specifications have been followed.

Thermal Stresses

Stresses caused by thermal cycling can eventually result in bond failure. Proper design and operating procedures can control this problem.

Permeation

Permeation through the lining to the substrate can cause debonding when corrosion products or fluids accumulate at the interface. In addition, deterioration of the substrate can also result. Proper selection of the lining material can help to control this problem. Refer to Chapter 1 and Tables 1.2 to 1.4 for more details on permeation.

Absorption

As with permeation, absorption of the corrodent by the liner material can result in swelling of the liner, cracking, and eventual penetration to the substrate. This can lead to high stresses and debonding. Refer to Chapter 1 for a discussion of absorption.

Welding Flaws

A welding flaw is a common cause of failure. Therefore, it is essential that only qualified personnel perform the welding and only qualified, experienced contractors be used to install linings.

Debonding

In addition to debonding occurring as a result of one of the preceding causes, it can also occur as the result of the use of the wrong bonding agent. Care should be taken that the proper bonding agent be employed for the specific lining being used.

Operation

It is important that those personnel operating lined equipment be thoroughly familiar with the allowable operating characteristics of the liner. Most failures from this cause usually result while vessels are being cleaned or repaired. Temperatures may be exceeded if live steam is used, or chemical attack may occur if the vessel is solvent cleaned.

SHELL DESIGN

The design of the vessel shell is critical if the lining is to perform satisfactorily. In order for the lining to be installed properly, the vessel must meet certain design configurations. Although these details may vary slightly depending upon the specific lining material being used, there are certain design principles that apply in all cases.

1. The vessel must be of butt welded construction.
2. All internal welds must be ground flush.
3. All weld spatter must be removed.
4. All sharp corners must be ground to a minimum of $1/8$ in. radius.
5. All outlets to be of flanged or pad type. Certain lining materials require that nozzles be not less than 2 in. (55 mm) in diameter.
6. No protrusions are permitted inside of the vessel.

Once the lining has been installed, there should be no welding permitted on the exterior of the vessel.

When a vessel shell is fabricated from a reinforced thermosetting plastic (RTP), several advantages are realized. The RTPs themselves generally have a wider range of corrosion resistance but relatively low allowable operating temperatures. When a fluoropolymer-type lining is applied to an RTP shell, the temperature to which the backup RTP is exposed is reduced, and the RTP is prevented from being exposed to the chemicals in the process system. An upper temperature limit for using RTP dual laminates is 350°F (177°C).

The dual laminate construction lessens the problem of permeation through liners. If there is permeation, it is believed to pass through the RTP structure at a rate equal to or greater than through the fluoropolymer itself, resulting in no potential for collection of permeate at the thermoplastic-thermoset interface. If delamination does not occur, permeation is not a problem.

SHELL PREPARATION

After the vessel shell is fabricated, it is essential that the surface be properly prepared to receive the lining. This is an extremely critical step. Unless the surface is properly prepared, proper bonding of the lining to the shell will not be achieved. The basic requirement is that the surface be absolutely clean. To achieve proper bonding, all surfaces to be lined should be abrasive blasted to white metal in accordance with SSPC specification SPS-63 or NACE specification NACE-1.

A white metal surface condition is defined as being one from which all rust, scale, paint, and the like have been removed and the surface has a uniform gray-white appearance. Streaks or stains of rust or other contaminants are not allowed. On occasion a near-white blast-cleaned finish equal to SSPC-SP-10 is allowed, which is a more economical finish. It is essential that the finish the lining contractor specifies be provided. In addition to installing the lining, some lining contractors will also undertake the fabrication and surface preparation of the vessel. Placing the total responsibility on the lining contractor can simplify the problem of coordination and usually will result in a better-quality product.

DESIGN CONSIDERATIONS

In addition to selecting a lining material that is resistant to the corrodent being handled, there are three other factors to be considered in the design: permeation, absorption, and environmental stress cracking. Permeation, and absorption can cause

1. Bond failure and blistering resulting from the accumulation of fluids at the bond interface when the substrate is less permeable than the liner, or from corrosion/reaction products if the substrate is attacked by the permeant.
2. Failure of the substrate from attack.
3. Loss of contents through the substrate and liner as a result of the eventual failure of the substrate. In unbonded linings it is important that the space between the liner and substrate be vented to the atmosphere, not only to allow minute quantities of permeant vapor to escape, but to prevent expansion of entrapped air from collapsing the liner. For a discussion of permeation, absorption, and environmental stress cracking, refer to Chapter 1.

Consideration must also be given to the adhesive used to bond the liner to the substrate and joining sheets of lining material (if used). This adhesive must be resistant to the corrodents that will come into contact with the liner.

BONDED LININGS

Bonded linings have several advantages over loose or unbonded linings.

1. They provide superior performance in vacuum service, resisting collapse to full vacuum if designed properly.
2. During thermal or pressure cycling, bonded linings follow the move-

ment of the structural wall, avoiding stress concentrations at nozzles and other anchor points.

3. There is less permeation for bonded linings than for unbonded linings, since permeate must pass through the liner and the substrate.

The construction of adhesive bonded sheet linings on steel substrate consists of fabric backed sheets bonded to steel vessel walls with neoprene or epoxy based adhesive. Liners may be installed in the field or in the shop.

In order to protect the liner from delamination, a differential expansion, or buffer layer, is used. This is the bonding layer between the vessel wall and the fabric backing of the liner. Delamination is due primarily to the difference in expansion rates between the structural wall and the lining.

The linear coefficient of a liner free to move would typically be approximately twice that of the structural wall. In any case the relative stresses set up by the differential expansion of the materials must be determined and the allowable physical properties not exceeded.

For process applications which require a lining that will resist harsh chemicals usually one of the fluoropolymers is selected. Since these linings are expensive, representing approximately 80% of the cost of the vessel, a fabricator must be selected that is experienced and knowledgeable in the handling, welding, and forming of the specific fluoropolymer sheet.

Special equipment is also required such as machines that fuse flat sheets in limited widths into massive sheets, heated forming tools for flanges and fittings, and forming machines that make heads with a minimum number of seams from large sheets. Table 2.1 lists the fluoropolymers and the thicknesses of sheets available.

TABLE 2.1 Fluoropolymer Sheet Linings (Fabric Backed)[a]

Fluoropolymer	Thickness
Fluorinated ethylene propylene (FEP)	1.5 mm (60 mils)
	2.3 mm (90 mils)
Perfluoroalkoxy (PFA)	1.5 mm (60 mils)
	2.3 mm (90 mils)
Ethylene trifluoroethylene copolymer (ETFE)	1.5 mm (60 mils)
	2.3 mm (90 mils)
Ethylene chlorotrifluoroethylene copolymer (ECTFE)	1.5 mm (60 mils)
	2.3 mm (90 mils)
Polyvinylidene fluoride (PVDF)	3 mm (120 mils)
	4 mm (160 mils)
	5 mm (200 mils)
	9 mm (360 mils)

[a] All fluoropolymers listed are 1 m wide.

Seams are fabricated by hot gas welding, extrusion welding, or by butt fusion. It requires considerable skill to hand fabricate using a rod and hot gas, so the vendor's welders should be checked as to their qualifications. Seams should be minimized. This can be done by using the widest sheet available (3 ft or wider) and by thermoforming heads instead of preformed sections.

UNBONDED LININGS

Loose linings are produced by welding fluoropolymer sheets into lining shapes, folding, and slipping into the housing. The lining is flared over body and nozzle flanges to keep it in place. Weep holes must be provided in the substrate to permit the release of permeants. Vacuum is permitted for diameters up to 12 in. (305 mm). Larger vessels cannot tolerate vacuum.

Performance of loose linings in large vessels has only been fair, but they have proven very successful in lined piping. Refer to Ref. 2.

INSPECTION OF LINING

Scheduling of the fabrication of the liner and installation in the vessel should be arranged so that timely inspections can be made at various stages. The following items should be inspected.

1. Liner fit-up and preparation for welding. Fusion welded seams may be made prior to application of the liner to the mold in the case of an RTP shell.
2. In the case of an RTP structural shell the liner welds should be inspected on the mold prior to application of the RTP shell. All welds should be checked both visually and with a spark tester. If the mold is not of metal construction, conductive backup material must be placed on the mold behind the seams.
3. If the structural shell is to be of RTP construction, it is necessary that conductive resin ground strips be placed over the back side of the seams so that the liner can be spark tested after the structural laminate has been applied. The resin ground strips are made from resin filled with carbon powder applied as a putty or are resin reinforced with carbon cloth laid up over the seams. The carbon provides sufficient electrical conductivity to create a ground for the spark. When a steel shell is to be used the strips are unnecessary since the steel shell will act as a ground for the spark.
4. Application of the prime coat to the backing of the liner.

5. With an RTP shell, fit-up of the components of the vessel. This includes head to shell, nozzles to heads and shell, and shell joints. Alignment is more critical in a lined vessel than in an unlined one.
6. When butt fusion joints are to be used inside the shell, the preparation for these inside welds should be inspected.
7. Hydrostatic and thermal cycling tests.
8. Halide tests.
9. Final spark testing of the seams prior to shipping.

Additional inspections may be required if the vessel has special features such as internal support ledges or pipe supports that may require hand welding of components.

When preparing for shipment, flange covers should be shipped in place. Flanges or pads that do not have covers should be protected with temporary covers securely fastened in place.

INSTALLATION AT THE SITE

No welding is to be done on the exterior of the vessel once the lining is in place. Lined vessels should be properly identified when installed with the allowable operating characteristics of the liner posted to avoid damage to the liner during cleaning or repair operations. Most failures from this cause result while vessels are being cleaned or repaired. If live steam is used to clean the vessel, allowable operating temperatures may be exceeded. If the vessel is solvent cleaned, chemical attack may occur. See Ref. 1.

Piping, valves, instruments, and the like should be supported independently and not hung off the tank nozzle. Alignment of piping to the tank nozzle is also critical. Misaligned piping can induce a stress on the nozzle lining.

If raised face flanges, or lined valves or piping are connected to the tank nozzles, there will be an unsupported area between the raised portion of these components and the full face lining of the tank nozzles. This provides an opportunity for overstressing the vessel flange and cracking it (in the case of an RTP vessel). This can be prevented by using a filler ring or blocking ring. The filler ring fills the open area between the two flanges, and its lack of compressibility prevents bending of the lined flange. The blocker ring is a plastic coated steel flange that provides a full face flange to the vessel side and has sufficient strength to take the mismatch on the nonfull face side. These rings are similar to those used in the installation of unlined RTP equipment.

The proper bolt torque should be used when making up flanged joints to prevent damage by overtorquing. Fluoropolymer coated bolts are useful in this situation since they provide consistent lubrication for accurate torques. They also permit easy disassembly at a later date since they provide an uncorroded surface. This prevents flange damage from excessive disassembly forces.

REFERENCES

1. Schweitzer, Philip A. *Corrosion Resistance Tables*, 4th ed., vols. 1–3, Marcel Dekker, New York, 1995.
2. Schweitzer, Philip A. *Corrosion Resistant Piping Systems*, Marcel Dekker, New York, 1994.

3

Specific Thermoplastic Sheet Lining Materials

Thermoplastic sheet linings are used more extensively than any other type of lining. Table 3.1 lists the most common thermoplasts used as linings. These materials are used to line piping as well as vessels. They may be installed as a loose lining or as a bonded lining.

When vessels are lined, the linings, with the exception of plasticized PVC, are fabricated from sheet stock that must be cut, shaped, and joined. Joining is usually accomplished by hot gas welding. Several problems exist when a thermoplastic lining is bonded to a metal shell, primarily the large differences in the coefficient of thermal expansion and the difficulty of adhesion. If the vessel is to be used under ambient temperature, such as a storage tank, the problem of thermal expansion differences is eliminated. However, the problem of adhesion is still present. Because of this, many linings are installed as loose linings.

Techniques have been developed to overcome the problem of adhesion, making use of an intermediate bond. One approach is to heat the thermoplastic sheet, then impress into one surface a fiber cloth or nonwoven web. This provides half of the bond. Bonding to the metal surface is accomplished by the use of an epoxy adhesive that will bond to both the fiber and the metal.

Welding techniques for joining the plastic sheets are critical. Each weld must be continuous, leak tight, and mechanically strong. Poor welding is a common cause of lining failure.

TABLE 3.1 Thermoplastic Lining Materials

Polyvinyl chloride (PVC)
Chlorinated polyvinyl chloride (CPVC)
Polyvinylidene chloride (Saran)
Polyvinylidene fluoride (PVDF)
Polypropylene (PP)
Polyethylene (PE)
Polytetrafluoroethylene (PTFE)
Fluorinated ethylene propylene (FEP)
Perfluoroalkoxy (PFA)
Ethylene tetrafluoroethylene (ETFE)
Ethylene chlorotrifluoroethylene (ECTFE)

These materials are capable of providing a wide range of corrosion resistance. Table 3.2 lists the general area of corrosion resistance for each thermoplast. This table is only a general guide. The resistance of a lining material to a specific corrodent should be checked.

Most of the lined pipe have loose fitting liners (the lining material is not bonded to the inside of the pipe). Although the lining provides the advantage of corrosion resistance to the material being conveyed, there are some disadvantages to these piping systems.

TABLE 3.2 General Corrosion Resistance of Thermoplastic Lining Materials

Material	Strong acids	Strong bases	Chlorinated solvents	Esters and ketones	Strong oxidants
PVC (type 1)	F	F	P	P	P
PVC (type 2)	F	G	P	P	P
CPVC	F	F	F	P	P
Saran	F	F		P	
PVDF	E	P	E	P	E
PP	G	E	P	F	P
PE	F	G	P	P	E
PTFE	E	E	E	E	E
FEP	E	E	E	E	E
PFA	E	E	E	E	E
ETFE	E	E	E	E	E
ECTFE	E	E	E	E	E

E = excellent; G = good; F = fair; P = poor.

1. Although the inside of the pipe is protected, the external portion is subject to corrosion. It may be necessary in certain atmospheres to apply a protective coating to the exterior of the carbon steel pipe.
2. Flanged joints are the common method of joining pipe sections, although some of the lining materials can be utilized with threaded joints. With the use of flanged joints, odd lengths must be fabricated by using special field fabricating techniques or be ordered from the factory.

More details can be found in Ref. 1.

POLYVINYL CHLORIDE (PVC)

Two types of PVC are produced: normal impact (type 1) and high impact (type 2). Type 1 is a rigid unplasticized PVC having normal impact with optimum chemical resistance. Type 2 is a plasticized PVC and has optimum impact resistance and reduced chemical resistance. Table 3.3 lists the physical and mechanical properties of both types of PVC. PVC is stronger and more rigid than other thermoplastic materials. It has a high tensile strength and modulus of elasticity. Additives are used to further specific end uses, such as thermal stability, lubricity, impact, and pigmentation.

TABLE 3.3 Physical and Mechanical Properties of PVC

Property	Type 1	Type 2
Specific gravity	1.45	1.38
Water absorption, 24 h at 73°F/23°C, %	0.04	0.05
Tensile strength at 73°F/23°C, psi	6,800	5,500
Modulus of elasticity in tension at 73°F/23°C $\times 10^5$	5.0	4.2
Compressive strength, psi	10,000	7,900
Flexural strength, psi	14,000	11,000
Izod impact strength, notched at 73°F/23°C, ft · lb/in.	0.88	12.15
Coefficient of thermal expansion		
in./in.-°F $\times 10^{-5}$	4.0	6.0
in./10°F/100 ft	0.40	0.60
Thermal conductivity Btu/h/ft^2/°F/in.	1.33	1.62
Heat distortion temperature, °F/°C		
at 66 psi	130/54	135/57
at 264 psi	155/68	160/71
Resistance to heat, °F/°C at continuous drainage	150/66	140/60
Limiting oxygen index, %		43
Flame spread		15–20
Underwriters lab rating (Sub 94)		94V-O

TABLE 3.4 Compatibility of Type 1 PVC with Selected Corrodents[a]

Chemical	Maximum temp. °F	Maximum temp. °C	Chemical	Maximum temp. °F	Maximum temp. °C
Acetaldehyde	x	x	Barium carbonate	140	60
Acetamide	x	x	Barium chloride	140	60
Acetic acid 10%	140	60	Barium hydroxide	140	60
Acetic acid 50%	140	60	Barium sulfate	140	60
Acetic acid 80%	140	60	Barium sulfide	140	60
Acetic acid, glacial	130	54	Benzaldehyde	x	x
Acetic anhydride	x	x	Benzene	x	x
Acetone	x	x	Benzene sulfonic acid 10%	140	60
Acetyl chloride	x	x	Benzoic acid	140	60
Acrylic acid	x	x	Benzyl alcohol	x	x
Acrylonitrile	x	x	Benzyl chloride	60	16
Adipic acid	140	60	Borax	140	60
Allyl alcohol	90	32	Boric acid	140	60
Allyl chloride	x	x	Bromine gas, dry	x	x
Alum	140	60	Bromine gas, moist	x	x
Aluminum acetate	100	38	Bromine liquid	x	x
Aluminum chloride, aqueous	140	60	Butadiene	140	60
Aluminum chloride, dry	140	60	Butyl acetate	x	x
Aluminum fluoride	140	60	Butyl alcohol	x	x
Aluminum hydroxide	140	60	n-Butylamine	x	x
Aluminum nitrate	140	60	Butyl phthalate	80	27
Aluminum oxychloride	140	60	Butyric acid	60	16
Aluminum sulfate	140	60	Calcium bisulfide	140	60
Ammonia gas	140	60	Calcium bisulfite	140	60
Ammonium bifluoride	90	32	Calcium carbonate	140	60
Ammonium carbonate	140	60	Calcium chlorate	140	60
Ammonium chloride 10%	140	60	Calcium chloride	140	60
Ammonium chloride 50%	140	60	Calcium hydroxide 10%	140	60
Ammonium chloride, sat.	140	60	Calcium hydroxide, sat.	140	60
Ammonium fluoride 10%	140	60	Calcium hypochlorite	140	60
Ammonium fluoride 25%	140	60	Calcium nitrate	140	60
Ammonium hydroxide 25%	140	60	Calcium oxide	140	60
Ammonium hydroxide, sat.	140	60	Calcium sulfate	140	60
Ammonium nitrate	140	60	Caprylic acid	120	49
Ammonium persulfate	140	60	Carbon bisulfide	x	x
Ammonium phosphate	140	60	Carbon dioxide, dry	140	60
Ammonium sulfate 10–40%	140	60	Carbon dioxide, wet	140	60
Ammonium sulfide	140	60	Carbon disulfide	x	x
Ammonium sulfite	120	49	Carbon monoxide	140	60
Amyl acetate	x	x	Carbon tetrachloride	x	x
Amyl alcohol	140	60	Carbonic acid	140	60
Amyl chloride	x	x	Cellosolve	x	x
Aniline	x	x	Chloracetic acid, 50% water	140	60
Antimony trichloride	140	60	Chloracetic acid	140	60
Aqua regia 3:1	x	x	Chlorine gas, dry	140	60

TABLE 3.4 Continued

Chemical	Maximum temp. °F	Maximum temp. °C	Chemical	Maximum temp. °F	Maximum temp. °C
Chlorine gas, wet	x	x	Magnesium chloride	140	60
Chlorine, liquid	x	x	Malic acid	140	60
Chlorobenzene	x	x	Manganese chloride	90	32
Chloroform	x	x	Methyl chloride	x	x
Chlorosulfonic acid	60	16	Methyl ethyl ketone	x	x
Chromic acid 10%	140	60	Methyl isobutyl ketone	x	x
Chromic acid 50%	140	60	Muriatic acid	140	60
Chromyl chloride	120	49	Nitric acid 5%	140	60
Citric acid 15%	140	60	Nitric acid 20%	140	60
Citric acid, conc.	140	60	Nitric acid 70%	140	60
Copper acetate	80	27	Nitric acid, anhydrous	x	x
Copper carbonate	140	60	Nitrous acid, concentrated	60	16
Copper chloride	140	60	Oleum	x	x
Copper cyanide	140	60	Pherchloric acid 10%	140	60
Copper sulfate	140	60	Perchloric acid 70%	60	16
Cresol	130	54	Phenol	x	x
Cupric chloride 5%	140	60	Phosphoric acid 50–80%	140	60
Cupric chloride 50%	140	60	Picric acid	x	x
Cyclohexane	80	27	Potassium bromide 30%	140	60
Cyclohexanol	x	x	Salicylic acid	140	60
Dichloroacetic acid	120	49	Silver bromide 10%	140	60
Dichloroethane	x	x	Sodium carbonate	140	60
(ethylene dichloride)			Sodium chloride	140	60
Ethylene glycol	140	60	Sodium hydroxide 10%	140	60
Ferric chloride	140	60	Sodium hydroxide 50%	140	60
Ferric chloride 50% in water	140	60	Sodium hydroxide, concentrated	140	60
Ferric nitrate 10–50%	140	60	Sodium hypochlorite 20%	140	60
Ferrous chloride	140	60	Sodium hypochlorite,	140	60
Ferrous nitrate	140	60	concentrated		
Fluorine gas, dry	x	x	Sodium sulfide to 50%	140	60
Fluorine gas, moist	x	x	Stannic chloride	140	60
Hydrobromic acid, dilute	140	60	Stannous chloride	140	60
Hydrobromic acid 20%	140	60	Sulfuric acid 10%	140	60
Hydrobromic acid 50%	140	60	Sulfuric acid 50%	140	60
Hydrochloric acid 20%	140	60	Sulfuric acid 70%	140	60
Hydrochloric acid 38%	140	60	Sulfuric acid 90%	140	60
Hydrocyanic acid 10%	140	60	Sulfuric acid 98%	x	x
Hydrofluoric acid 30%	120	49	Sulfuric acid 100%	x	x
Hydrofluoric acid 70%	68	20	Sulfuric acid, fuming	x	x
Hydrofluoric acid 100%	x	x	Sulfurous acid	140	60
Hypochlorous acid	140	60	Thionyl chloride	x	x
Iodine solution 10%	100	38	Toluene	x	x
Ketones, general	x	x	Trichloroacetic acid	90	32
Lactic acid 25%	140	60	White liquor	140	60
Lactic acid, concentrated	80	27	Zinc chloride	140	60

[a] The chemicals listed are in the pure state or in a saturated solution unless otherwise indicated. Compatibility is shown to the maximum allowable temperature for which data are available. Incompatibility is shown by an x. A blank space indicates that data are unavailable.
Source: Ref. 2.

TABLE 3.5 Compatibility of Type 2 PVC with Selected Corrodents[a]

Chemical	Maximum temp. °F	°C	Chemical	Maximum temp. °F	°C
Acetaldehyde	x	x	Barium chloride	140	60
Acetamide	x	x	Barium hydroxide	140	60
Acetic acid 10%	100	38	Barium sulfate	140	60
Acetic acid 50%	90	32	Barium sulfide	140	60
Acetic acid 80%	x	x	Benzaldehyde	x	x
Acetic acid, glacial	x	x	Benzene	x	x
Acetic anhydride	x	x	Benzene sulfonic acid 10%	140	60
Acetone	x	x	Benzoic acid	140	60
Acetyl chloride	x	x	Benzyl alcohol	x	x
Acrylic acid	x	x	Borax	140	60
Acrylonitrile	x	x	Boric acid	140	60
Adipic acid	140	60	Bromine gas, dry	x	x
Allyl alcohol	90	32	Bromine gas, moist	x	x
Allyl chloride	x	x	Bromine liquid	x	x
Alum	140	60	Butadiene	60	16
Aluminum acetate	100	38	Butyl acetate	x	x
Aluminum chloride, aqueous	140	60	Butyl alcohol	x	x
Aluminum fluoride	140	60	n-Butylamine	x	x
Aluminum hydroxide	140	60	Butyric acid	x	x
Aluminum nitrate	140	60	Calcium bisulfide	140	60
Aluminum oxychloride	140	60	Calcium bisulfite	140	60
Aluminum sulfate	140	60	Calcium carbonate	140	60
Ammonia gas	140	60	Calcium chlorate	140	60
Ammonium bifluoride	90	32	Calcium chloride	140	60
Ammonium carbonate	140	60	Calcium hydroxide 10%	140	60
Ammonium chloride 10%	140	60	Calcium hydroxide, sat.	140	60
Ammonium chloride 50%	140	60	Calcium hypochlorite	140	60
Ammonium chloride, sat.	140	60	Calcium nitrate	140	60
Ammonium fluoride 10%	90	32	Calcium oxide	140	60
Ammonium fluoride 25%	90	32	Calcium sulfate	140	60
Ammonium hydroxide 25%	140	60	Calcium bisulfide	x	x
Ammonium hydroxide, sat.	140	60	Carbon dioxide, dry	140	60
Ammonium nitrate	140	60	Carbon dioxide, wet	140	60
Ammonium persulfate	140	60	Carbon disulfide	x	x
Ammonium phosphate	140	60	Carbon monoxide	140	60
Ammonium sulfate 10–40%	140	60	Carbon tetrachloride	x	x
Ammonium sulfide	140	60	Carbonic acid	140	60
Amyl acetate	x	x	Cellosolve	x	x
Amyl alcohol	x	x	Chloracetic acid	105	40
Amyl chloride	x	x	Chlorine gas, dry	140	60
Aniline	x	x	Chlorine gas, wet	x	x
Antimony trichloride	140	60	Chlorine, liquid	x	x
Aqua regia 3:1	x	x	Chlorobenzene	x	x
Barium carbonate	140	60	Chloroform	x	x

TABLE 3.5   Continued

Chemical	Maximum temp. °F	Maximum temp. °C	Chemical	Maximum temp. °F	Maximum temp. °C
Chlorosulfonic acid	60	16	Nitric acid 5%	100	38
Chromic acid 10%	140	60	Nitric acid 20%	140	60
Chromic acid 50%	x	x	Nitric acid 70%	70	23
Citric acid 15%	140	60	Nitric acid, anhydrous	x	x
Citric acid, con.	140	60	Nitrous acid, concentrated	60	16
Copper carbonate	140	60	Oleum	x	x
Copper chloride	140	60	Perchloric acid 10%	60	16
Copper cyanide	140	60	Perchloric acid 70%	60	16
Copper sulfate	140	60	Phenol	x	x
Cresol	x	x	Phosphoric acid 50–80%	140	60
Cyclohexanol	x	x	Picric acid	x	x
Dichloroacetic acid	120	49	Potassium bromide 30%	140	60
Dichloroethane (ethylene dichloride)	x	x	Salicylic acid	x	x
			Silver bromide 10%	105	40
Ethylene glycol	140	60	Sodium carbonate	140	60
Ferric chloride	140	60	Sodium chloride	140	60
Ferric nitrate 10–50%	140	60	Sodium hydroxide 10%	140	60
Ferrous chloride	140	60	Sodium hydroxide 50%	140	60
Ferrous nitrate	140	60	Sodium hydroxide, concen-	140	60
Fluorine gas, dry	x	x	trated		
Fluorine gas, moist	x	x	Sodium hypochlorite 20%	140	60
Hydrobromic acid, dilute	140	60	Sodium hypochlorite, concen-	140	60
Hydrobromic acid 20%	140	60	trated		
Hydrobromic acid 50%	140	60	Sodium sulfide to 50%	140	60
Hydrochloric acid 20%	140	60	Stannic chloride	140	60
Hydrochloric acid 38%	140	60	Stannous chloride	140	60
Hydrocyanic acid 10%	140	60	Sulfuric acid 10%	140	60
Hydrofluoric acid 30%	120	49	Sulfuric acid 50%	140	60
Hydrofluoric acid 70%	68	20	Sulfuric acid 70%	140	60
Hypochlorous acid	140	60	Sulfuric acid 90%	x	x
Ketones, general	x	x	Sulfuric acid 98%	x	x
Lactic acid 25%	140	60	Sulfuric acid 100%	x	x
Lactic acid, concentrated	80	27	Sulfuric acid, fuming	x	x
Magnesium chloride	140	60	Sulfurous acid	140	60
Malic acid	140	60	Thionyl chloride	x	x
Methyl chloride	x	x	Toluene	x	x
Methyl ethyl ketone	x	x	Trichloroacetic acid	x	x
Methyl isobutyl ketone	x	x	White liquor	140	60
Muriatic acid	140	60	Zinc chloride	140	60

[a] The chemicals listed are in the pure state or in a saturated solution unless otherwise indicated. Compatibility is shown to the maximum allowable temperature for which data are available. Incompatibility is shown by an x. A blank space indicates that the data are unavailable.
Source: Ref. 2.

Unplasticized PVC cannot be bonded directly to a metal substrate, while plastisols (type 2 PVC) can be bonded directly. Because of this, many PVC sheet linings are dual laminates with an unplasticized PVC sheet bonded to a plastisol. This dual laminate can then be bonded to the metal substrate, leaving the unplasticized PVC sheet exposed to the corrodent.

Polyvinyl chloride linings can also be applied by dipping in a plastisol solution or by spray coating. More details will be found in Chapter 5: Liquid Applied Linings.

Because plasticized PVC is compounded of a polyvinyl chloride dispersion of high molecular weight vinyl chloride polymers in a suitable liquid plasticizer, formulations can be made for special applications. By selective compounding, both physical and corrosion resistant properties can be modified. For certain applications this feature can be most advantageous.

Sheet linings with either material must have the joints joined by hot-gas welding. The plastisol material is flexible enough to withstand thermal stresses and can be stretched to cover minor irregularities on the vessel surface. It has a maximum allowable temperature rating of 140°F/60°C.

Type 1 PVC (unplasticized) resists attack by most acids and strong alkalies, gasoline, kerosene, aliphatic alcohols, and hydrocarbons. It is particularly useful in the handling of hydrochloric acid. The chemical resistance of type 2 PVC to oxidizing and highly alkaline materials is reduced.

PVC may be attacked by aromatics, chlorinated organic compounds, and lacquer solvents.

In addition to handling highly corrosive and abrasive chemicals, these linings have found many applications in marine environments. Table 3.4 lists the compatibility of type 1 PVC with selected corrodents, while Table 3.5 does the same for type 2 PVC. Additional data can be found in Ref. 2.

CHLORINATED POLYVINYL CHLORIDE (CPVC)

Chlorinated polyvinyl chloride is very similar in properties to PVC. Table 3.6 lists the physical and mechanical properties. CPVC has an allowable operating temperature of 200°F/93°C which is somewhat higher than the temperature rating for PVC.

It can be installed as a sheet lining and the joints hot gas welded. A word of caution: It is extremely difficult to hot-gas-weld CPVC. This factor should be considered when selecting this material.

CPVC is resistant to most acids, alkalies, salts, halogens, and many corrosive waters. It has a better resistance to chlorinated solvents than PVC. In general, it cannot be used in contact with most organic materials including chlorinated

TABLE 3.6 Physical and Mechanical Properties of CPVC

Specific gravity	1.55
Water absorption, 24 h at 73°F/23°C, %	0.03
Tensile strength at 73°F/23°C, psi	8000
Modulus of elasticity in tension at 73°F/23°C × 10^5	4.15
Compressive strength at 73°F/23°C, psi	9000
Flexural strength, psi	15,100
Izod impact strength at 73°F/23°C, ft · lb/in.	1.5
Coefficient of thermal expansion	
in./in.-°F × 10^{-5}	3.4
in./10°F/100 ft	0.034
Thermal conductivity Btu/h/ft²/°F/in.	0.95
Heat distortion temperature, °F/°C	
at 66 psi	238/114
at 264 psi	217/102
Resistance to heat at continuous drainage, °F/°C	200/93
Limiting oxygen index, %	60
Flame spread	15
Underwriters lab rating (U.L. 94)	VO; 5VA; 5VB

Source: Courtesy of B. F. Goodrich, Specialty Polymers and Chemical Division.

or aromatic esters, hydrocarbons, and ketones. Table 3.7 lists the compatibility of CPVC with selected corrodents. Reference 2 provides a more detailed listing.

POLYPROPYLENE (PP)

Polypropylene is used to line vessels and piping. The material may be joined by thermal fusion. In order to be bonded to a substrate, polypropylene sheets must have a backing imbedded since PP by itself can not be bonded. If no backing is imbedded, the liner will be loose. Piping is lined with extruded material and installed in the pipe as a loose liner.

Polypropylene is produced either as a homopolymer or a copolymer with polyethylene. The copolymer is less brittle than the homopolymer and is able to withstand impact forces down to −20°F/−29°C, while the homopolymer is extremely brittle below 40°F/4°C.

The homopolymers, being long-chain high molecular weight molecules with a minimum of random orientation, have optimum thermal, chemical, and physical properties. For this reason homopolymer material is preferred for difficult chemical, thermal, and physical conditions. The maximum allowable operating temperature of polypropylene is 225°F/107°C, however this may have

TABLE 3.7 Compatibility of CPVC with Selected Corrodents[a]

Chemical	Maximum temp. °F	°C	Chemical	Maximum temp. °F	°C
Acetaldehyde	x	x	Barium chloride	180	82
Acetic acid 10%	90	32	Barium hydroxide	180	82
Acetic acid 50%	x	x	Barium sulfate	180	82
Acetic acid 80%	x	x	Barium sulfide	180	82
Acetic acid, glacial	x	x	Benzaldehyde	x	x
Acetic anhydride	x	x	Benzene	x	x
Acetone	x	x	Benzene sulfonic acid 10%	180	82
Acetyl chloride	x	x	Benzoic acid	200	93
Acrylic acid	x	x	Benzyl alcohol	x	x
Acrylonitrile	x	x	Benzyl chloride	x	x
Adipic acid	200	93	Borax	200	93
Allyl alcohol 96%	200	93	Boric acid	210	99
Allyl chloride	x	x	Bromine gas, dry	x	x
Alum	200	93	Bromine gas, moist	x	x
Aluminum acetate	100	38	Bromine liquid	x	x
Aluminum chloride, aqueous	200	93	Butadiene	150	66
Aluminum chloride, dry	180	82	Butyl acetate	x	x
Aluminum fluoride	200	93	Butyl alcohol	140	60
Aluminum hydroxide	200	93	n-Butylamine	x	x
Aluminum nitrate	200	93	Butyric acid	140	60
Aluminum oxychloride	200	93	Calcium bisulfide	180	82
Aluminum sulfate	200	93	Calcium bisulfite	210	99
Ammonia gas, dry	200	93	Calcium carbonate	210	99
Ammonium bifluoride	140	60	Calcium chlorate	180	82
Ammonium carbonate	200	93	Calcium chloride	180	82
Ammonium chloride 10%	180	82	Calcium hydroxide 10%	170	77
Ammonium chloride 50%	180	82	Calcium hydroxide, sat.	210	99
Ammonium chloride, sat.	200	93	Calcium hypochlorite	200	93
Ammonium fluoride 10%	200	93	Calcium nitrate	180	82
Ammonium fluoride 25%	200	93	Calcium oxide	180	82
Ammonium hydroxide 25%	x	x	Calcium sulfate	180	82
Ammonium hydroxide, sat.	x	x	Caprylic acid	180	82
Ammonium nitrate	200	93	Carbon bisulfide	x	x
Ammonium persulfate	200	93	Carbon dioxide, dry	210	99
Ammonium phosphate	200	93	Carbon dioxide, wet	160	71
Ammonium sulfate 10–40%	200	93	Carbon disulfide	x	x
Ammonium sulfide	200	93	Carbon monoxide	210	99
Ammonium sulfite	160	71	Carbon tetrachloride	x	x
Amyl acetate	x	x	Carbonic acid	180	82
Amyl alcohol	130	54	Cellosolve	180	82
Amyl chloride	x	x	Chloracetic acid, 50% water	100	38
Aniline	x	x	Chloracetic acid	x	x
Antimony trichloride	200	93	Chlorine gas, dry	140	60
Aqua regia 3:1	80	27	Chlorine gas, wet	x	x
Barium carbonate	200	93	Chlorine, liquid	x	x

TABLE 3.7 Continued

Chemical	°F	°C	Chemical	°F	°C
Chlorobenzene	x	x	Manganese chloride	180	82
Chloroform	x	x	Methyl chloride	x	x
Chlorosulfonic acid	x	x	Methyl ethyl ketone	x	x
Chromic acid 10%	210	99	Methyl isobutyl ketone	x	x
Chromic acid 50%	210	99	Muriatic acid	170	77
Chromyl chloride	180	82	Nitric acid 5%	180	82
Citric acid 15%	180	82	Nitric acid 20%	160	71
Citric acid, conc.	180	82	Nitric acid 70%	180	82
Copper acetate	80	27	Nitric acid, anhydrous	x	x
Copper carbonate	180	82	Nitrous acid, concentrated	80	27
Copper chloride	210	99	Oleum	x	x
Copper cyanide	180	82	Perchloric acid 10%	180	82
Copper sulfate	210	99	Perchloric acid 70%	180	82
Cresol	x	x	Phenol	140	60
Cupric chloride 5%	180	82	Phosphoric acid 50–80%	180	82
Cupric chloride 50%	180	82	Picric acid	x	x
Cyclohexane	x	x	Potassium bromide 30%	180	82
Cyclohexanol	x	x	Salicylic acid	x	x
Dichloroacetic acid, 20%	100	38	Silver bromide 10%	170	77
Dichloroethane (ethylene di-chloride)	x	x	Sodium carbonate	210	99
			Sodium chloride	210	99
Ethylene glycol	210	99	Sodium hydroxide 10%	190	88
Ferric chloride	210	99	Sodium hydroxide 50%	180	82
Ferric chloride 50% in water	180	82	Sodium hydroxide, concentrated	190	88
Ferric nitrate 10–50%	180	82			
Ferrous chloride	210	99	Sodium hypochlorite 20%	190	88
Ferrous nitrate	180	82	Sodium hypochlorite, concentrated	180	82
Fluorine gas, dry	x	x			
Fluorine gas, moist	80	27	Sodium sulfide to 50%	180	82
Hydrobromic acid, dilute	130	54	Stannic chloride	180	82
Hydrobromic acid 20%	180	82	Stannous chloride	180	82
Hydrobromic acid 50%	190	88	Sulfuric acid 10%	180	82
Hydrochloric acid 20%	180	82	Sulfuric acid 50%	180	82
Hydrochloric acid 38%	170	77	Sulfuric acid 70%	200	93
Hydrocyanic acid 10%	80	27	Sulfuric acid 90%	x	x
Hydrofluoric acid 30%	x	x	Sulfuric acid 98%	x	x
Hydrofluoric acid 70%	90	32	Sulfuric acid 100%	x	x
Hydrofluoric acid 100%	x	x	Sulfuric acid, fuming	x	x
Hypochlorous acid	180	82	Sulfurous acid	180	82
Ketones, general	x	x	Thionyl chloride	x	x
Lactic acid 25%	180	82	Toluene	x	x
Lactic acid, concentrated	100	38	Trichloroacetic acid, 20%	140	60
Magnesium chloride	230	110	White liquor	180	82
Malic acid	180	82	Zinc chloride	180	82

[a] The chemicals listed are in the pure state or in a saturated solution unless otherwise indicated. Compatibility is shown to the maximum allowable temperature for which data are available. Incompatibility is shown by an x.
Source: Ref. 2.

TABLE 3.8 Physical and Mechanical Properties of Copolymer and
Homopolymer PP

Property	Homopolymer	Copolymer
Specific gravity	0.905	0.91
Water absorption, 24 h at 73°F/23°C, %	0.02	0.03
Tensile strength at 73°F/23°C, psi	5000	4000
Modulus of elasticity in tension at 73°F/23°C × 10^5 psi	1.7	1.5
Compressive strength, psi	9243	8500
Flexural strength, psi	7000	—
Izod impact strength, notched at 73°F/23°C, ft-lb/in.	1.3	8
Coefficient of thermal expansion in./in. °F × 10^{-5}	5.0	6.1
Thermal conductivity, Btu/h/ft^2/°F/in.	1.2	1.3
Heat distortion temperature, °F/°C		
at 66 psi	220/107	220/107
at 264 psi	140/60	124/49
Limiting oxygen index, %	17	—
Flame spread	Slow burning	
Underwriters lab rating (Sub 94)	94 HB	

to be limited depending upon the corrodent being handled. Table 3.8 provides
the physical and mechanical properties of polypropylene.

Polypropylene is not affected by most inorganic chemicals, except halogens
and severe oxidizing conditions. It can be used with sulfur bearing compounds,
caustics, solvents, acids, and other organic chemicals. Polypropylene has FDA
approval for handling of food products. It should not be used with oxidizing
type acids, detergents, low boiling hydrocarbons, alcohols, aromatics, and some
chlorinated organic materials.

Polypropylene may be subject to environmental stress cracking (see Chapter 1). The occurrence is difficult to predict. It is dependent upon the process
chemistry, operating conditions, and quality of fabrication. In table 3.9, which
gives the compatibility of polypropylene with selected corrodents, certain
chemical/liner ratings are identified by an "e," indicating that the liner in contact
with that specific chemical may be susceptible to ESC. Reference 2 provides a
more comprehensive compatibility listing.

POLYETHYLENE (PE)

Polyethylene is the least expensive of all the plastic materials. Large quantities are used as thin sheet or film liners in drums and other packages. It can be

TABLE 3.9   Compatibility of PP with Selected Corrodents[a]

Chemical	Maximum temp. °F	Maximum temp. °C	Chemical	Maximum temp. °F	Maximum temp. °C
Acetaldehyde	120	49	Aqua regia 3:1	x	x
Acetamide	110	43	Barium carbonate	200	93
Acetic acid 10%	220	104	Barium chloride	220	104
Acetic acid 50%	200	93	Barium hydroxide	200	93
Acetic acid 80%	200	93	Barium sulfate	200	93
Acetic acid, glacial	190	88	Barium sulfide	200	93
Acetic anhydride	100	38	Benzaldehyde	80	27
Acetone	220	104	Benzene	140	60
Acetyl chloride	x	x	Benzene sulfonic acid 10%	180	82
Acrylic acid	x	x	Benzoic acid	190	88
Acrylonitrile	90	32	Benzyl alcohol	140	60
Adipic acid	100	38	Benzyl chloride	80	27
Allyl alcohol	140	60	Borax	210	99
Allyl chloride	140	60	Boric acid	220	104
Alum	220	104	Bromine gas, dry	x	x
Aluminum acetate	100	38	Bromine gas, moist	x	x
Aluminum chloride, aqueous	200	93	Bromine liquid	x	x
Aluminum chloride, dry	220	104	Butadiene	x	x
Aluminum fluoride	200	93	Butyl acetate	x	x
Aluminum hydroxide	200	93	Butyl alcohol	200	93
Aluminum nitrate	200	93	n-Butylamine	90	32
Aluminum oxychloride	220	104	Butyl phthalate	180	82
Aluminum sulfate			Butyric acid	180	82
Ammonia gas	150	66	Calcium bisulfide	210	99
Ammonium bifluoride	200	93	Calcium bisulfite	210	99
Ammonium carbonate	220	104	Calcium carbonate	210	99
Ammonium chloride 10%	180	82	Calcium chlorate	220	104
Ammonium chloride 50%	180	82	Calcium chloride	220	104
Ammonium chloride, sat.	200	93	Calcium hydroxide 10%	200	93
Ammonium fluoride 10%	210	99	Calcium hydroxide, sat.	220	104
Ammonium fluoride 25%	200	93	Calcium hypochlorite	210	99
Ammonium hydroxide 25%	200	93	Calcium nitrate	210	99
Ammonium hydroxide, sat.	200	93	Calcium oxide	220	104
Ammonium nitrate	200	93	Calcium sulfate	220	104
Ammonium persulfate	220	104	Caprylic acid	140	60
Ammonium phosphate	200	93	Carbon bisulfide	x	x
Ammonium sulfate 10–40%	200	93	Carbon dioxide, dry	220	104
Ammonium sulfide	220	104	Carbon dioxide, wet	140	60
Ammonium sulfite	220	104	Carbon disulfide	x	x
Amyl acetate	x	x	Carbon monoxide	220	104
Amyl alcohol	200	93	Carbon tetrachloride	x	x
Amyl chloride	x	x	Carbonic acid	220	104
Aniline	180	82	Cellosolve	200	93
Antimony trichloride	180	82	Chloracetic acid, 50% water	80	27

TABLE 3.9 Continued

Chemical	Maximum temp. °F	Maximum temp. °C	Chemical	Maximum temp. °F	Maximum temp. °C
Chloracetic acid	180	82	Iodine solution 10%	x	x
Chlorine gas, dry	x	x	Ketones, general	110	43
Chlorine gas, wet	x	x	Lactic acid 25%	150	66
Chlorine, liquid	x	x	Lactic acid, concentrated	150	66
Chlorobenzene	x	x	Magnesium chloride	210	99
Chloroform	x	x	Malic acid	130	54
Chlorosulfonic acid	x	x	Manganese chloride	120	49
Chromic acid 10%	140	60	Methyl chloride	x	x
Chromic acid 50%	e150	66	Methyl ethyl ketone	x	x
Chromyl chloride	140	60	Methyl isobutyl ketone	80	27
Citric acid 15%	220	104	Muriatic acid	200	93
Citric acid, conc.	220	104	Nitric acid 5%	140	60
Copper acetate	80	27	Nitric acid 20%	140	60
Copper carbonate	200	93	Nitric acid 70%	x	x
Copper chloride	200	93	Nitric acid, anhydrous	x	x
Copper cyanide	200	93	Nitrous acid, concentrated	x	x
Copper sulfate	200	93	Oleum	x	x
Cresol	x	x	Perchloric acid 10%	140	60
Cupric chloride 5%	140	60	Perchloric acid 70%	x	x
Cupric chloride 50%	140	60	Phenol	180	82
Cyclohexane	x	x	Phosphoric acid 50–80%	210	99
Cyclohexanol	150	66	Picric acid	140	60
Dichloroacetic acid	100	38	Potassium bromide 30%	210	99
Dichloroethane (ethylene di-chloride)	80	27	Salicylic acid	130	54
			Silver bromide 10%	170	77
Ethylene glycol	210	99	Sodium carbonate	220	104
Ferric chloride	e210	99	Sodium chloride	200	93
Ferric chloride 50% in water	210	99	Sodium hydroxide 10%	220	104
			Sodium hydroxide 50%	220	104
Ferric nitrate 10–50%	210	99	Sodium hydroxide, concentrated	140	60
Ferrous chloride	e210	99			
Ferrous nitrate	210	99	Sodium hypochlorite 20%	120	49
Fluorine gas, dry	x	x	Sodium hypochlorite, con-centrated	110	43
Fluorine gas, moist	x	x			
Hydrobromic acid, dilute	230	110	Sodium sulfide to 50%	190	88
Hydrobromic acid 20%	200	93	Stannic chloride	150	66
Hydrobromic acid 50%	190	88	Stannous chloride	200	93
Hydrochloric acid 20%	220	104	Sulfuric acid 10%	200	93
Hydrochloric acid 38%	200	93	Sulfuric acid 50%	200	93
Hydrocyanic acid 10%	150	66	Sulfuric acid 70%	180	82
Hydrofluoric acid 30%	180	82	Sulfuric acid 90%	180	82
Hydrofluoric acid 70%	200	93	Sulfuric acid 98%	120	49
Hydrofluoric acid 100%	200	93	Sulfuric acid 100%	x	x
Hypochlorous acid	140	60	Sulfuric acid, fuming	x	x

TABLE 3.9 Continued

Chemical	Maximum temp. °F	°C	Chemical	Maximum temp. °F	°C
Sulfurous acid	180	82	Trichloroacetic acid	150	66
Thionyl chloride	100	38	White liquor	220	104
Toluene	x	x	Zinc chloride	200	93

[a] The chemicals listed are in the pure state or in a saturated solution unless otherwise indicated. Compatibility is shown to the maximum allowable temperature for which data are available. Incompatibility is shown by an x. A blank space indicates that the data are unavailable. e indicates that the liner in contact with this material may be subject to environmental stress cracking.
Source: Ref. 2.

readily joined by heat sealing and fusion welding, when sheet linings are installed in vessels. The maximum allowable temperature at continuous contact is 120°F/49°C.

Physical and mechanical properties differ by density and molecular weight. The three main classifications of density are low, medium, and high. These specific gravity ranges are 0.91–0.925, 0.925–0.940, and 0.940–0.965. These grades are sometimes referred to as types 1, 11, and 111.

Industry practice breaks the molecular weight of polyethylenes into four distinct classifications:

Medium molecular weight: less than 100,000
High molecular weight: 110,000–250,000
Extra high molecular weight: 250,000–1,500,000
Ultra high molecular weight: 1,500,000 and higher

Usually the ultra high molecular weight material has a molecular weight of at least 3.1 million.

The two varieties of polyethylene used for corrosive applications are EHMW and UHMW. The key properties of these two varieties are:

Good abrasion resistance
Excellent impact resistance
Lightweight
Easily heat fused
High tensile strength
Low moisture absorption
Nontoxic
Nonstaining
Corrosion resistant

TABLE 3.10 Physical and Mechanical Properties of PE UHMW and EHMW

	UHMW	EHMW
Specific gravity	0.94–0.96	0.947–0.955
Water absorption, 24 h at 73°F/23°C, %	<0.01	
Tensile strength at 73°F/23°C, psi	3100–3500	2500–4300
Modulus of elasticity in tension at 73°F/23°C × 10^5	1.18	1.36
Flexural modulus, psi × 10^5	1.33	1.25–1.75
Izod impact strength, notched at 73°F/23°C, ft · lb/in.	0.4–6.0	3.2–4.5
Coefficient of thermal expansion		
in./in.-°F × 10^{-5}	11.1	7.0–11
in./10°F/100 ft	0.111	0.007–0.11
Thermal conductivity, Btu/h/ft²/°F/in.	0.269	0.269
Heat distortion temperature, °F/°C		
at 66 psi	150/66	154/68
at 264 psi	250/121	
Resistance to heat at continuous drainage, °F/°C	180/82	180/82
Flame spread	Slow burning	Slow burning

Their physical and mechanical properties are shown in Table 3.10.

Polyethylene exhibits a wide range of corrosion resistance—ranging from potable water to corrosive wastes. It is resistant to most mineral acids, including sulfuric up to 70% concentration, inorganic salts including chlorides, alkalies, and many organic acids.

It is not resistant to bromine, aromatics, or chlorinated hydrocarbons. Refer to Table 3.11 for the compatibility of EHMW PE with selected corrodents and Table 3.12 for HMW PE's compatibility. Reference 2 provides a more detailed listing.

POLYVINYLIDENE CHLORIDE (SARAN) PVDC

Saran is the trademark of Dow Chemical for their proprietary polyvinylidene chloride resin. It has a maximum allowable operating temperature of 175°F/81°C. It is used to line piping as well as vessels. Refer to Table 3.13 for the physical and mechanical properties of Saran.

PVDC has found wide application in the plating industry and for handling deionized water, pharmaceuticals, food processing, and other applications where stream purity protection is critical. The material complies with FDA regulations for food processing and potable water and also with regulations prescribed by

TABLE 3.11 Compatibility of EHMW PE with Selected Corrodents[a]

Chemical	Maximum temp.		Chemical	Maximum temp.	
	°F	°C		°F	°C
Acetaldehyde 40%	90	32	Aqua regia 3:1	130	54
Acetamide			Barium carbonate	140	60
Acetic acid 10%	140	60	Barium chloride	140	60
Acetic acid 50%	140	60	Barium hydroxide	140	60
Acetic acid 80%	80	27	Barium sulfate	140	60
Acetic acid, glacial			Barium sulfide	140	60
Acetic anhydride	x	x	Benzaldehyde	x	x
Acetone	120	49	Benzene	x	x
Acetyl chloride			Benzene sulfonic acid 10%	140	60
Acrylic acid			Benzoic acid	140	60
Acrylonitrile	150	66	Benzyl alcohol	170	77
Adipic acid	140	60	Benzyl chloride		
Allyl alcohol	140	60	Borax	140	60
Allyl chloride	80	27	Boric acid	140	60
Alum	140	60	Bromine gas, dry	x	x
Aluminum acetate			Bromine gas, moist	x	x
Aluminum chloride, aqueous	140	60	Bromine liquid	x	x
Aluminum chloride, dry	140	60	Butadiene	x	x
Aluminum fluoride	140	60	Butyl acetate	90	32
Aluminum hydroxide	140	60	Butyl alcohol	140	60
Aluminum nitrate			n-Butylamine	x	x
Aluminum oxychloride			Butyl phthalate	80	27
Aluminum sulfate	140	60	Butyric acid	130	54
Ammonia gas	140	60	Calcium bisulfide	140	60
Ammonium bifluoride			Calcium bisulfite	80	27
Ammonium carbonate	140	60	Calcium carbonate	140	60
Ammonium chloride 10%	140	60	Calcium chlorate	140	60
Ammonium chloride 50%	140	60	Calcium chloride	140	60
Ammonium chloride, sat.	140	60	Calcium hydroxide 10%	140	60
Ammonium fluoride 10%	140	60	Calcium hydroxide, sat.	140	60
Ammonium fluoride 25%	140	60	Calcium hypochlorite	140	60
Ammonium hydroxide 25%	140	60	Calcium nitrate	140	60
Ammonium hydroxide, sat.	140	60	Calcium oxide	140	60
Ammonium nitrate	140	60	Calcium sulfate	140	60
Ammonium persulfate	140	60	Caprylic acid		
Ammonium phosphate	80	27	Carbon bisulfide	x	x
Ammonium sulfate 10–40%	140	60	Carbon dioxide, dry	140	60
Ammonium sulfide	140	60	Carbon dioxide, wet	140	60
Ammonium sulfite			Carbon disulfide	x	x
Amyl acetate	140	60	Carbon monoxide	140	60
Amyl alcohol	140	60	Carbon tetrachloride	x	x
Amyl chloride	x	x	Carbonic acid	140	60
Aniline	130	54	Cellosolve		
Antimony trichloride	140	60	Chloracetic acid, 50% in water	x	x

TABLE 3.11   Continued

Chemical	Maximum temp. °F	°C	Chemical	Maximum temp. °F	°C
Chloracetic acid	x	x	Iodine solution 10%	80	27
Chlorine gas, dry	80	27	Ketones, general	x	x
Chlorine gas, wet, 10%	120	49	Lactic acid 25%	140	60
Chlorine, liquid	x	x	Lactic acid, concentrated	140	60
Chlorobenzene	x	x	Magnesium chloride	140	60
Chloroform	80	27	Malic acid	100	38
Chlorosulfonic acid	x	x	Manganese chloride	80	27
Chromic acid 10%	140	60	Methyl chloride	x	x
Chromic acid 50%	90	32	Methyl ethyl ketone	x	x
Chromyl chloride			Methyl isobutyl ketone	80	27
Citric acid 15%	140	60	Muriatic acid	140	60
Citric acid, conc.	140	60	Nitric acid 5%	140	60
Copper acetate			Nitric acid 20%	140	60
Copper carbonate			Nitric acid 70%	x	x
Copper chloride	140	60	Nitric acid, anhydrous	x	x
Copper cyanide	140	60	Nitrous acid, concentrated		
Copper sulfate	140	60	Oleum		
Cresol	80	27	Perchloric acid 10%	140	60
Cupric chloride 5%	80	27	Perchloric acid 70%	x	x
Cupric chloride 50%			Phenol	100	38
Cyclohexane	130	54	Phosphoric acid 50–80%	100	38
Cyclohexanol	170	77	Picric acid	100	38
Dichloroacetic acid	73	23	Potassium bromide 30%	140	60
Dichloroethane (ethylene di-chloride)	x	x	Salicylic acid		
			Silver bromide 10%		
Ethylene glycol	140	60	Sodium carbonate	140	60
Ferric chloride	140	60	Sodium chloride	140	60
Ferric chloride 50% in water	140	60	Sodium hydroxide 10%	170	77
Ferric nitrate 10–50%	140	60	Sodium hydroxide 50%	170	77
Ferrous chloride	140	60	Sodium hydroxide, concen-trated		
Ferrous nitrate	140	60			
Fluorine gas, dry	x	x	Sodium hypochlorite 20%	140	60
Fluorine gas, moist	x	x	Sodium hypochlorite, concen-trated	140	60
Hydrobromic acid, dilute	140	60			
Hydrobromic acid 20%	140	60	Hydrobromic acid, dilute	140	60
Hydrobromic acid 50%	140	60	Sodium sulfide to 50%	140	60
Hydrochloric acid 20%	140	60	Stannic chloride	140	60
Hydrochloric acid 38%	140	60	Stannous chloride	140	60
Hydrocyanic acid 10%	140	60	Sulfuric acid 10%	140	60
Hydrofluoric acid 30%	80	27	Sulfuric acid 50%	140	60
Hydrofluoric acid 70%	x	x	Sulfuric acid 70%	80	27
Hydrofluoric acid 100%	x	x	Sulfuric acid 90%	x	x
Hypochlorous acid			Sulfuric acid 98%	x	x

TABLE 3.11 Continued

Chemical	Maximum temp.		Chemical	Maximum temp.	
	°F	°C		°F	°C
Sulfuric acid 100%	x	x	Toluene	x	x
Sulfuric acid, fuming	x	x	Trichloroacetic acid	140	60
Sulfurous acid	140	60	White liquor		
Thionyl chloride	x	x	Zinc chloride	140	60

[a] The chemicals listed are in the pure state or in a saturated solution unless otherwise indicated. Compatibility is shown to the maximum allowable temperature for which data are available. Incompatibility is shown by an x. A blank space indicates that the data are unavailable. *Source*: Ref. 2.

the Meat Inspection Division of the Department of Agriculture for transporting fluids used in meat production. In applications such as plating solutions, chlorines, and certain other chemicals, Saran is superior to polypropylene and finds many applications in the handling of municipal water supplies and waste waters. Refer to Table 3.14 for the compatibility of PVDC with selected corrodents. Reference 2 provides a more comprehensive listing.

POLYVINYLIDENE FLUORIDE (PVDF)

Polyvinylidene fluoride is one of the most popular lining materials for both vessels and piping because of its range of corrosion resistance and high allowable operating temperature (275°F/135°C). Because of this factor it is much more difficult to weld the PVDF than lower melting materials such as PVC. PVDF cannot be bonded directly to a metal substrate. A dual laminate must be used or the lining will be loose. The material is used to line vessels as well as piping. Piping liners are loose. Refer to Ref. (1).

PVDF is manufactured under various trade names:

Trade Name	Manufacturer
Kynar	Elf Atochem
Solef	Solvay
Hylar	Ausimont USA

Other manufacturers produce PVDF as a solid piping system.

TABLE 3.12 Compatibility of HMWPE with Selected Corrodents[a]

Chemical	Maximum temp.		Chemical	Maximum temp.	
	°F	°C		°F	°C
Acetaldehyde	x	x	Barium sulfate	140	60
Acetamide	140	60	Barium sulfide	140	60
Acetic acid 10%	140	60	Benzaldehyde	x	x
Acetic acid 50%	140	60	Benzene	x	x
Acetic acid 80%	80	27	Benzoic acid	140	60
Acetic anhydride	x	x	Benzyl alcohol	x	x
Acetone	80	27	Borax	140	60
Acetyl chloride	x	x	Boric acid	140	60
Acrylonitrile	150	66	Bromine gas, dry	x	x
Adipic acid	140	60	Bromine gas, moist	x	x
Allyl alcohol	140	60	Bromine, liquid	x	x
Allyl chloride	110	43	Butadiene	x	x
Alum	140	60	Butyl acetate	90	32
Aluminum chloride, aqueous	140	60	Butyl alcohol	140	60
Aluminum chloride, dry	140	60	n-Butylamine	x	x
Aluminum fluoride	140	60	Butyric acid	x	x
Aluminum hydroxide	140	60	Calcium bisulfide	140	60
Aluminum nitrate	140	60	Calcium bisulfite	140	60
Aluminum sulfate	140	60	Calcium carbonate	140	60
Ammonium gas	140	60	Calcium chlorate	140	60
Ammonium bifluoride	140	60	Calcium chloride	140	60
Ammonium carbonate	140	60	Calcium hydroxide 10%	140	60
Ammonium chloride 10%	140	60	Calcium hydroxide, sat.	140	60
Ammonium chloride 50%	140	60	Calcium hypochlorite	140	60
Ammonium chloride, sat.	140	60	Calcium nitrate	140	60
Ammonium fluoride 10%	140	60	Calcium oxide	140	60
Ammonium fluoride 25%	140	60	Calcium sulfate	140	60
Ammonium hydroxide 25%	140	60	Carbon bisulfide	x	x
Ammonium hydroxide, sat.	140	60	Carbon dioxide, dry	140	60
Ammonium nitrate	140	60	Carbon dioxide, wet	140	60
Ammonium persulfate	150	66	Carbon disulfide	x	x
Ammonium phosphate	80	27	Carbon monoxide	140	60
Ammonium sulfate to 40%	140	60	Carbon tetrachloride	x	x
Ammonium sulfide	140	60	Carbonic acid	140	60
Ammonium sulfite	140	60	Cellosolve	x	x
Amyl acetate	140	60	Chloroacetic acid	x	x
Amyl alcohol	140	60	Chlorine gas, dry	x	x
Amyl chloride	x	x	Chlorine gas, wet	x	x
Aniline	130	44	Chlorine, liquid	x	x
Antimony trichloride	140	60	Chlorobenzene	x	x
Aqua regia 3:1	130	44	Chloroform	x	x
Barium carbonate	140	60	Chlorosulfonic acid	x	x
Barium chloride	140	60	Chromic acid 10%	140	60
Barium hydroxide	140	60	Chromic acid 50%	90	32

TABLE 3.12   Continued

Chemical	Maximum temp. °F	°C	Chemical	Maximum temp. °F	°C
Citric acid 15%	140	60	Methyl isobutyl ketone	80	27
Citric acid, concd	140	60	Nitric acid 5%	140	60
Copper chloride	140	60	Nitric acid 20%	140	60
Copper cyanide	140	60	Nitric acid 70%	x	x
Copper sulfate	140	60	Nitric acid, anhydrous	x	x
Cresol	x	x	Nitrous acid, concd	120	49
Cupric chloride 5%	140	60	Perchloric acid 10%	140	60
Cupric chloride 50%	140	60	Perchloric acid 70%	x	x
Cyclohexane	80	27	Phenol	100	38
Cyclohexanol	80	27	Phosphoric acid 50–80%	100	38
Dibutyl phthalate	80	27	Picric acid	100	38
Dichloroethane	80	27	Potassium bromide 30%	140	60
Ethylene glycol	140	60	Salicylic acid	140	60
Ferric chloride	140	60	Sodium carbonate	140	60
Ferrous chloride	140	60	Sodium chloride	140	60
Ferrous nitrate	140	60	Sodium hydroxide 10%	150	66
Fluorine gas, dry	x	x	Sodium hydroxide 50%	150	66
Fluorine gas, moist	x	x	Sodium hypochlorite 20%	140	60
Hydrobromic acid, dil	140	60	Sodium hypochlorite, concd	140	60
Hydrobromic acid 20%	140	60	Sodium sulfide to 50%	140	60
Hydrobromic acid 50%	140	60	Stannic chloride	140	60
Hydrochloric acid 20%	140	60	Stannous chloride	140	60
Hydrochloric acid 38%	140	60	Sulfuric acid 10%	140	60
Hydrocyanic acid 10%	140	60	Sulfuric acid 50%	140	60
Hydrofluoric acid 30%	140	60	Sulfuric acid 70%	80	27
Hydrofluoric acid 70%	x	x	Sulfuric acid 90%	x	x
Hydrochlorous acid	150	66	Sulfuric acid 98%	x	x
Iodine solution 10%	80	27	Sulfuric acid 100%	x	x
Ketones, general	80	27	Sulfuric acid, fuming	x	x
Lactic acid 25%	150	66	Sufurous acid	140	60
Magnesium chloride	140	60	Thionyl chloride	x	x
Malic acid	140	60	Toluene	x	x
Manganese chloride	80	27	Trichloroacetic acid	80	27
Methyl chloride	x	x	Zinc chloride	140	60
Methyl ethyl ketone	x	x			

[a] The chemicals listed are in the pure state or in a saturated solution unless otherwise indicated. Compatibility is shown to the maximum allowable temperature for which data are available. Incompatibility is shown by an x.
Source: Ref. 2.

TABLE **3.13**   Physical and Mechanical Properties of
Polyvinylidene Chloride

Specific gravity	1.75–1.85
Water absorption, 24 h at 73°F/23°C, %	nil
Tensile strength at 73°F/23°C, psi	2700–3700
Coefficient of thermal expansion	
in./in.- °F/°C $\times$ 10^{-5}	3.9–5
in./10°F/100 ft	0.039–0.05
Thermal conductivity, Btu/h/ft^2/°F/in.	1.28
Flame spread	Self-extinguishing

Refer to Table 3.15 for the physical and mechanical properties of PVDF. It is a fluoropolymer resin (one that is not fully fluorinated). As such it is resistant to permeation by gases.

PVDF is chemically resistant to most acids, bases, and organic solvents. It is also resistant to wet or dry chlorine, bromine, and other halogens. Approval has been granted by the Food and Drug Administration for repeated use in contact with food, as in food handling and processing equipment.

It should not be used with strong alkalies, fuming acids, polar solvents, amines, ketones, and esters. When used with strong alkalies, it stress cracks. Refer to Table 3.16 for the compatibility of PVDF with selected corrodents. Reference 2 provides a more comprehensive listing.

POLYTETRAFLUOROETHYLENE (PTFE)

Extruded tubing is used to line pipe, and sheet material to line vessels. This is one of the most expensive linings, but also the one having the greatest range of corrosion resistance. PTFE cannot be bonded directly to a metal substrate. Unless a fiberglass backing sheet is impressed, the lining will be loose. Lining thicknesses vary from 60 to 120 mils. The material tends to creep under stress at elevated temperatures. When the vessel is designed, this should be taken into account and provisions made to contain the material. PTFE is also a relatively weak material. Since PTFE is relatively difficult to work with, perfluoroalkoxy (PFA), also a fully fluorinated material, is often used in conjunction with PTFE. PFA is readily fabricated.

PTFE has a maximum operating temperature of 500°F/260°C, and can be fusion welded. It is subject to permeation by some corrodents. Refer to Table 1.2 for the permeation of vapors in PTFE. Refer to Table 3.17 for the physical and mechanical properties of PTFE.

PTFE is chemically inert in the presence of most corrodents. There are very few chemicals that will attack it within normal use temperatures. These

TABLE 3.14 Compatibility of Polyvinylidene Chloride (Saran) with Selected Corrodents[a]

Chemical	Maximum temp. °F	Maximum temp. °C	Chemical	Maximum temp. °F	Maximum temp. °C
Acetaldehyde	150	66	Benzene	x	x
Acetic acid 10%	150	66	Benzene sulfonic acid 10%	120	49
Acetic acid 50%	130	54	Benzoic acid	120	49
Acetic acid 80%	130	54	Benzyl chloride	80	27
Acetic acid, glacial	140	60	Boric acid	170	77
Acetic anhydride	90	32	Bromine liquid	x	x
Acetone	90	32	Butadiene	x	x
Acetyl chloride	130	54	Butyl acetate	120	49
Acrylonitrile	90	32	Butyl alcohol	150	66
Adipic acid	150	66	Butyl phthalate	180	82
Allyl alcohol	80	27	Butyric acid	80	27
Alum	180	82	Calcium bisulfite	80	27
Aluminum chloride, aqueous	150	66	Calcium carbonate	180	82
Aluminum fluoride	150	66	Calcium chlorate	160	71
Aluminum hydroxide	170	77	Calcium chloride	180	82
Aluminum nitrate	180	82	Calcium hydroxide 10%	160	71
Aluminum oxychloride	140	60	Calcium hydroxide, sat.	180	82
Aluminum sulfate	180	82	Calcium hypochlorite	120	49
Ammonia gas	x	x	Calcium nitrate	150	66
Ammonium bifluoride	140	60	Calcium oxide	180	82
Ammonium carbonate	180	82	Calcium sulfate	180	82
Ammonium chloride, sat.	160	71	Caprylic acid	90	32
Ammonium fluoride 10%	90	32	Carbon bisulfide	90	32
Ammonium fluoride 25%	90	32	Carbon dioxide, dry	180	82
Ammonium hydroxide 25%	x	x	Carbon dioxide, wet	80	27
Ammonium hydroxide, sat.	x	x	Carbon disulfide	80	27
Ammonium nitrate	120	49	Carbon monoxide	180	82
Ammonium persulfate	90	32	Carbon tetrachloride	140	60
Ammonium phosphate	150	66	Carbonic acid	180	82
Ammonium sulfate 10–40%	120	49	Cellosolve	80	27
Ammonium sulfide	80	27	Chloracetic acid, 50% water	120	49
Amyl acetate	120	49	Chloracetic acid	120	49
Amyl alcohol	150	66	Chlorine gas, dry	80	27
Amyl chloride	80	27	Chlorine gas, wet	80	27
Aniline	x	x	Chlorine, liquid	x	x
Antimony trichloride	150	66	Chlorobenzene	80	27
Aqua regia 3:1	120	49	Chloroform	x	x
Barium carbonate	180	82	Chlorosulfonic acid	x	x
Barium chloride	180	82	Chromic acid 10%	180	82
Barium hydroxide	180	82	Chromic acid 50%	180	82
Barium sulfate	180	82	Citric acid 15%	180	82
Barium sulfide	150	66	Citric acid, concentrated	180	82
Benzaldehyde	x	x	Copper carbonate	180	82

TABLE 3.14 Continued

Chemical	°F	°C	Chemical	°F	°C
Copper chloride	180	82	Nitric acid 5%	90	32
Copper cyanide	130	54	Nitric acid 20%	150	66
Copper sulfate	180	82	Nitric acid 70%	x	x
Cresol	150	66	Nitric acid, anhydrous	x	x
Cupric chloride 5%	160	71	Oleum	x	x
Cupric chloride 50%	170	77	Perchloric acid 10%	130	54
Cyclohexane	120	49	Perchloric acid 70%	120	49
Cyclohexanol	90	32	Phenol	x	x
Dichloroacetic acid	120	49	Phosphoric acid 50–80%	130	54
Dichloroethane (ethylene di-chloride)	80	27	Picric acid	120	49
			Potassium bromide 30%	110	43
Ethylene glycol	180	82	Salicylic acid	130	54
Ferric chloride	140	60	Sodium carbonate	180	82
Ferric chloride 50% in water	140	60	Sodium chloride	180	82
Ferric nitrate 10–50%	130	54	Sodium hydroxide 0%	90	32
Ferrous chloride	130	54	Sodium hydroxide 50%	150	66
Ferrous nitrate	80	27	Sodium hydroxide, concen-trated	x	x
Fluorine gas, dry	x	x			
Fluorine gas, moist	x	x	Sodium hypochlorite 10%	130	54
Hydrobromic acid, dilute	120	49	Sodium hypochlorite, con-centrated	120	49
Hydrobromic acid 20%	120	49			
Hydrobromic acid 50%	130	54	Sodium sulfide to 50%	140	60
Hydrochloric acid 20%	180	82	Stannic chloride	180	82
Hydrochloric acid 38%	180	82	Stannous chloride	180	82
Hydrocyanic acid 10%	120	49	Sulfuric acid 10%	120	49
Hydrofluoric acid 30%	160	71	Sulfuric acid 50%	x	x
Hydrofluoric acid 100%	x	x	Sulfuric acid 70%	x	x
Hypochlorous acid	120	49	Sulfuric acid 90%	x	x
Ketones, general	90	32	Sulfuric acid 98%	x	x
Lactic acid, concentrated	80	27	Sulfuric acid 100%	x	x
Magnesium chloride	180	82	Sulfuric acid, fuming	x	x
Malic acid	80	27	Sulfurous acid	80	27
Methyl chloride	80	27	Thionyl chloride	x	x
Methyl ethyl ketone	x	x	Toluene	80	27
Methyl isobutyl ketone	80	27	Trichloroacetic acid	80	27
Muriatic acid	180	82	Zinc chloride	170	77

[a] The chemicals listed are in the pure state or in a saturated solution unless otherwise indicated. Compatibility is shown to the maximum allowable temperature for which data are available. Incompatibility is shown by an x. A blank space indicates that the data are unavailable. *Source*: Ref. 2.

TABLE 3.15 Physical and Mechanical Properties of PVDF

Specific gravity	1.76
Water absorption, 24 h at 73°/23°C, %	<0.04
Tensile strength at 73°F/23°C, psi	6000
Modulus of elasticity in tension at 73°F/23°C × 10^5	2.1
Compressive strength, psi	11600
Flexural strength, psi	10750
Izod impact strength, notched at 73°F/23°C, ft · lb/in.	3.8
Coefficient of thermal expansion	
in./in.-°F × 10^{-5}	7.9
in./10°F/100 ft	0.079
Thermal conductivity, Btu/h/ft²/°F/in.	0.79
Heat distortion temperature, °F/°C	
at 66 psi	284/140
at 264 psi	194/90
Resistance to heat at continuous drainage, °F/°C	280/138
Limiting oxygen index, %	44
Flame spread	0
Underwriters lab rating (Sub 94)	94V-O

reactants are among the most violent oxidizers and reducing agents known. Elemental sodium in intimate contact with fluorocarbons removes fluorine from the polymer molecule. The other alkali metals (potassium, lithium, etc.) react in a similar manner.

Fluorine and related compounds (e.g., chlorine trifluoride) are absorbed into the PTFE resin with such intimate contact that the mixture becomes sensitive to a source of ignition such as impact.

The handling of 80% sodium hydroxide, aluminum chloride, ammonia, and certain amines at high temperatures may produce the same effect as elemental sodium. Also, slow oxidative attack can be produced by 70% nitric acid under pressure at 480°F/250°C.

Refer to Table 3.18 for the compatibility of PTFE with selected corrodents. Refer to Ref. 2 for a more comprehensive listing. Reference 1 provides details of PTFE lined pipe.

PERFLUOROALKOXY (PFA)

PFA can be used up to 500°F/260°C. It lacks the physical strength of PTFE at elevated temperatures but has somewhat better physical and mechanical properties than FEP above 300°F/149°C. While PFA matches the hardness and impact strength of PTFE, it sustains only one quarter of the life of PTFE in flexibility tests. The physical and mechanical properties are shown in Table 3.19. PFA also

TABLE 3.16 Compatibility of PVDF with Selected Corrodents[a]

Chemical	Maximum temp. °F	Maximum temp. °C	Chemical	Maximum temp. °F	Maximum temp. °C
Acetaldehyde	150	66	Barium sulfate	280	138
Acetamide	90	32	Barium sulfide	280	138
Acetic acid 10%	300	149	Benzaldehyde	120	49
Acetic acid 50%	300	149	Benzene	150	66
Acetic acid 80%	190	88	Benzene sulfonic acid 10%	100	38
Acetic acid, glacial	190	88	Benzoic acid	250	121
Acetic anhydride	100	38	Benzyl alcohol	280	138
Acetone	x	x	Benzyl chloride	280	138
Acetyl chloride	120	49	Borax	280	138
Acrylic acid	150	66	Boric acid	280	138
Acrylonitrile	130	54	Bromine gas, dry	210	99
Adipic acid	280	138	Bromine gas, moist	210	99
Allyl alcohol	200	93	Bromine liquid	140	60
Allyl chloride	200	93	Butadiene	280	138
Alum	180	82	Butyl acetate	140	60
Aluminum acetate	250	121	Butyl alcohol	280	138
Aluminum chloride, aqueous	300	149	n-Butylamine	x	x
Aluminum chloride, dry	270	132	Butyl phthalate	80	27
Aluminum fluoride	300	149	Butyric acid	230	110
Aluminum hydroxide	260	127	Calcium bisulfide	280	138
Aluminum nitrate	300	149	Calcium bisulfite	280	138
Aluminum oxychloride	290	143	Calcium carbonate	280	138
Aluminum sulfate	300	149	Calcium chlorate	280	138
Ammonia gas	270	132	Calcium chloride	280	138
Ammonium bifluoride	250	121	Calcium hydroxide 10%	270	132
Ammonium carbonate	280	138	Calcium hydroxide, sat.	280	138
Ammonium chloride 10%	280	138	Calcium hypochlorite	280	138
Ammonium chloride 50%	280	138	Calcium nitrate	280	138
Ammonium chloride, sat.	280	138	Calcium oxide	250	121
Ammonium fluoride 10%	280	138	Calcium sulfate	280	138
Ammonium fluoride 25%	280	138	Caprylic acid	220	104
Ammonium hydroxide 25%	280	138	Carbon bisulfide	80	27
Ammonium hydroxide, sat.	280	138	Carbon dioxide, dry	280	138
Ammonium nitrate	280	138	Carbon dioxide, wet	280	138
Ammonium persulfate	280	138	Carbon disulfide	80	27
Ammonium phosphate	280	138	Carbon monoxide	280	138
Ammonium sulfate 10–40%	280	138	Carbon tetrachloride	280	138
Ammonium sulfide	280	138	Carbonic acid	280	138
Ammonium sulfite	280	138	Cellosolve	280	138
Amyl acetate	190	88	Chloracetic acid, 50% water	210	99
Amyl alcohol	280	138	Chloracetic acid	200	93
Amyl chloride	280	138	Chlorine gas, dry	210	99
Aniline	200	93	Chlorine gas, wet, 10%	210	99
Antimony trichloride	150	66	Chlorine, liquid	210	99
Aqua regia 3:1	130	54	Chlorobenzene	220	104
Barium carbonate	280	138	Chloroform	250	121
Barium chloride	280	138	Chlorosulfonic acid	110	43
Barium hydroxide	280	138	Chromic acid 10%	220	104

TABLE 3.16 Continued

Chemical	Maximum temp. °F	Maximum temp. °C	Chemical	Maximum temp. °F	Maximum temp. °C
Chromic acid 50%	250	121	Methyl ethyl ketone	x	x
Chromyl chloride	110	43	Methyl isobutyl ketone	110	43
Citric acid 15%	250	121	Muriatic acid	280	138
Citric acid, concentrated	250	121	Nitric acid 5%	200	93
Copper acetate	250	121	Nitric acid 20%	180	82
Copper carbonate	250	121	Nitric acid 70%	120	49
Copper chloride	280	138	Nitric acid, anhydrous	150	66
Copper cyanide	280	138	Nitrous acid, concentrated	210	99
Copper sulfate	280	138	Oleum	x	x
Cresol	210	99	Perchloric acid 10%	210	99
Cupric chloride 5%	270	132	Perchloric acid 70%	120	49
Cupric chloride 50%	270	132	Phenol	200	93
Cyclohexane	250	121	Phosphoric acid 50–80%	220	104
Cyclohexanol	210	99	Picric acid	80	27
Dichloroacetic acid	120	49	Potassium bromide 30%	280	138
Dichlorethane (ethylene di-chloride)	280	138	Salicylic acid	220	104
			Silver bromide 10%	250	121
Ethylene glycol	280	138	Sodium carbonate	280	138
Ferric chloride	280	138	Sodium chloride	280	138
Ferric chloride 50% in water	280	138	Sodium hydroxide 10%	230	110
Ferrous nitrate 10–50%	280	138	Sodium hydroxide 50%	220	104
Ferrous chloride	280	138	Sodium hydroxide, concen-trated[b]	150	66
Ferrous nitrate	280	138			
Fluorine gas, dry	80	27	Sodium hypochlorite 20%	280	138
Fluorine gas, moist	80	27	Sodium hypochlorite, concen-trated	280	138
Hydrobromic acid, dilute	260	127			
Hydrobromic acid 20%	280	138	Sodium sulfide to 50%	280	138
Hydrobromic acid 50%	280	138	Stannic chloride	280	138
Hydrochloric acid 20%	280	138	Stannous chloride	280	138
Hydrochloric acid 38%	280	138	Sulfuric acid 10%	250	121
Hydrocyanic acid 10%	280	138	Sulfuric acid 50%	220	104
Hydrofluoric acid 30%	260	127	Sulfuric acid 70%	220	104
Hydrofluoric acid 70%	200	93	Sulfuric acid 90%	210	99
Hydrofluoric acid 100%	200	93	Sulfuric acid 98%	140	60
Hypochlorous acid	280	138	Sulfuric acid 100%	x	x
Iodine solution 10%	250	121	Sulfuric acid, fuming	x	x
Ketones, general	110	43	Sulfurous acid	220	104
Lactic acid 25%	130	54	Thionyl chloride	x	x
Lactic acid, concentrated	110	43	Toluene	x	x
Magnesium chloride	280	138	Trichloroacetic acid	130	54
Malic acid	250	121	White liquor	80	27
Manganese chloride	280	138	Zinc chloride	260	127
Methyl chloride	x	x			

[a] The chemicals listed are in the pure state or in a saturated solution unless otherwise indicated. Compatibility is shown to the maximum allowable temperature for which data are available. Incompatibility is shown by an x. A blank space indicates that the data are unavailable.
[b] Subject to stress corrosion cracking.
Source: Ref. 2.

TABLE 3.17 Physical and Mechanical Properties of PTFE

Specific gravity	2.13–2.2
Water absorption, 24 h at 73°F/23°C, %	0.01
Tensile strength at 73°F/23°C, psi	2000–6500
Compressive strength, psi	1700
Flexural strength, psi	No break
Flexural modulus, psi $\times$ 10^{-5}	0.7–1.1
Izod impact strength, notched at 73°F/23°C, ft-lb/in.	3
Coefficient of thermal expansion, in./in. °F $\times$ 10^{-5}	5.5
Heat distortion temperature at 66 psi, °F/°C	250/121
Low-temperature embrittlement, °F/°C	−450/−268

performs well under cryogenic conditions. Refer to Table 3.20 for a comparison of mechanical properties of PFA at room temperature and cryogenic temperature.

Like PTFE, PFA is subject to permeation by certain gases and will absorb selected chemicals. Table 1.4 shows the permeation of certain gases in PFA while Table 3.21 shows the absorption of representative liquids in PFA.

PFA has the advantage over PTFE in that it can be thermoformed, heat sealed, welded, and heat laminated. This makes the material much easier to produce a lining from than that of PTFE. PFA can be bonded to a metal substrate. It is also used to line pipe.

Perfluoroalkoxy is inert to strong mineral acids, organic bases, inorganic oxidizers, aromatics, some aliphatic hydrocarbons, alcohols, aldehydes, ketones, ethers, esters, chlorocarbons, fluorocarbons, and mixtures of these.

PFA will be attacked by certain halogenated complexes containing fluorine. This includes chlorine trifluoride, bromine trifluoride, iodine pentafluoride, and fluorine. It can also be attacked by such metals as sodium or potassium, particularly in their molten state. Refer to Table 3.22 for the compatibility of PFA with selected corrodents. Reference 2 provides a more comprehensive listing.

FLUORINATED ETHYLENE PROPYLENE (FEP)

FEP is another fluorinated thermoplast. It is considerably easier to fabricate and is less expensive than PTFE while having basically the same corrosion resistant properties.

After prolonged exposure at 400°F/204°C, FEP exhibits changes in physical strength and is therefore limited to a maximum operating temperature of 375°F/190°C. Table 3.23 lists the physical and mechanical properties of FEP. It is a relatively soft plastic with lower tensile strength, wear resistance, and creep resistance than other plastics.

TABLE 3.18 Compatibility of PTFE with Selected Corrodents[a]

Chemical	°F	°C	Chemical	°F	°C
Acetaldehyde	450	232	Barium sulfide	450	232
Acetamide	450	232	Benzaldehyde	450	232
Acetic acid 10%	450	232	Benzene[b]	450	232
Acetic acid 50%	450	232	Benzene sulfonic acid 10%	450	232
Acetic acid 80%	450	232	Benzoic acid	450	232
Acetic acid, glacial	450	232	Benzyl alcohol	450	232
Acetic anhydride	450	232	Benzyl chloride	450	232
Acetone	450	232	Borax	450	232
Acetyl chloride	450	232	Boric acid	450	232
Acrylonitrile	450	232	Bromine gas, dry[b]	450	232
Adipic acid	450	232	Bromine liquid[b]	450	232
Allyl alcohol	450	232	Butadiene[b]	450	232
Allyl chloride	450	232	Butyl acetate	450	232
Alum	450	232	Butyl alcohol	450	232
Aluminum chloride, aqueous	450	232	n-Butylamine	450	232
Aluminum fluoride	450	232	Butyl phthalate	450	232
Aluminum hydroxide	450	232	Butyric acid	450	232
Aluminum nitrate	450	232	Calcium bisulfide	450	232
Aluminum oxychloride	450	232	Calcium bisulfite	450	232
Aluminum sulfate	450	232	Calcium carbonate	450	232
Ammonia gas[b]	450	232	Calcium chlorate	450	232
Ammonium bifluoride	450	232	Calcium chloride	450	232
Ammonium carbonate	450	232	Calcium hydroxide 10%	450	232
Ammonium chloride 10%	450	232	Calcium hydroxide, sat.	450	232
Ammonium chloride 50%	450	232	Calcium hypochlorite	450	232
Ammonium chloride, sat.	450	232	Calcium nitrate	450	232
Ammonium fluoride 10%	450	232	Calcium oxide	450	232
Ammonium fluoride 25%	450	232	Calcium sulfate	450	232
Ammonium hydroxide 25%	450	232	Caprylic acid	450	232
Ammonium hydroxide, sat.	450	232	Carbon bisulfide[b]	450	232
Ammonium nitrate	450	232	Carbon dioxide, dry	450	232
Ammonium persulfate	450	232	Carbon dioxide, wet	450	232
Ammonium phosphate	450	232	Carbon disulfide	450	232
Ammonium sulfate 10–40%	450	232	Carbon monoxide	450	232
Ammonium sulfide	450	232	Carbon tetrachloride[c]	450	232
Amyl acetate	450	232	Carbonic acid	450	232
Amyl alcohol	450	232	Chloracetic acid, 50% water	450	232
Amyl chloride	450	232	Chloracetic acid	450	232
Aniline	450	232	Chlorine gas, dry	x	x
Antimony trichloride	450	232	Chlorine gas, wet[b]	450	232
Aqua regia 3:1	450	232	Chlorine, liquid	x	x
Barium carbonate	450	232	Chlorobenzene[b]	450	232
Barium chloride	450	232	Chloroform[b]	450	232
Barium hydroxide	450	232	Chlorosulfonic acid	450	232
Barium sulfate	450	232	Chromic acid 10%	450	232

TABLE 3.18 Continued

Chemical	°F	°C	Chemical	°F	°C
Chromic acid 50%	450	232	Methyl ethyl ketone[b]	450	232
Chromyl chloride	450	232	Methyl isobutyl ketone[c]	450	232
Citric acid 15%	450	232	Muriatic acid[b]	450	232
Citric acid, concentrated	450	232	Nitric acid 5%[b]	450	232
Copper carbonate	450	232	Nitric acid 20%[b]	450	232
Copper chloride	450	232	Nitric acid 70%[b]	450	232
Copper cyanide 10%	450	232	Nitric acid, anhydrous[b]	450	232
Copper sulfate	450	232	Nitrous acid 10%	450	232
Cresol	450	232	Oleum	450	232
Cupric chloride 5%	450	232	Perchloric acid 10%	450	232
Cupric chloride 50%	450	232	Perchloric acid 70%	450	232
Cyclohexane	450	232	Phenol[b]	450	232
Cyclohexanol	450	232	Phosphoric acid 50–80%	450	232
Dichloroacetic acid	450	232	Picric acid	450	232
Dichloroethane (ethylene di-chloride)[b]	450	232	Potassium bromide 30%	450	232
			Salicylic acid	450	232
Ethylene glycol	450	232	Sodium carbonate	450	232
Ferric chloride	450	232	Sodium chloride	450	232
Ferric chloride 50% in water	450	232	Sodium hydroxide 10%	450	232
Ferric nitrate 10–50%	450	232	Sodium hydroxide 50%	450	232
Ferrous chloride	450	232	Sodium hydroxide, concen-trated	450	232
Ferrous nitrate	450	232			
Fluorine gas, dry	x	x	Sodium hypochlorite 20%	450	232
Fluorine gas, moist	x	x	Sodium hypochlorite, concen-trated	450	232
Hydrobromic acid, dilute[bc]	450	232			
Hydrobromic acid 20%[c]	450	232	Sodium sulfide to 50%	450	232
Hydrobromic acid 50%[c]	450	232	Stannic chloride	450	232
Hydrochloric acid 20%[c]	450	232	Stannous chloride	450	232
Hydrochloric acid 38%[c]	450	232	Sulfuric acid 10%	450	232
Hydrocyanic acid 10%	450	232	Sulfuric acid 50%	450	232
Hydrofluoric acid 30%[b]	450	232	Sulfuric acid 70%	450	232
Hydrofluoric acid 70%[b]	450	232	Sulfuric acid 90%	450	232
Hydrofluoric acid 100%[b]	450	232	Sulfuric acid 98%	450	232
Hypochlorous acid	450	232	Sulfuric acid 100%	450	232
Iodine solution 10%[b]	450	232	Sulfuric acid, fuming[b]	450	232
Ketones, general	450	232	Sulfurous acid	450	232
Lactic acid 25%	450	232	Thionyl chloride	450	232
Lactic acid, concentrated	450	232	Toluene[b]	450	232
Magnesium chloride	450	232	Trichloroacetic acid	450	232
Malic acid	450	232	White liquor	450	232
Methyl chloride[b]	450	232	Zinc chloride[d]	450	232

[a] The chemicals listed are in the pure state or in a saturated solution unless otherwise indicated. Compatibility is shown to the maximum allowable temperature for which data are available. Incompatibility is shown by an x. A blank space indicates that the data are unavailable.
[b] Material will permeate.
[c] Material will cause stress cracking.
[d] Material will be absorbed.
Source: Ref. 2.

TABLE 3.19 Physical and Mechanical Properties of PFA

Specific gravity	2.12–2.17
Water absorption, 24 h at 73°F/23°C, %	<0.03
Tensile strength	
at 73°F/23°C psi	4000
at 482°F/250°C psi	2000
Modulus of elasticity in tension, psi	
at 73°F/23°C	40,000
at 482°F/250°C	6000
Compressive strength, psi	
at 73°F/23°C psi	3500
at 320°F/196°C	60,000
Flexural modulus, psi	
at 73°F/23°C	90,000
at 482°F/250°C	10,000
Izod impact, notched at 73°F/23°C, ft-lb/in.	No break
Coefficient of linear thermal expansion, in./in. °F	
at 70–212°F/20–100°C	7.8×10^{-5}
at 212–300°F/100–150°C	9.8×10^{-5}
at 300–480°F/150–210°C	12.1×10^{-5}
Heat distortion temperature, °F/°C	
at 66 psi	164/73
at 264 psi	118/48
Limiting oxygen index, %	<95
Flame spread	10
Underwriters lab rating (Sub 94)	94-V-O

TABLE 3.20 Comparison of Mechanical Properties of PFA at Room Temperature and Cryogenic Temperature

	Temperature	
Property	73°F/23°C	−320°F/−190°C
Yield strength, psi	2100	No yield
Ultimate tensile strength, psi	2600	18,700
Elongation, %	260	8
Flexural modulus, psi	81,000	840,000
Izod impact strength, notched, ft-lb/in.	No break	12
Compressive strength, psi	3500	60,000
Compressive strain, %	20	35
Modulus of elasticity, psi	10,000	680,000

TABLE 3.21 Absorption of Representative Liquids in PFA

Liquid[a]	Temperature, °F/°C	Range of weight gains, %
Aniline	365/185	0.3–0.4
Acetophenone	394/201	0.6–0.8
Benzaldehyde	354/179	0.4–0.5
Benzyl alcohol	400/204	0.3–0.4
n-Butylamine	172/78	0.3–0.4
Carbon tetrachloride	172/78	2.3–2.4
Dimethyl sulfoxide	372/190	0.1–0.2
Freon 113	117/47	1.2
Isooctane	210/99	0.7–0.8
Nitrobenzene	410/210	0.7–0.9
Perchlorethylene	250/121	2.0–2.3
Sulfuryl chloride	154/68	1.7–2.7
Toluene	230/110	0.7–0.8
Tributyl phosphate	392/200[b]	1.8–2.0
Bromine, anhydrous	−5/−22	0.5
Chlorine, anhydrous	248/120	0.5–0.6
Chlorosulfonic acid	302/150	0.7–0.8
Chromic acid 50%	248/120	0.00–0.01
Ferric chloride	212/100	0.00–0.01
Hydrochloric acid 37%	248/120	0.00–0.03
Phosphoric acid, concentrated	212/100	0.00–0.01
Zinc chloride	212/100	0.00–0.03

[a] Liquids were exposed for 168 hours at the boiling point of the solvents. The acidic reagents were exposed for 168 hours.
[b] Not boiling.

FEP extruded tubing is used to line pipe while sheets are used to produce vessel linings. Reference 1 provides details regarding FEP lined pipe.

As with other fluorinated thermoplasts FEP is subject to permeation by certain gases, refer to Table 1.3. It is also subject to absorption of certain liquids. Refer to Table 3.24.

FEP, with a few exceptions, exhibits the same corrosion resistance as PTFE but at a lower temperature. It is resistant to practically all chemicals, the exceptions being extremely potent oxidizers such as chlorine trifluoride and related compounds. Some chemicals will attack FEP when present in high concentrations at or near the service temperature limit. Refer to Table 3.25 for the compatibility of FEP with selected corrodents. Reference 2 provides a more detailed listing.

TABLE 3.22 Compatibility of PFA with Selected Corrodents[a]

Chemical	°F	°C	Chemical	°F	°C
Acetaldehyde	450	232	Benzaldehyde[c]	450	232
Acetamide	450	232	Benzene[b]	450	232
Acetic acid 10%	450	232	Benzene sulfonic acid 10%	450	232
Acetic acid 50%	450	232	Benzoic acid	450	232
Acetic acid 80%	450	232	Benzyl alcohol[c]	450	232
Acetic acid, glacial	450	232	Benzyl chloride[b]	450	232
Acetic anhydride	450	232	Borax	450	232
Acetone	450	232	Boric acid	450	232
Acetyl chloride	450	232	Bromine gas, dry[b]	450	232
Acrylonitrile	450	232	Bromine liquid[b,c]	450	232
Adipic acid	450	232	Butadiene[b]	450	232
Allyl alcohol	450	232	Butyl acetate	450	232
Allyl chloride	450	232	Butyl alcohol	450	232
Alum	450	232	n-Butylamine[c]	450	232
Aluminum chloride, aqueous	450	232	Butyl phthalate	450	232
Aluminum fluoride	450	232	Butyric acid	450	232
Aluminum hydroxide	450	232	Calcium bisulfide	450	232
Aluminum nitrate	450	232	Calcium bisulfite	450	232
Aluminum oxychloride	450	232	Calcium carbonate	450	232
Aluminum sulfate	450	232	Calcium chlorate	450	232
Ammonia gas[b]	450	232	Calcium chloride	450	232
Ammonium bifluoride[b]	450	232	Calcium hydroxide 10%	450	232
Ammonium carbonate	450	232	Calcium hydroxide, sat.	450	232
Ammonium chloride 10%	450	232	Calcium hypochlorite	450	232
Ammonium chloride 50%	450	232	Calcium nitrate	450	232
Ammonium chloride, sat.	450	232	Calcium oxide	450	232
Ammonium fluoride 10%[b]	450	232	Calcium sulfate	450	232
Ammonium fluoride 25%[b]	450	232	Caprylic acid	450	232
Ammonium hydroxide 25%	450	232	Carbon bisulfide[b]	450	232
Ammonium hydroxide, sat.	450	232	Carbon dioxide, dry	450	232
Ammonium nitrate	450	232	Carbon dioxide, wet	450	232
Ammonium persulfate	450	232	Carbon disulfide[b]	450	232
Ammonium phosphate	450	232	Carbon monoxide	450	232
Ammonium sulfate 10–40%	450	232	Carbon tetrachloride[b,c,d]	450	232
Ammonium sulfide	450	232	Carbonic acid	450	232
Amyl acetate	450	232	Chloracetic acid, 50% water	450	232
Amyl alcohol	450	232	Chloracetic acid	450	232
Amyl chloride	450	232	Chlorine gas, dry	x	x
Aniline[c]	450	232	Chlorine gas, wet[b]	450	232
Antimony trichloride	450	232	Chlorine, liquid[c]	x	x
Aqua regia 3:1	450	232	Chlorobenzene[b]	450	232
Barium carbonate	450	232	Chloroform[b]	450	232
Barium chloride	450	232	Chlorosulfonic acid[c]	450	232
Barium hydroxide	450	232	Chromic acid 10%	450	232
Barium sulfate	450	232	Chromic acid 50%[c]	450	232
Barium sulfide	450	232	Chromyl chloride	450	232

TABLE 3.22 Continued

Chemical	Maximum temp. °F	°C	Chemical	Maximum temp. °F	°C
Citric acid 15%	450	232	Methyl isobutyl ketone[b]	450	232
Citric acid, concentrated	450	232	Muriatic acid[b]	450	232
Copper carbonate	450	232	Nitric acid 5%[b]	450	232
Copper chloride	450	232	Nitric acid 20%[b]	450	232
Copper cyanide	450	232	Nitric acid 70%[b]	450	232
Copper sulfate	450	232	Nitric acid, anhydrous[b]	450	232
Cresol	450	232	Nitrous acid 10%	450	232
Cupric chloride 5%	450	232	Oleum	450	232
Cupric chloride 50%	450	232	Perchloric acid 10%	450	232
Cyclohexane	450	232	Perchloric acid 70%	450	232
Cyclohexanol	450	232	Phenol[b]	450	232
Dichloroacetic acid	450	232	Phosphoric acid 50–80%[c]	450	232
Dichloroethane (ethylene di-chloride)[b]	450	232	Picric acid	450	232
			Potassium bromide 30%	450	232
Ethylene glycol	450	232	Salicylic acid	450	232
Ferric chloride	450	232	Sodium carbonate	450	232
Ferric chloride 50% in water[c]	450	232	Sodium chloride	450	232
Ferric nitrate 10–50%	450	232	Sodium hydroxide 10%	450	232
Ferrous chloride	450	232	Sodium hydroxide 50%	450	232
Ferrous nitrate	450	232	Sodium hydroxide, concen-trated	450	232
Fluorine gas, dry	x	x	Sodium hypochlorite 20%	450	232
Fluorine gas, moist	x	x	Sodium hypochlorite, concen-trated	450	232
Hydrobromic acid, dilute[b,d]	450	232			
Hydrobromic acid 20%[b,d]	450	232	Sodium sulfide to 50%	450	232
Hydrobromic acid 50%[b,d]	450	232	Stannic chloride	450	232
Hydrochloric acid 20%[b,d]	450	232	Stannous chloride	450	232
Hydrochloric acid 38%[b,d]	450	232	Sulfuric acid 10%	450	232
Hydrocyanic acid 10%	450	232	Sulfuric acid 50%	450	232
Hydrofluoric acid 30%[b]	450	232	Sulfuric acid 70%	450	232
Hydrofluoric acid 70%[b]	450	232	Sulfuric acid 90%	450	232
Hydrofluoric acid 100%[b]	450	232	Sulfuric acid 98%	450	232
Hypochlorous acid	450	232	Sulfuric acid 100%	450	232
Iodine solution 10%[b]	450	232	Sulfuric acid, fuming[b]	450	232
Ketones, general	450	232	Sulfurous acid	450	232
Lactic acid 25%	450	232	Thionyl chloride[b]	450	232
Lactic acid, concentrated	450	232	Toluene[b]	450	232
Magnesium chloride	450	232	Trichloroacetic acid	450	232
Malic acid	450	232	White liquor	450	232
Methyl chloride[b]	450	232	Zinc chloride[c]	450	232
Methyl ethyl ketone[b]	450	232			

[a] The chemicals listed are in the pure state or in a saturated solution unless otherwise indicated. Compatibility is shown to the maximum allowable temperature for which data are available. Incompatibility is shown by an x. A blank space indicates that the data are unavailable.
[b] Material will permeate.
[c] Material will be absorbed.
[d] Material will cause stress cracking.
Source: Ref. 2.

TABLE **3.23** Physical and Mechanical Properties of FEP

Specific gravity	2.15
Water absorption, 24 h at 73°F/23°C, %	<0.01
Tensile strength at 73°F/23°C, psi	2700–3100
Modulus of elasticity in tension at 73°F/23°C $\times$ 10^5 psi	0.9
Compressive strength, psi	16,000
Flexural strength, psi	3000
Izod impact strength, notched at 73°F/23°C, ft · lb/in.	No break
Coefficient of thermal expansion, in./in. °F $\times$ 10^{-5}	8.3–10.5
Thermal conductivity, Btu/h/ft^2/°F/in.	0.11
Heat distortion temperature, at 66 psi °F/°C	158/70
Resistance to heat at continuous drainage, °F/°C	400/204
Limiting oxygen index, %	95
Flame spread	Nonflammable

CHLOROTRIFLUOROETHYLENE (ECTFE)

ECTFE is sold under the tradename of Halar by Ausimont USA, Inc. Halar has many of the desirable properties of PTFE without some of the disadvantages. ECTFE can be welded with ordinary thermoplastic welding equipment. It is used to line vessels and pipe. Liner thicknesses are usually 0.160 inch thick, which resists permeation. The water absorption rate of ECTFE is low, being less than 0.1%.

TABLE **3.24** Absorption of Selected Liquids[a] by FEP

Chemical	Temperature, °F/°C	Range of weight gains, %
Aniline	365/185	0.3–0.4
Acetophenone	394/201	0.6–0.8
Benzaldehyde	354/179	0.4–0.5
Benzyl alcohol	400/204	0.3–0.4
n-Butylamine	172/78	0.3–0.4
Carbon tetrachloride	172/78	2.3–2.4
Dimethyl sulfoxide	372/190	0.1–0.2
Nitrobenzene	410/210	0.7–0.9
Perchlorethylene	250/121	2.0–2.3
Sulfuryl chloride	154/68	1.7–2.7
Toluene	230/110	0.7–0.8
Tributyl phosphate	392/200[b]	1.8–2.0

[a] 168 hour exposure at their boiling points.
[b] Not boiling.

TABLE 3.25 Compatibility of FEP with Selected Corrodents[a]

Chemical	Maximum temp. °F	Maximum temp. °C	Chemical	Maximum temp. °F	Maximum temp. °C
Acetaldehyde	200	93	Barium sulfate	400	204
Acetamide	400	204	Barium sulfide	400	204
Acetic acid 10%	400	204	Benzaldehyde[b]	400	204
Acetic acid 50%	400	204	Benzene[b,c]	400	204
Acetic acid 80%	400	204	Benzene sulfonic acid 10%	400	204
Acetic acid, glacial	400	204	Benzoic acid	400	204
Acetic anhydride	400	204	Benzyl alcohol	400	204
Acetone[b]	400	204	Benzyl chloride	400	204
Acetyl chloride	400	204	Borax	400	204
Acrylic acid	200	93	Boric acid	400	204
Acrylonitrile	400	204	Bromine gas, dry[c]	200	93
Adipic acid	400	204	Bromine gas, moist[c]	200	93
Allyl alcohol	400	204	Bromine liquid[b,c]	400	204
Allyl chloride	400	204	Butadiene[c]	400	204
Alum	400	204	Butyl acetate	400	204
Aluminum acetate	400	204	Butyl alcohol	400	204
Aluminum chloride, aqueous	400	204	n-Butylamine[b]	400	204
Aluminum chloride, dry	300	149	Butyl phthalate	400	204
Aluminum fluoride[c]	400	204	Butyric acid	400	204
Aluminum hydroxide	400	204	Calcium bisulfide	400	204
Aluminum nitrate	400	204	Calcium bisulfite	400	204
Aluminum oxychloride	400	204	Calcium carbonate	400	204
Aluminum sulfate	400	204	Calcium chlorate	400	204
Ammonia gas[c]	400	204	Calcium chloride	400	204
Ammonium bifluoride[c]	400	204	Calcium hydroxide 10%	400	204
Ammonium carbonate	400	204	Calcium hydroxide, sat.	400	204
Ammonium chloride 10%	400	204	Calcium hypochlorite	400	204
Ammonium chloride 50%	400	204	Calcium nitrate	400	204
Ammonium chloride, sat.	400	204	Calcium oxide	400	204
Ammonium fluoride 10%[c]	400	204	Calcium sulfate	400	204
Ammonium fluoride 25%[c]	400	204	Caprylic acid	400	204
Ammonium hydroxide 25%	400	204	Carbon bisulfide[c]	400	204
Ammonium hydroxide, sat.	400	204	Carbon dioxide, dry	400	204
Ammonium nitrate	400	204	Carbon dioxide, wet	400	204
Ammonium persulfate	400	204	Carbon disulfide	400	204
Ammonium phosphate	400	204	Carbon monoxide	400	204
Ammonium sulfate 10–40%	400	204	Carbon tetrachloride[b,c,d]	400	204
Ammonium sulfide	400	204	Carbonic acid	400	204
Ammonium sulfite	400	204	Cellosolve	400	204
Amyl acetate	400	204	Chloracetic acid, 50% water	400	204
Amyl alcohol	400	204	Chloracetic acid	400	204
Amyl chloride	400	204	Chlorine gas, dry	x	x
Aniline[b]	400	204	Chlorine gas, wet[c]	400	204
Antimony trichloride	250	121	Chlorine, liquid[b]	400	204
Aqua regia 3:1	400	204	Chlorobenzene[c]	400	204
Barium carbonate	400	204	Chloroform[c]	400	204
Barium chloride	400	204	Chlorosulfonic acid[b]	400	204
Barium hydroxide	400	204	Chromic acid 10%	400	204

TABLE 3.25 Continued

Chemical	Maximum temp. °F	Maximum temp. °C	Chemical	Maximum temp. °F	Maximum temp. °C
Chromic acid 50%[b]	400	204	Methyl chloride[c]	400	204
Chromyl chloride	400	204	Methyl ethyl ketone[c]	400	204
Citric acid 15%	400	204	Methyl isobutyl ketone[c]	400	204
Citric acid, concentrated	400	204	Muriatic acid[c]	400	204
Copper acetate	400	204	Nitric acid 5%[c]	400	204
Copper carbonate	400	204	Nitric acid 20%[c]	400	204
Copper chloride	400	204	Nitric acid 70%[c]	400	204
Copper cyanide	400	204	Nitric acid, anhydrous[c]	400	204
Copper sulfate	400	204	Nitrous acid, concentrated	400	204
Cresol	400	204	Oleum	400	204
Cupric chloride 5%	400	204	Perchloric acid 10%	400	204
Cupric chloride 50%	400	204	Perchloric acid 70%	400	204
Cyclohexane	400	204	Phenol[c]	400	204
Cyclohexanol	400	204	Phosphoric acid 50–80%	400	204
Dichloroacetic acid	400	204	Picric acid	400	204
Dichloroethane (ethylene di-chloride)[c]	400	204	Potassium bromide 30%	400	204
			Salicylic acid	400	204
Ethylene glycol	400	204	Silver bromide 10%	400	204
Ferric chloride	400	204	Sodium carbonate	400	204
Ferric chloride 50% in water[b]	260	127	Sodium chloride	400	204
Ferric nitrate 10–50%	260	127	Sodium hydroxide 10%[b]	400	204
Ferrous chloride	400	204	Sodium hydroxide 50%	400	204
Ferrous nitrate	400	204	Sodium hydroxide, concentrated	400	204
Fluorine gas, dry	200	93	Sodium hypochlorite 20%	400	204
Fluorine gas, moist	x	x	Sodium hypochlorite, concen-trated	400	204
Hydrobromic acid, dilute	400	204			
Hydrobromic acid 20%[c,d]	400	204	Sodium sulfide to 50%	400	204
Hydrobromic acid 50%[c,d]	400	204	Stannic chloride	400	204
Hydrochloric acid 20%[c,d]	400	204	Stannous chloride	400	204
Hydrochloric acid 38%[c,d]	400	204	Sulfuric acid 10%	400	204
Hydrocyanic acid 10%	400	204	Sulfuric acid 50%	400	204
Hydrofluoric acid 30%[c]	400	204	Sulfuric acid 70%	400	204
Hydrofluoric acid 70%[c]	400	204	Sulfuric acid 90%	400	204
Hydrofluoric acid 100%[c]	400	204	Sulfuric acid 98%	400	204
Hypochlorous acid	400	204	Sulfuric acid 100%	400	204
Iodine solution 10%[c]	400	204	Sulfuric acid, fuming[c]	400	204
Ketones, general	400	204	Sulfurous acid	400	204
Lactic acid 25%	400	204	Thionyl chloride[c]	400	204
Lactic acid, concentrated	400	204	Toluene[c]	400	204
Magnesium chloride	400	204	Trichloroacetic acid	400	204
Malic acid	400	204	White liquor	400	204
Manganese chloride	300	149	Zinc chloride[d]	400	204

[a] The chemicals listed are in the pure state or in a saturated solution unless otherwise indicated. Compatibility is shown to the maximum allowable temperature for which data are available. Incompatibility is shown by an x. A blank space indicates that data are unavailable.
[b] Material will be absorbed.
[c] Material will permeate.
[d] Material can cause stress cracking.
Source: Ref. 2.

Halar has a broad use temperature range from cryogenic to 340°F/171°C with continuous service to 300°F/149°C. It exhibits excellent impact strength and abrasion resistance over its entire operating range, and good tensile, flexural, and wear related properties. The physical and mechanical properties are shown in Table 3.26.

Halar is very similar in its corrosion resistance to PTFE, but does not have the permeation problem associated with PTFE, PFA, and FEP. It is resistant to strong mineral and oxidizing acids, alkalies, metal etchants, liquid oxygen, and practically all organic solvents except hot amines, aniline, dimethylamine, etc. ECTFE is not subject to chemically induced stress cracking from strong acids, bases, or solvents. Some halogenated solvents can cause ECTFE to become slightly plasticized when it comes into contact with them. Under normal circumstances this does not affect the usefulness of the polymer since upon removal of the solvent from contact, and upon drying its mechanical properties return to their original values, indicating that no chemical attack has taken place.

Like other fluoropolymers, ECTFE will be attacked by metallic sodium and potassium. Table 3.27 provides the compatibility of ECTFE with selected corrodents.

Reference 1 provides details of Halar lined pipe, and Ref. 2 provides a more comprehensive listing of corrosion resistance.

TABLE 3.26 Physical and Mechanical Properties of ECTFE

Specific gravity	1.68
Water absorption, 24 h at 73°F/23°C, %	<0.01
Tensile strength at 73°F/23°C, psi	4500
Modulus of elasticity in tension at 73°F/23°C × 10^5 psi	2.4
Flexural strength, psi	7000
Izod impact strength, notched at 73°F/23°C, ft-lb/in.	No break
Linear coefficient of thermal expansion, in./in. °F	
at −22 to 122°F/−30 to 50°C	4.4×10^{-5}
at 122 to 185°F/50 to 80°C	5.6×10^{-5}
at 185 to 257°F/80 to 125°C	7.5×10^{-5}
at 257 to 356°F/125 to 180°C	9.2×10^{-5}
Thermal conductivity, Btu /h/ft²/°F/in.	1.07
Heat distortion temperature, °F/°C	
at 66 psi	195/91
at 264 psi	151/66
Limiting oxygen index, %	60
Underwriters lab rating (Sub 94)	V-O

TABLE 3.27 Compatibility of ECTFE with Selected Corrodents[a]

Chemical	Maximum temp. °F	Maximum temp. °C	Chemical	Maximum temp. °F	Maximum temp. °C
Acetic acid 10%	250	121	Benzaldehyde	150	66
Acetic acid 50%	250	121	Benzene	150	66
Acetic acid 80%	150	66	Benzene sulfonic acid 10%	150	66
Acetic acid, glacial	200	93	Benzoic acid	250	121
Acetic anhydride	100	38	Benzyl alcohol	300	149
Acetone	150	66	Benzyl chloride	300	149
Acetyl chloride	150	66	Borax	300	149
Acrylonitrile	150	66	Boric acid	300	149
Adipic acid	150	66	Bromine gas, dry	x	x
Allyl chloride	300	149	Bromine liquid	150	66
Alum	300	149	Butadiene	250	121
Aluminum chloride, aqueous	300	149	Butyl acetate	150	66
Aluminum chloride, dry			Butyl alcohol	300	149
Aluminum fluoride	300	149	Butyric acid	250	121
Aluminum hydroxide	300	149	Calcium bisulfide	300	149
Aluminum nitrate	300	149	Calcium bisulfite	300	149
Aluminum oxychloride	150	66	Calcium carbonate	300	149
Aluminum sulfate	300	149	Calcium chlorate	300	149
Ammonia gas	300	149	Calcium chloride	300	149
Ammonium bifluoride	300	149	Calcium hydroxide 10%	300	149
Ammonium carbonate	300	149	Calcium hydroxide, sat.	300	149
Ammonium chloride 10%	290	143	Calcium hypochlorite	300	149
Ammonium chloride 50%	300	149	Calcium nitrate	300	149
Ammonium chloride, sat.	300	149	Calcium oxide	300	149
Ammonium fluoride 10%	300	149	Calcium sulfate	300	149
Ammonium fluoride 25%	300	149	Caprylic acid	220	104
Ammonium hydroxide 25%	300	149	Carbon bisulfide	80	27
Ammonium hydroxide, sat.	300	149	Carbon dioxide, dry	300	149
Ammonium nitrate	300	149	Carbon dioxide, wet	300	149
Ammonium persulfate	150	66	Carbon disulfide	80	27
Ammonium phosphate	300	149	Carbon monoxide	150	66
Ammonium sulfate 10–40%	300	149	Carbon tetrachloride	300	149
Ammonium sulfide	300	149	Carbonic acid	300	149
Amyl acetate	160	71	Cellosolve	300	149
Amyl alcohol	300	149	Chloracetic acid, 50% water	250	121
Amyl chloride	300	149	Chloracetic acid	250	121
Aniline	90	32	Chlorine gas, dry	150	66
Antimony trichloride	100	38	Chlorine gas, wet	250	121
Aqua regia 3:1	250	121	Chlorine, liquid	250	121
Barium carbonate	300	149	Chlorobenzene	150	66
Barium chloride	300	149	Chloroform	250	121
Barium hydroxide	300	149	Chlorosulfonic acid	80	27
Barium sulfate	300	149	Chromic acid 10%	250	121
Barium sulfide	300	149	Chromic acid 50%	250	121

TABLE **3.27** Continued

Chemical	°F	°C	Chemical	°F	°C
	Maximum temp.			Maximum temp.	
Citric acid 15%	300	149	Nitric acid 5%	300	149
Citric acid, conc.	300	149	Nitric acid 20%	250	121
Copper carbonate	150	66	Nitric acid 70%	150	66
Copper chloride	300	149	Nitric acid, anhydrous	150	66
Copper cyanide	300	149	Nitrous acid, concentrated	250	121
Copper sulfate	300	149	Oleum	x	x
Cresol	300	149	Perchloric acid 10%	150	66
Cupric chloride 5%	300	149	Perchloric acid 70%	150	66
Cupric chloride 50%	300	149	Phenol	150	66
Cyclohexane	300	149	Phosphoric acid 50–80%	250	121
Cyclohexanol	300	149	Picric acid	80	27
Ethylene glycol	300	149	Potassium bromide 30%	300	149
Ferric chloride	300	149	Salicylic acid	250	121
Ferric chloride 50% in water	300	149	Sodium carbonate	300	149
Ferric nitrate 10–50%	300	149	Sodium chloride	300	149
Ferrous chloride	300	149	Sodium hydroxide 10%	300	149
Ferrous nitrate	300	149	Sodium hydroxide 50%	250	121
Fluorine gas, dry	x	x	Sodium hydroxide, concentrated	150	66
Fluorine gas, moist	80	27	Sodium hypochlorite 20%	300	149
Hydrobromic acid, dilute	300	149	Sodium hypochlorite, concentrated	300	149
Hydrobromic acid 20%	300	149			
Hydrobromic acid 50%	300	149	Sodium sulfide to 50%	300	149
Hydrochloric acid 20%	300	149	Stannic chloride	300	149
Hydrochloric acid 38%	300	149	Stannous chloride	300	149
Hydrocyanic acid 10%	300	149	Sulfuric acid 10%	250	121
Hydrofluoric acid 30%	250	121	Sulfuric acid 50%	250	121
Hydrofluoric acid 70%	240	116	Sulfuric acid 70%	250	121
Hydrofluoric acid 100%	240	116	Sulfuric acid 90%	150	66
Hypochlorous acid	300	149	Sulfuric acid 98%	150	66
Iodine solution 10%	250	121	Sulfuric acid 100%	80	27
Lactic acid 25%	150	66	Sulfuric acid, fuming	300	149
Lactic acid, concentrated	150	66	Sulfurous acid	250	121
Magnesium chloride	300	149	Thionyl chloride	150	66
Malic acid	250	121	Toluene	150	66
Methyl chloride	300	149	Trichloroacetic acid	150	66
Methyl ethyl ketone	150	66	White liquor	250	121
Methyl isobutyl ketone	150	66	Zinc chloride	300	149
Muriatic acid	300	149			

[a] The chemicals listed are in the pure state or in a saturated solution unless otherwise indicated. Compatibility is shown to the maximum allowable temperature for which data are available. Incompatibility is shown by an x. A blank space indicates that the data are unavailable. *Source*: Ref. 2.

TABLE **3.28** Physical and Mechanical Properties of ETFE

Specific gravity	1.70
Tensile strength, psi	6500
Modulus of elasticity, psi $\times 10^5$	2.17
Elongation, %	300
Flexural modulus, psi $\times 10^5$	1.7
Impact strength, ft-lb/in.	No break
Hardness, Shore D	67
Water absorption, 24 h at 73°F/23°C, %	<0.03
Thermal conductivity, Btu/h/ft^2/°F/in.	1.6
Heat distortion temperature, °F/°C	
at 66 psi	220/104
at 264 psi	160/71
Limiting oxygen index, %	30
Underwriters lab rating (Sub 94)	V-O

ETHYLENE TETRAFLUOROETHYLENE (ETFE)

ETFE is sold under the trade name of Tefzel by DuPont. It is a partially fluorinated copolymer of ethylene and tetrafluoroethylene with a maximum service temperature of 300°F/149°C. Tefzel can be melt bonded to untreated aluminum, steel, and copper. It can also be melt bonded to itself. In order to adhesive bond Tefzel with polyester or epoxy compounds, the surface must be chemically etched or subjected to corona or flame treatments.

ETFE is a rugged thermoplastic. It is less dense, tougher, stiffer, and exhibits a higher tensile strength and creep resistance than PTFE and FEP fluorocarbon resins. The physical and mechanical properties are shown in Table 3.28.

Vapor permeation of ETFE is as follows:

Gas	Permeation (cm^3 mil thickness/100 in.^{2} 24 h atm)
Carbon dioxide	250
Nitrogen	30
Oxygen	100
Helium	900

Tefzel is inert to strong mineral acids, inorganic bases, halogens, and metal salt solutions. Even carboxylic acids, anhydrides, aromatic and aliphatic hydro-

TABLE 3.29 Compatibility of ETFE with Selected Corrodents[a]

Chemical	°F	°C	Chemical	°F	°C
Acetaldehyde	200	93	Barium sulfate	300	149
Acetamide	250	121	Barium sulfide	300	149
Acetic acid 10%	250	121	Benzaldehyde	210	99
Acetic acid 50%	250	121	Benzene	210	99
Acetic acid 80%	230	110	Benzene sulfonic acid 10%	210	99
Acetic acid, glacial	230	110	Benzoic acid	270	132
Acetic anhydride	300	149	Benzyl alcohol	300	149
Acetone	150	66	Benzyl chloride	300	149
Acetyl chloride	150	66	Borax	300	149
Acrylonitrile	150	66	Boric acid	300	149
Adipic acid	280	138	Bromine gas, dry	150	66
Allyl alcohol	210	99	Bromine water 10%	230	110
Allyl chloride	190	88	Butadiene	250	121
Alum	300	149	Butyl acetate	230	110
Aluminum chloride, aqueous	300	149	Butyl alcohol	300	149
Aluminum chloride, dry	300	149	n-Butylamine	120	49
Aluminum fluoride	300	149	Butyl phthalate	150	66
Aluminum hydroxide	300	149	Butyric acid	250	121
Aluminum nitrate	300	149	Calcium bisulfide	300	149
Aluminum oxychloride	300	149	Calcium carbonate	300	149
Aluminum sulfate	300	149	Calcium chlorate	300	149
Ammonium bifluoride	300	149	Calcium chloride	300	149
Ammonium carbonate	300	149	Calcium hydroxide 10%	300	149
Ammonium chloride 10%	300	149	Calcium hydroxide, sat.	300	149
Ammonium chloride 50%	290	143	Calcium hypochlorite	300	149
Ammonium chloride, sat.	300	149	Calcium nitrate	300	149
Ammonium fluoride 10%	300	149	Calcium oxide	260	127
Ammonium fluoride 25%	300	149	Calcium sulfate	300	149
Ammonium hydroxide 25%	300	149	Caprylic acid	210	99
Ammonium hydroxide, sat.	300	149	Carbon bisulfide	150	66
Ammonium nitrate	230	110	Carbon dioxide, dry	300	149
Ammonium persulfate	300	149	Carbon dioxide, wet	300	149
Ammonium phosphate	300	149	Carbon disulfide	150	66
Ammonium sulfate 10–40%	300	149	Carbon monoxide	300	149
Ammonium sulfide	300	149	Carbon tetrachloride	270	132
Amyl acetate	250	121	Carbonic acid	300	149
Amyl alcohol	300	149	Cellosolve	300	149
Amyl chloride	300	149	Chloracetic acid, 50% water	230	110
Aniline	230	110	Chloracetic acid 50%	230	110
Antimony trichloride	210	99	Chlorine gas, dry	210	99
Aqua regia 3:1	210	99	Chlorine gas, wet	250	121
Barium carbonate	300	149	Chlorine, water	100	38
Barium chloride	300	149	Chlorobenzene	210	99
Barium hydroxide	300	149	Chloroform	230	110

TABLE 3.29 Continued

Chemical	Maximum temp. °F	Maximum temp. °C	Chemical	Maximum temp. °F	Maximum temp. °C
Chlorosulfonic acid	80	27	Methyl isobutyl ketone	300	149
Chromic acid 10%	150	66	Muriatic acid	300	149
Chromic acid 50%	150	66	Nitric acid 5%	150	66
Chromyl chloride	210	99	Nitric acid 20%	150	66
Citric acid 15%	120	49	Nitric acid 70%	80	27
Copper chloride	300	149	Nitric acid, anhydrous	x	x
Copper cyanide	300	149	Nitrous acid, concentrated	210	99
Copper sulfate	300	149	Oleum	150	66
Cresol	270	132	Perchloric acid 10%	230	110
Cupric chloride 5%	300	149	Perchloric acid 70%	150	66
Cyclohexane	300	149	Phenol	210	99
Cyclohexanol	250	121	Phosphoric acid 50–80%	270	132
Dichloroacetic acid	150	66	Picric acid	130	54
Ethylene glycol	300	149	Potassium bromide 30%	300	149
Ferric chloride 50% in water	300	149	Salicylic acid	250	121
Ferric nitrate 10–50%	300	149	Sodium carbonate	300	149
Ferrous chloride	300	149	Sodium chloride	300	149
Ferrous nitrate	300	149	Sodium hydroxide 10%	230	110
Fluorine gas, dry	100	38	Sodium hydroxide 50%	230	110
Fluorine gas, moist	100	38	Sodium hypochlorite 20%	300	149
Hydrobromic acid, dilute	300	149	Sodium hypochlorite, concen-trated	300	149
Hydrobromic acid 20%	300	149			
Hydrobromic acid 50%	300	149	Sodium sulfide to 50%	300	149
Hydrochloric acid 20%	300	149	Stannic chloride	300	149
Hydrochloric acid 38%	300	149	Stannous chloride	300	149
Hydrocyanic acid 10%	300	149	Sulfuric acid 10%	300	149
Hydrofluoric acid 30%	270	132	Sulfuric acid 50%	300	149
Hydrofluoric acid 70%	250	121	Sulfuric acid 70%	300	149
Hydrofluoric acid 100%	230	110	Sulfuric acid 90%	300	149
Hypochlorous acid	300	149	Sulfuric acid 98%	300	149
Lactic acid 25%	250	121	Sulfuric acid 100%	300	149
Lactic acid, concentrated	250	121	Sulfuric acid, fuming	120	49
Magnesium chloride	300	149	Sulfurous acid	210	99
Malic acid	270	132	Thionyl chloride	210	99
Manganese chloride	120	49	Toluene	250	121
Methyl chloride	300	149	Trichloroacetic acid	210	99
Methyl ethyl ketone	230	110	Zinc chloride	300	149

[a] The chemicals listed are in the pure state or in a saturated solution unless otherwise indicated. Compatibility is shown to the maximum allowable temperature for which data are available. Incompatibility is shown by an x. A blank space indicates that the data are unavailable.
Source: Material extracted from Ref. 2.

carbons, alcohols, ketones, ethers, esters, chlorocarbons, and classic polymer solvents have little effect on ETFE.

Very strong oxidizing acids near their boiling points, such as nitric acid, at high concentration will affect ETFE in varying degrees, as well as organic bases such as amines and sulfonic acids. Refer to Table 3.29 for the compatibility of ETFE with selected corrodents. Reference 2 provides a more comprehensive listing.

REFERENCES

1. Schweitzer, Philip A. *Corrosion Resistant Piping Systems*, Marcel Dekker, New York, 1994.
2. Schweitzer, Philip A. *Corrosion Resistance Tables*, 4th ed., vols. 1–3, Marcel Dekker, New York, 1995.

4

Specific Elastomeric Sheet Linings

An elastomer is generally considered to be any material, either natural or synthetic, that is elastic or resilient and in general resembles natural rubber in feeling and appearance. A more technical definition is provided by ASTM, which states,

> An elastomer is a polymeric material which at room temperature can be stretched to at least twice its original length and upon immediate release of the stress will return quickly to its original length.

These materials are sometimes referred to as rubbers.

Elastomers are primarily composed of large molecules that tend to form spiral threads, similar to a coiled spring, which are attached to each other at infrequent intervals. These coils tend to stretch or compress when a small stress is applied, but exert an increasing resistance to the application of addition stresses. This phenonenon is illustrated by the reaction of rubber to the application of additional stress.

In the raw state elastomers tend to be soft and sticky when hot, and hard and brittle when cold. Compounding increases the utility of rubber and synthetic elastomers. Vulcanization extends the temperature range within which they are flexible and elastic. In addition to vulcanizing agents, ingredients are added to make elastomers stronger, tougher, or harder, to make them age better, to color them, and in general to improve specific properties to meet specific needs. Properties required for a lining material, depending upon the application may include:

Resistance to high temperature
Resistance to cold
Long life
Impermeability
Resistance to sunlight
Resistance to chemicals
Resistance to oils
Lack of odor or taste

Elastomers, both natural and synthetic, are used for the lining of vessels and piping. These linings have provided many years of service in the protection of steel vessels from corrosion. They are sheet applied and bonded to a steel substrate. The bonding agent used is dependent upon the specific elastomer.

Bonding of vulcanized elastomers to themselves or metal substrates is generally completed by using pressure sensitive adhesive, derived from an elastomer similar to the one being bonded. Flexible thermosetting adhesives such as epoxy-polyamide or polyurethane also offer excellent bond strength to most elastomers. Surface treatment consists of washing with a solvent, abrading, or acid cyclizing.

Unvulcanized elastomers may be bonded to metals by priming the substrate with a suitable air- or heat-drying adhesive before the elastomer is molded against the substrate. The most common elastomers to be bonded in this way include nitrile, neoprene, urethane, natural rubber, and butyl rubber.

Elastomers are also used extensively as membrane linings in acid-brick lined vessels to protect the steel shell from corrosive attack, while the acid-brick lining in turn protects the elastomer from abrasion and excessive temperature.

Another major use is as an impermeable liner for settling ponds or basins. These materials are employed to prevent pond contaminents from seeping into the soil and causing pollution of groundwater and contamination of the soil.

The most common elastomers used for lining applications, along with their operating temperature range are shown in Table 4.1.

Several factors must be taken into account when selecting an elastomer for a lining application. First and foremost is the compatibility of the elastomer with the medium at the temperature and concentration to which it will be exposed. It should also be remembered that each of the elastomeric materials may be formulated to improve certain of its properties. However, an improvement in one property may have an adverse affect on another property, such as corrosion resistance. Because of the ability to change formulation of many of these elastomers, the wisest policy is to permit a competent manufacturer to make the selection of the elastomer to satisfy the application.

Elastomeric materials can fail as the result of chemical action and/or mechanical damage. Chemical deterioration occurs as a result of a chemical reaction

TABLE 4.1 Elastomers Used as Liners

	Temperature range			
	°F		°C	
Elastomer	Min	Max	Min	Max
Natural rubber NR	−59	175	−50	80
Butyl rubber IIR	−30	300	−34	149
Chlorobutyl rubber CIIR	−30	300	−34	149
Neoprene CR	−13	203	−25	95
Hypalon CSM	−20	250	−30	121
Urethane rubber AU	−65	250	−54	121
EPDM rubber	−65	300	−54	149
Nitrile rubber, NBR, Buna-N	−40	250	−40	121
Polyester elastomer PE	−40	302	−40	150
Perfluoroelastomers FPM	−58	600	−50	316
Fluoroelastomers FKM	−10	400	−18	204

between an elastomer and the medium or by the absorption of the medium into the elastomer. This attack results in a swelling of the elastomer and a reduction in its tensile strength.

The degree of deterioration is a function of the temperature and concentration of the corrodent. In general, the higher the temperature and the higher the concentration of the corrodent, the greater will be the chemical attack. Elastomers, unlike metals, absorb varying quantities of the material they are in contact with, especially organic liquids. This can result in swelling, cracking, and penetration to the substrate of an elastomer lined vessel. Swelling can cause softening of the elastomer, and in a lined vessel introduce high stresses and failure of the bond. If an elastomeric lining has high absorption, permeation will probably result. Some elastomers, such as the fluorocarbons, are easily permeated but have very little absorption. An approximation of the expected permeation and/or absorption of an elastomer can be based on the absorption of water. These data are usually available. See Chapter 1 for a discussion of permeation.

Permeation and absorption can result in:

1. Bond failure and blistering. These are caused by accumulation of fluids at the bond when the substrate is less permeable than the lining or from the formation of corrosion or reaction products if the substrate is attacked by the permeant.
2. Failure of the substrate due to corrosive attack.
3. Loss of contents through lining and substrate as the result of eventual failure of the substrate.

Thickness of lining is a factor affecting permeation. For general corrosion resistance, thicknesses of 0.010–0.020 in. are usually satisfactory, depending upon the combination of elastomerc material and specific corrodent. When mechanical factors such as thinning due to cold flow, mechanical abuse, and permeation rates are a consideration, thicker linings may be required.

Although elastomers can be damaged by mechanical means alone, this is not usually the case. When in good physical condition, an elastomer will exhibit abrasion resistance superior to that of metal. The actual size, shape, and hardness of the particles and their velocity are the determining factors as to how well a particular rubber resists mechanical damage from the medium. Hard, sharp objects, including those foreign to the normal medium, may cut or gouge the elastomer. Mechanical damage occurs as the result of chemical deterioration of the elastomer. When the elastomer is in a deteriorated condition, the material is weakened, and consequently it is more susceptible to mechanical damage from flowing or agitated media.

NATURAL RUBBER (NR)

The degree of curing to which natural rubber is subjected will determine whether it is classified as soft, semihard, or hard. Soft rubber is the form primarily used as a lining material although some hard rubber linings are produced. Natural rubber is also used for lining pipelines.

The maximum temperature for continuous use is 175°F/80°C which will vary depending upon the grade of rubber. The physical and mechanical properties of natural rubber are shown in Table 4.2.

Natural rubber offers excellent resistance to most inorganic salt solutions, alkalies, and nonoxidizing acids. Hydrochloric acid will react with soft rubber to form rubber hydrochloride, and therefore is not recommended to be used for this application. Hard rubber is resistant to hydrochloric acid. Strong oxidizing media, such as nitric acid, concentrated sulfuric acid, permanganates, dichromates, chlorine dioxide, and sodium hypochlorite will severely attack rubber. Mineral and vegetable oils, gasoline, benzene, toluene, and chlorinated hydrocarbons also affect rubber. Natural rubber offers good resistance to radiation and alcohols.

Soft Natural Rubber Linings

These linings have the advantages of:

Ease of application, cure, and repair unaffected by mechanical stresses or rapid temperature changes, making it ideal for outdoor tanks.
Good general chemical resistance within temperature limitation. See Table 4.3.

TABLE 4.2 Physical and Mechanical Properties of Natural Rubber[a]

Specific gravity	0.92
Refractive index	1.52
Specific heat, cal/g	0.452
Swelling, % by volume	
in kerosene at 77°F/25°C	200
in benzene at 77°F/25°C	200
in acetone at 77°F/25°C	25
in mineral oil at 100°F/38°C	120
Brittle point	$-68°F/-56°C$
Relative permeability to hydrogen	50
Relative permeability to air	11
Insulation resistance, ohms/cm	10
Resilience, %	90
Tear resistance, psi	1640
Coefficient of linear expansion at 32–140°F, in./in.-°F	0.000036
Coefficient of heat conduction K, B tu/ft^2-in.-°F	1.07
Tensile strength, psi	3000–4500
Elongation, % at break	775–780
Hardness, Shore A	40–100
Abrasion resistance	Excellent
Maximum temperature, continuous use	175°F/80°C
Impact resistance	Excellent
Compression set	Good
Machining qualities	Can be ground
Effect of sunlight	Deteriorates
Effect of aging	Moderately resistant
Effect of heat	Softens

[a] These are representative values since they may be altered by compounding.

Excellent to superior physical properties. Maximum tensile, elongation, abrasion and tear resistance.

Low cost.

Limitations of these linings include:

Not oil or flame resistant.

Not ozone, sunlight and weather resistant.

Temperature limited to 140°F/60°C.

Cannot be used with dilute hydrochloric acid (5 to 10%) or spent acids.

Typical applications for soft natural rubber linings include chemical storage tanks to handle phosphoric acid (37%), hydrofluosilicic acid, alum, chlorides and sulfates of ammonia, cadmium, iron phosphorus and sodium, inorganic salts in gen-

TABLE 4.3 Compatibility of Soft Natural Rubber with Selected Corrodents[a]

Chemical	°F	°C	Chemical	°F	°C
Acetaldehyde	x	x	Barium hydroxide	140	60
Acetamide	x	x	Barium sulfate	140	60
Acetic acid 10%	150	66	Barium sulfide	140	60
Acetic acid 50%	x	x	Benzaldehyde	x	x
Acetic acid 80%	x	x	Benzene	x	x
Acetic acid, glacial	x	x	Benzene sulfonic acid 10%	x	x
Acetic anhydride	x	x			
Acetone	140	60	Benzoic acid	140	60
Acetyl chloride	x	x	Benzyl alcohol	x	x
Alum	140	60	Benzyl chloride	x	x
Aluminum chloride, aqueous	140	60	Borax	140	60
			Boric acid	140	60
Aluminum chloride, dry	160	71	Butyl acetate	x	x
Aluminum fluoride	x	x	Butyl alcohol	140	60
Aluminum nitrate	x	x	Butyric acid	x	x
Aluminum sulfate	140	60	Calcium bisulfite	140	60
Ammonium carbonate	140	60	Calcium carbonate	140	60
Ammonium chloride 10%	140	60	Calcium chlorate	140	60
			Calcium chloride	140	60
Ammonium chloride 50%	140	60	Calcium hydroxide 10%	140	60
			Calcium hydroxide, sat.	140	60
Ammonium chloride, sat.	140	60	Calcium hypochlorite	x	x
Ammonium fluoride 10%	x	x	Calcium nitrate	x	x
Ammonium fluoride 25%	x	x	Calcium oxide	140	60
Ammonium hydroxide 25%	140	60	Calcium sulfate	140	60
			Carbon bisulfide	x	x
Ammonium hydroxide, sat.	140	60	Carbon disulfide	x	x
			Carbon monoxide	x	x
Ammonium nitrate	140	60	Carbon tetrachloride	x	x
Ammonium phosphate	140	60	Carbonic acid	140	60
Ammonium sulfate 10–40%	140	60	Cellosolve	x	x
			Chloracetic acid, 50% water	x	x
Ammonium sulfide	140	60	Chloracetic acid	x	x
Amyl acetate	x	x	Chlorine gas, dry	x	x
Amyl alcohol	140	60	Chlorine gas, wet	x	x
Amyl chloride	x	x	Chlorine, liquid	x	x
Aniline	x	x	Chlorobenzene	x	x
Aqua regia 3:1	x	x	Chloroform	x	x
Barium carbonate	140	60	Chlorosulfonic acid	x	x
Barium chloride	140	60			

TABLE 4.3 Continued

Chemical	Maximum temp. °F	Maximum temp. °C	Chemical	Maximum temp. °F	Maximum temp. °C
Chromic acid 10%	x	x	Methyl ethyl ketone	x	x
Chromic acid 50%	x	x	Methyl isobutyl ketone	x	x
Citric acid 15%	140	60	Muriatic acid	140	60
Citric acid, concentrated	x	x	Nitric acid 5%	x	x
Copper carbonate	x	x	Nitric acid 20%	x	x
Copper chloride	x	x	Nitric acid 70%	x	x
Copper cyanide	140	60	Nitric acid, anhydrous	x	x
Copper sulfate	140	60	Nitrous acid, concen-	x	x
Cresol	x	x	trated		
Cupric chloride 5%	x	x	Phenol	x	x
Cupric chloride 50%	x	x	Phosphoric acid 50–	140	60
Cyclohexane	x	x	80%		
Dichloroethane (ethylene	x	x	Potassium bromide 30%	140	60
dichloride)			Sodium carbonate	140	60
Ethylene glycol	140	60	Sodium chloride	140	60
Ferric chloride	140	60	Sodium hydroxide 10%	140	60
Ferric chloride 50% in	140	60	Sodium hydroxide 50%	x	x
water			Sodium hydroxide, con-	x	x
Ferric nitrate 10–50%	x	x	centrated		
Ferrous chloride	140	60	Sodium hypochlorite	x	x
Ferrous nitrate	x	x	20%		
Fluorine gas, dry	x	x	Sodium hypochlorite,	x	x
Hydrobromic acid, dilute	140	60	concentrated		
Hydrobromic acid 20%	140	60	Sodium sulfide to 50%	140	60
Hydrobromic acid 50%	140	60	Stannic chloride	140	60
Hydrochloric acid 20%	x	x	Stannous chloride	140	60
Hydrochloric acid 38%	140	60	Sulfuric acid 10%	140	60
Hydrofluoric acid 30%	x	x	Sulfuric acid 50%	x	x
Hydrofluoric acid 70%	x	x	Sulfuric acid 70%	x	x
Hydrofluoric acid 100%	x	x	Sulfuric acid 90%	x	x
Lactic acid 25%	x	x	Sulfuric acid 98%	x	x
Lactic acid, concentrated	x	x	Sulfuric acid 100%	x	x
Magnesium chloride	140	60	Sulfuric acid, fuming	x	x
Malic acid	x	x	Sulfurous acid	x	x
Methyl chloride	x	x	Zinc chloride	140	60

[a] The chemicals listed are in the pure state or in a saturated solution unless otherwise indicated. Compatibility is shown to the maximum allowable temperature for which data are available. Incompatibility is shown by an x. A blank space indicates that data are unavailable.
Source: Ref. 1.

TABLE 4.4 Compatibility of Multiple Ply (Soft/Hard/Soft) Natural Rubber with Selected Corrodents[a]

Chemical	Maximum temp.		Chemical	Maximum temp.	
	°F	°C		°F	°C
Acetaldehyde	x	x	Barium hydroxide	160	71
Acetamide	x	x	Barium sulfate	160	71
Acetic acid 10%	x	x	Barium sulfide	160	71
Acetic acid 50%	x	x	Benzaldehyde	x	x
Acetic acid 80%	x	x	Benzene	x	x
Acetic acid, glacial	x	x	Benzene sulfonic acid 10%	x	x
Acetic anhydride	x	x			
Acetone	140	60	Benzoic acid	160	71
Alum	160	71	Benzyl alcohol	x	x
Aluminum chloride, aqueous	160	71	Benzyl chloride	x	x
			Borax	160	71
Aluminum chloride, dry	160	71	Boric acid	140	60
Aluminum fluoride	x	x	Butyl acetate	x	x
Aluminum nitrate	x	x	Butyl alcohol	160	71
Aluminum oxychloride			Butyric acid	x	x
Aluminum sulfate	160	71	Calcium bisulfite	160	71
Ammonium carbonate	160	71	Calcium carbonate	160	71
Ammonium chloride 10%	160	71	Calcium chlorate	140	60
Ammonium chloride 50%	160	71	Calcium chloride	140	60
Ammonium chloride, sat.	160	71	Calcium hydroxide 10%	160	71
Ammonium fluoride 10%	x	x	Calcium hydroxide, sat.	160	71
Ammonium fluoride 25%	x	x	Calcium hypochlorite	x	x
Ammonium hydroxide 25%	100	38	Calcium nitrate	x	x
			Calcium oxide	160	71
Ammonium hydroxide, sat.	100	38	Calcium sulfate	160	71
			Carbon bisulfide	x	x
Ammonium nitrate	160	71	Carbon disulfide	x	x
Ammonium persulfate			Carbon monoxide	x	x
Ammonium phosphate	160	71	Carbon tetrachloride	x	x
Ammonium sulfate 10–40%	160	71	Carbonic acid	160	71
			Cellosolve	x	x
Ammonium sulfide	160	71	Chloracetic acid, 50% water	x	x
Amyl acetate	x	x			
Amyl alcohol	100	38	Chloracetic acid	x	x
Amyl chloride	x	x	Chlorine gas, dry	x	x
Aniline	x	x	Chlorine gas, wet	x	x
Aqua regia 3:1	x	x	Chlorine, liquid	x	x
Barium carbonate	160	71	Chlorobenzene	x	x
Barium chloride	160	71	Chloroform	x	x

TABLE 4.4 Continued

Chemical	Maximum temp. °F	°C	Chemical	Maximum temp. °F	°C
Chlorosulfonic acid	x	x	Methyl chloride	x	x
Chromic acid 10%	x	x	Methyl ethyl ketone	x	x
Chromic acid 50%	x	x	Methyl isobutyl ketone	x	x
Citric acid 15%	x	x	Muriatic acid	140	60
Citric acid, concentrated	x	x	Nitric acid 5%	x	x
Copper carbonate	x	x	Nitric acid 20%	x	x
Copper chloride	x	x	Nitric acid 70%	x	x
Copper cyanide	160	71	Nitric acid, anhydrous	x	x
Copper sulfate	160	71	Nitrous acid, concen-	x	x
Cresol	x	x	trated		
Cupric chloride 5%	x	x	Phenol	x	x
Cupric chloride 50%	x	x	Phosphoric acid 50–	160	71
Cyclohexane	x	x	80%		
Dichloroethane (ethylene	x	x	Potassium bromide 30%	160	71
dichloride)			Sodium carbonate	160	71
Ethylene glycol	160	71	Sodium chloride	160	71
Ferric chloride	160	71	Sodium hydroxide 10%	160	71
Ferric chloride 50% in	160	71	Sodium hydroxide 50%	x	x
water			Sodium hydroxide, con-	x	x
Ferric nitrate 10–50%	x	x	centrated		
Ferrous chloride	140	60	Sodium hypochlorite 20%	x	x
Ferrous nitrate	x	x	Sodium hypochlorite,	x	x
Fluorine gas, dry	x	x	concentrated		
Hydrobromic acid, dilute	160	71	Sodium sulfide to 50%	160	71
Hydrobromic acid 20%	160	71	Stannic chloride	160	71
Hydrobromic acid 50%	160	71	Stannous chloride	160	71
Hydrochloric acid 20%	x	x	Sulfuric acid 10%	160	71
Hydrochloric acid 38%	160	71	Sulfuric acid 50%	x	x
Hydrofluoric acid 30%	x	x	Sulfuric acid 70%	x	x
Hydrofluoric acid 70%	x	x	Sulfuric acid 90%	x	x
Hydrofluoric acid 100%	x	x	Sulfuric acid 98%	x	x
Lactic acid 25%	x	x	Sulfuric acid 100%	x	x
Lactic acid, concentrated	x	x	Sulfuric acid, fuming	x	x
Magnesium chloride	160	71	Sulfurous acid	x	x
Malic acid	100	38	Zinc chloride	160	71

[a] The chemicals listed are in the pure state or in a saturated solution unless otherwise indicated. Compatibility is shown to the maximum allowable temperature for which data are available. Incompatibility is shown by an x. A blank space indicates that data are unavailable.
Source: Ref. 1.

TABLE 4.5 Compatibility of Semihard Natural Rubber with Selected Corrodents[a]

Chemical	Maximum temp.		Chemical	Maximum temp.	
	°F	°C		°F	°C
Acetamide	x	x	Calcium hydroxide 10%	180	82
Acetic acid, glacial	x	x	Calcium hydroxide, sat.	180	82
Acetic anhydride	x	x	Calcium hypochlorite	x	x
Acetone	x	x	Chloracetic acid, 50%	100	38
Alum	180	82	water		
Aluminum chloride,	180	82	Chloracetic acid	100	38
aqueous			Chromic acid 10%	x	x
Aluminum fluoride	x	x	Chromic acid 50%	x	x
Aluminum nitrate	100	38	Citric acid 15%	100	38
Aluminum sulfate	180	82	Citric acid, concentrated	100	38
Ammonium carbonate	180	82	Copper carbonate	180	82
Ammonium chloride	180	82	Copper chloride	x	x
10%			Copper cyanide	180	82
Ammonium chloride	180	82	Copper sulfate	180	82
50%			Cupric chloride 5%	x	x
Ammonium chloride, sat.	180	82	Cupric chloride 50%	x	x
Ammonium fluoride 10%	x	x	Ethylene glycol	180	82
Ammonium fluoride 25%	x	x	Ferric chloride	180	82
Ammonium nitrate	180	82	Ferric chloride 50% in	180	82
Ammonium phosphate	180	82	water		
Ammonium sulfate 10–	180	82	Ferric nitrate 10–50%	100	38
40%			Ferrous chloride	180	82
Ammonium sulfide	180	82	Ferrous nitrate	100	38
Amyl alcohol	180	82	Hydrobromic acid, dilute	180	82
Aniline	x	x	Hydrobromic acid 20%	180	82
Barium carbonate	180	82	Hydrobromic acid 50%	180	82
Barium chloride	180	82	Hydrochloric acid 20%	180	82
Barium hydroxide	180	82	Hydrochloric acid 38%	180	82
Barium sulfate	180	82	Hydrofluoric acid 70%	x	x
Benzene	x	x	Hydrofluoric acid 100%	x	x
Benzoic acid	180	82	Lactic acid 25%	100	38
Borax	180	82	Lactic acid, concentrated	100	38
Boric acid	180	82	Magnesium chloride	180	82
Butyl acetate	100	38	Malic acid	100	38
Butyl alcohol	180	82	Methyl chloride	100	38
Butyric acid	100	38	Nitric acid 5%	100	38
Calcium bisulfite	180	82	Nitric acid 20%	x	x
Calcium carbonate	180	82	Nitric acid 70%	x	x
Calcium chloride	180	82	Nitric acid, anhydrous	x	x

TABLE 4.5   Continued

Chemical	Maximum temp.		Chemical	Maximum temp.	
	°F	°C		°F	°C
Nitrous acid, concentrated	100	38	Sodium sulfide to 50%	180	82
Phenol	x	x	Stannic chloride	180	82
Phosphoric acid 50–80%	180	82	Stannous chloride	180	82
Potassium bromide 30%	180	82	Sulfuric acid 10%	180	82
Sodium carbonate	180	82	Sulfuric acid 50%	100	38
Sodium chloride	180	82	Sulfuric acid 70%	x	x
Sodium hydroxide 10%	180	82	Sulfuric acid 90%	x	x
Sodium hydroxide 50%	100	38	Sulfuric acid 98%	x	x
Sodium hydroxide, con- centrated	100	38	Sulfuric acid 100%	x	x
			Sulfuric acid, fuming	x	x
Sodium hypochlorite 20%	x	x	Sulfurous acid	150	66
Sodium hypochlorite, concentrated	x	x	Zinc chloride	180	82

ᵃThe chemicals listed are in the pure state or in a saturated solution unless otherwise indicated. Compatibility is shown to the maximum allowable temperature for which data are available. Incompatibility is shown by an x. A blank space indicates that data are unavailable.
Source: Ref. 1.

eral. This material is also suitable to line trailer tanks for transporting the above chemicals. Soft natural rubber is also used to line pipe and fittings.

Multiply Natural Rubber Linings

Multiply natural rubber linings consist of three layers of rubber: soft natural rubber/ hard natural rubber/soft natural rubber. The advantages of these linings are:

> Better permeation resistance than soft natural rubber
> Heat resistance to 160°F/71°C.
> Ease of application, cure, and repair.
> Good flexibility, reducing danger of cracking in cold weather, rapid temperature changes, or mechanical stresses.
> Moderate cost.

The limitations of these linings are:

> Not oil or flame resistant.
> Cannot be used in transport trucks.

> Refer to Table 4.4 for the compatibility of multiply linings with selected

TABLE 4.6 Compatibility of Hard Natural Rubber with Selected Corrodents[a]

Chemical	Maximum temp. °F	°C	Chemical	Maximum temp. °F	°C
Acetamide	x	x	Benzaldehyde	x	x
Acetic acid 10%	200	93	Benzene	x	x
Acetic acid 50%	200	93	Benzoic acid	200	93
Acetic acid 80%	150	66	Borax	200	93
Acetic acid, glacial	100	38	Boric acid	200	93
Acetic anhydride	100	38	Butyl acetate	x	x
Acetone	x	x	Butyl alcohol	160	71
Acrylonitrile	90	32	Butyric acid	150	66
Adipic acid	80	27	Calcium bisulfite	200	93
Allyl alcohol	x	x	Calcium carbonate	200	93
Alum	200	93	Calcium chloride	200	93
Aluminum chloride, aqueous	200	93	Calcium hydroxide 10%	200	93
			Calcium hydroxide, sat.	200	93
Aluminum fluoride	x	x	Calcium hypochlorite	x	x
Aluminum hydroxide	200	93	Calcium nitrate	200	93
Aluminum nitrate	190	88	Calcium oxide	200	93
Aluminum sulfate	200	93	Calcium sulfate	200	93
Ammonium bifluoride	200	93	Carbon bisulfide	x	x
Ammonium carbonate	200	93	Carbon disulfide	x	x
Ammonium chloride 10%	200	93	Carbon monoxide	x	x
Ammonium chloride 50%	200	93	Carbon tetrachloride	x	x
Ammonium chloride, sat.	200	93	Carbonic acid	200	93
Ammonium fluoride 10%	x	x	Chloracetic acid, 50% water	120	49
Ammonium fluoride 25%	x	x			
Ammonium hydroxide 25%	x	x	Chloracetic acid	120	49
			Chlorine gas, dry	x	x
Ammonium hydroxide, sat.	x	x	Chlorine gas, wet	190	88
Ammonium nitrate	150	66	Chlorine, liquid	x	x
Ammonium persulfate	200	93	Chlorobenzene	x	x
Ammonium phosphate	200	93	Chloroform	x	x
Ammonium sulfate 10–40%	200	93	Chlorosulfonic acid	x	x
			Chromic acid 10%	x	x
Ammonium sulfide	200	93	Chromic acid 50%	x	x
Amyl alcohol	200	93	Citric acid 15%	150	66
Aniline	x	x	Citric acid, concentrated	150	66
Aqua regia 3:1	x	x	Copper carbonate	200	93
Barium carbonate	200	93	Copper chloride	100	38
Barium chloride	200	93	Copper cyanide	200	93
Barium hydroxide	x	x	Copper sulfate	200	93
Barium sulfate	200	93	Ethylene glycol	200	93
Barium sulfide	200	93	Ferric chloride	200	93

TABLE 4.6 Continued

Chemical	Maximum temp.		Chemical	Maximum temp.	
	°F	°C		°F	°C
Ferric chloride 50% in water	200	93	Nitric acid 70%	x	x
			Nitric acid, anhydrous	x	x
Ferric nitrate 10–50%	150	66	Nitrous acid, concen-	150	66
Ferrous chloride	200	93	trated		
Ferrous nitrate	150	66	Phenol	x	x
Hydrobromic acid, dilute	200	93	Phosphoric acid 50–80%	200	93
Hydrobromic acid 20%	200	93	Potassium bromide 30%	200	93
Hydrobromic acid 50%	200	93	Sodium carbonate	200	93
Hydrochloric acid 20%	200	93	Sodium chloride	200	93
Hydrochloric acid 38%	200	93	Sodium hydroxide 10%	200	93
Hydrocyanic acid 10%	200	93	Sodium hydroxide 50%	150	66
Hydrofluoric acid 30%	x	x	Sodium hydroxide, con-	150	66
Hydrofluoric acid 70%	x	x	centrated		
Hydrofluoric acid 100%	x	x	Sodium hypochlorite 20%	x	x
Hypochlorous acid	150	66	Sodium hypochlorite,	x	x
Lactic acid 25%	150	66	concentrated		
Lactic acid, concentrated	150	66	Sodium sulfide to 50%	200	93
Magnesium chloride	200	93	Stannic chloride	200	93
Malic acid	150	66	Stannous chloride	200	93
Methyl ethyl ketone	x	x	Sulfuric acid 10%	200	93
Muriatic acid	200	93	Sulfuric acid 70%	x	x
Nitric acid 5%	150	66	Sulfurous acid	200	93
Nitric acid 20%	x	x	Zinc chloride	200	93

[a] The chemicals listed are in the pure state or in a saturated solution unless otherwise indicated. Compatibility is shown to the maximum allowable temperature for which data are available. Incompatibility is shown by an x. A blank space indicates that data are unavailable.
Source: Ref. 1.

corrodents. Applications for multiply linings include those for soft natural rubber linings with the exceptions noted above.

Semihard Natural Rubber Linings

Semihard natural rubber linings provide:

Ease of application, cure, and repair.
Excellent chemical and permeation resistance. Refer to Table 4.5.
Heat resistance to 180°F/82°C.
Moderate cost.

These linings are limited by their inability to resist oil or flames and the fact that they are subject to damage by cold weather, sudden extreme temperature changes, or mechanical stresses. In addition, they are not abrasion resistant.

Applications include chemical process tanks, pumps, fans, pipe and fittings.

Hard Natural Rubber Linings

Hard natural rubber linings have the advantages of:

Ease of application, cure, and repair.
Better chemical, heat, and permeation resistance than soft natural rubber. Refer to Table 4.6.
Heat resistance to 200°F/93°C.
Moderate cost.

Limitations of these linings are:

They are not oil or flame resistant.
Subject to damage by cold weather, exposure, sudden extreme temperature changes or mechanical stresses.
Not abrasion resistant.
Not to be used in transport vehicles.

Typical applications include chemical storage tanks, pickling tanks, pipe and fittings, plating tanks and various accessories such as agitators, pumps, etc.

BUTYL RUBBER (IIR)

Butyl rubber contains isobutylene as its parent material with small proportions of butadiene or isoprene added. Commercial butyl rubber may contain 5% butadiene as a copolymer.

The single outstanding physical property of butyl rubber is its impermeability. Its abrasion resistance, tear resistance, and adhesion to fabrics and metals is good. The flame resistance of butyl rubber is poor. Table 4.7 lists the physical and mechanical properties of butyl rubber.

Butyl rubber is very nonpolar. It has exceptional resistance to dilute mineral acids, alkalies, phosphate ester oils, ethylene, ethylene glycol, and water. It is also resistant to oxygenated solvents such as methyl ethyl ketone, ethyl acetate, and acetone. Resistance to concentrated acids, except nitric and sulfuric, is good. Unlike natural rubber it is very resistant to swelling by vegetable and animal oils. It has poor resistance to petroleum oils, gasoline, and most solvents, except oxygenated solvents. Refer to Table 4.8 for the compatibility of

TABLE **4.7** Physical and Mechanical Properties of Butyl Rubber
(IIR) [a]

Specific gravity	0.91
Dielectric strength, V/mm	25,000
Tensile strength, psi	500–3000
Hardness, Shore A	15–90
Abrasion resistance	Excellent
Maximum temperature, continuous use	250–300°F/120–148°C
Machining qualities	Can be ground
Resistance to sunlight	Excellent
Effect of aging	Highly resistant
Resistance to heat	Stiffens slightly

[a] These are representative values since they may be altered by compounding.

butyl rubber with selected corrodents. Reference 1 provides a more detailed listing.

Butyl rubber linings have the following advantages:

Unaffected by cold water or rapid temperature changes.
Heat resistant to 200°F/93°C.
Resistant to ozone, sunlight, and aging.
Possessed of good chemical and gaseous permeation resistance.
Resistant to water and oxygenated solvents.
Resistant to sliding abrasion.
Able to be used for hydrochloric acid.

Limitations of butyl rubber:

Not flame or oil resistant.
More costly and difficult to apply than natural rubber.

Applications include linings for hydrochloric acid storage and processing vessels, and mixed acid wastes. Rubber is also used to line pipe and fittings.

CHLOROBUTYL RUBBER (CIIR)

Chlorobutyl rubber is chlorinated isobutylene-isoprene. It has the same general properties as butyl rubber. Refer to Table 4.7.

CIIR exhibits the same general corrosion resistance as natural rubber, but can be used at higher temperatures. Unlike butyl rubber, CIIR cannot be used with hydrochloric acid. Refer to Table 4.9 for the compatibility of chlorobutyl rubber with selected corrodents and Ref. 1 for additional listings.

TABLE **4.8** Compatibility of Butyl Rubber with Selected Corrodents[a]

Chemical	Maximum temp.		Chemical	Maximum temp.	
	°F	°C		°F	°C
Acetaldehyde	80	27	Aniline	150	66
Acetic acid 10%	150	66	Antimony trichloride	150	66
Acetic acid 50%	110	43	Barium chloride	150	66
Acetic acid 80%	110	43	Barium hydroxide	190	88
Acetic acid, glacial	x	x	Barium sulfide	190	88
Acetic anhydride	x	x	Benzaldehyde	90	32
Acetone	100	38	Benzene	x	x
Acrylonitrile	x	x	Benzene sulfonic acid	90	32
Adipic acid	x	x	10%		
Allyl alcohol	190	88	Benzoic acid	150	66
Allyl chloride	x	x	Benzyl alcohol	190	88
Alum	200	93	Benzyl chloride	x	x
Aluminum acetate	200	93	Borax	190	88
Aluminum chloride,	200	93	Boric acid	150	66
aqueous			Butyl acetate	x	x
Aluminum chloride, dry	200	93	Butyl alcohol	140	60
Aluminum fluoride	180	82	Butyric acid	x	x
Aluminum hydroxide	100	38	Calcium bisulfite	120	49
Aluminum nitrate	100	38	Calcium carbonate	150	66
Aluminum sulfate	200	93	Calcium chlorate	190	88
Ammonium bifluoride	x	x	Calcium chloride	190	88
Ammonium carbonate	190	88	Calcium hydroxide 10%	190	88
Ammonium chloride	200	93	Calcium hydroxide, sat.	190	88
10%			Calcium hypochlorite	x	x
Ammonium chloride	200	93	Calcium nitrate	190	88
50%			Calcium sulfate	100	38
Ammonium chloride, sat.	200	93	Carbon dioxide, dry	190	88
Ammonium fluoride 10%	150	66	Carbon dioxide, wet	190	88
Ammonium fluoride 25%	150	66	Carbon disulfide	190	88
Ammonium hydroxide	190	88	Carbon monoxide	x	x
25%			Carbon tetrachloride	90	32
Ammonium hydroxide,	190	88	Carbonic acid	150	66
sat.			Cellosolve	150	66
Ammonium nitrate	200	93	Chloracetic acid, 50%	150	66
Ammonium persulfate	190	88	water		
Ammonium phosphate	150	66	Chloracetic acid	100	38
Ammonium sulfate 10–	150	66	Chlorine gas, dry	x	x
40%			Chlorine, liquid	x	x
Amyl acetate	x	x	Chlorobenzene	x	x
Amyl alcohol	150	66	Chloroform	x	x

TABLE 4.8 Continued

Chemical	Maximum temp. °F	Maximum temp. °C	Chemical	Maximum temp. °F	Maximum temp. °C
Chlorosulfonic acid	x	x	Methyl isobutyl ketone	80	27
Chromic acid 10%	x	x	Muriatic acid	x	x
Chromic acid 50%	x	x	Nitric acid 5%	200	93
Citric acid 15%	190	88	Nitric acid 20%	150	66
Citric acid, concentrated	190	88	Nitric acid 70%	x	x
Copper chloride	150	66	Nitric acid, anhydrous	x	x
Copper sulfate	190	88	Nitrous acid, concen-	125	52
Cresol	x	x	trated		
Cupric chloride 5%	150	66	Oleum	x	x
Cupric chloride 50%	150	66	Perchloric acid 10%	150	66
Cyclohexane	x	x	Phenol	150	66
Dichloroethane (ethylene	x	x	Phosphoric acid 50–	150	66
dichloride)			80%		
Ethylene glycol	200	93	Salicylic acid	80	27
Ferric chloride	175	79	Sodium chloride	200	93
Ferric chloride 50% in	160	71	Sodium hydroxide 10%	150	66
water			Sodium hydroxide 50%	150	66
Ferric nitrate 10–50%	190	88	Sodium hydroxide, con-	150	66
Ferrous chloride	175	79	centrated		
Ferrous nitrate	190	88	Sodium hypochlorite 20%	x	x
Fluorine gas, dry	x	x	Sodium hypochlorite,	x	x
Hydrobromic acid, dilute	125	52	concentrated		
Hydrobromic acid 20%	125	52	Sodium sulfide to 50%	150	66
Hydrobromic acid 50%	125	52	Stannic chloride	150	66
Hydrochloric acid 20%	125	52	Stannous chloride	150	66
Hydrochloric acid 38%	125	52	Sulfuric acid 10%	200	93
Hydrocyanic acid 10%	140	60	Sulfuric acid 50%	150	66
Hydrofluoric acid 30%	150	66	Sulfuric acid 70%	x	x
Hydrofluoric acid 70%	150	66	Sulfuric acid 90%	x	x
Hydrofluoric acid 100%	150	66	Sulfuric acid 98%	x	x
Hypochlorous acid	x	x	Sulfuric acid 100%	x	x
Lactic acid 25%	125	52	Sulfuric acid, fuming	x	x
Lactic acid, concentrated	125	52	Sulfurous acid	200	93
Magnesium chloride	200	93	Thionyl chloride	x	x
Malic acid	x	x	Toluene	x	x
Methyl chloride	90	32	Trichloroacetic acid	x	x
Methyl ethyl ketone	100	38	Zinc chloride	200	93

[a]The chemicals listed are in the pure state or in a saturated solution unless otherwise indicated. Compatibility is shown to the maximum allowable temperature for which data are available. Incompatibility is shown by an x. A blank space indicates that data are unavailable.
Source: Ref. 1.

TABLE **4.9** Compatibility of Chlorobutyl Rubber with Selected Corrodents[a]

Chemical	°F	°C	Chemical	°F	°C
Acetic acid 10%	150	60	Cupric chloride 5%	150	66
Acetic acid 50%	150	60	Cupric chloride 50%	150	66
Acetic acid 80%	150	60	Ethylene glycol	200	93
Acetic acid, glacial	x	x	Ferric chloride	175	79
Acetic anhydride	x	x	Ferric chloride 50% in	100	38
Acetone	100	38	water		
Alum	200	93	Ferric nitrate 10–50%	160	71
Aluminum chloride,	200	93	Ferrous chloride	175	79
aqueous			Hydrobromic acid, dilute	125	52
Aluminum nitrate	190	88	Hydrobromic acid 20%	125	52
Aluminum sulfate	200	93	Hydrobromic acid 50%	125	52
Ammonium carbonate	200	93	Hydrochloric acid 20%	x	x
Ammonium chloride 10%	200	93	Hydrochloric acid 38%	x	x
Ammonium chloride 50%	200	93	Hydrofluoric acid 70%	x	x
Ammonium chloride, sat.	200	93	Hydrofluoric acid 100%	x	x
Ammonium nitrate	200	93	Lactic acid 25%	125	52
Ammonium phosphate	150	66	Lactic acid, concentrated	125	52
Ammonium sulfate 10–	150	66	Magnesium chloride	200	93
40%			Nitric acid 5%	200	93
Amyl alcohol	150	66	Nitric acid 20%	150	66
Aniline	150	66	Nitric acid 70%	x	x
Antimony trichloride	150	66	Nitric acid, anhydrous	x	x
Barium chloride	150	66	Nitrous acid, concen-	125	52
Benzoic acid	150	66	trated		
Boric acid	150	66	Phenol	150	66
Calcium chloride	160	71	Phosphoric acid 50–80%	150	66
Calcium nitrate	160	71	Sodium chloride	200	93
Calcium sulfate	160	71	Sodium hydroxide 10%	150	66
Carbon monoxide	100	38	Sodium sulfide to 50%	150	66
Carbonic acid	150	66	Sulfuric acid 10%	200	93
Chloracetic acid	100	38	Sulfuric acid 70%	x	x
Chromic acid 10%	x	x	Sulfuric acid 90%	x	x
Chromic acid 50%	x	x	Sulfuric acid 98%	x	x
Citric acid 15%	90	32	Sulfuric acid 100%	x	x
Copper chloride	150	66	Sulfuric acid, fuming	x	x
Copper cyanide	160	71	Sulfurous acid	200	93
Copper sulfate	160	71	Zinc chloride	200	93

[a] The chemicals listed are in the pure state or in a saturated solution unless otherwise indicated. Compatibility is shown to the maximum allowable temperature for which data are available. Incompatibility is shown by an x. A blank space indicates that data are unavailable.
Source: Ref. 1.

Chlorobutyl rubber has the advantages of:

Heat resistance of 200°F/93°C.
Unaffected by cold weather or rapid temperature changes.
Good chemical and permeation resistance.
Good resistance to ozone, sunlight, and weathering.
Easier to apply than butyl rubber.

Limitations of CIIR include its inability to be used with hydrochloric acid or oils. It is also more expensive than natural rubber.

NEOPRENE (POLYCHLOROPRENE) (CR)

Neoprene is one of the oldest and most versatile of the synthetic rubbers. It was originally introduced as an oil resistant substitute for natural rubber.

Neoprene can be ignited by an open flame but will stop burning when the flame is removed. Natural rubber and many other synthetic rubber will continue to burn once ignited even if the flame is removed. In an actual fire situation neoprene will burn.

Compared with natural rubber, neoprene is relatively impermeable to gases. At the maximum operating temperature of 200°F/93°C, neoprene continues to maintain good physical properties and has excellent resistance to long-term heat degradation. Heat failure of neoprene results from the hardening of the product and lack of resilience, rather than softening or melting as with other elastomers. Neoprene can be used in temperatures as low as 0°F/−18°C. As temperatures decrease further, the material stiffens until the brittle point is reached at approximately −40°F/−40°C, depending upon the compounding. Table 4.10 lists the physical and mechanical properties of neoprene.

Neoprene will form an extremely strong mechanical bond with cotton fabric. It can also be molded in contact with metals, particularly carbon and alloy steels, stainless steels, aluminum and aluminum alloys, brass and copper, using any of the commercially available bonding agents. If suitable treatments or additives are provided, it can also be made to adhere to such manmade fibers as glass, nylon, rayon, acrylic, and polyester.

Neoprene displays excellent resistance to sun, weather, and ozone. Because of its low rate of oxidation, neoprene has a high resistance to both outdoor and indoor aging. One of the outstanding properties of neoprene is its resistance to attack from solvents, waxes, fats, oils, greases, and many other petroleum based products. Excellent service is also experienced when it is in contact with aliphatic compounds (methyl and ethyl alcohols, ethylene glycols, etc.) aliphatic hydrocarbons, and most freon refrigerants. When exposed to dilute mineral acids, inorganic salt solutions, or alkalies, neoprene shows little if any change in appearance or change in properties.

TABLE 4.10 Physical and Mechanical Properties of Neoprene (CR)[a]

Specific gravity	1.4
Refractive index	1.56
Specific heat, cal/g	0.40
Volumetric coefficient of thermal expansions	
at 77°F	$403 \times 10^{-6}/°F$
at 25°F	$725 \times 10^{-6}/°C$
Thermal conductivity	
Btu/h-ft²-in. °F	1.45
g-cal/h-cm²cm °C	1.80
Brittle point	−40°F/−40°C
DC resistivity, ohm-cm	2×10^{13}
Dielectric strength, V/mil	600
Permeability (cm³/cm²-cm-sec-atm) at 77°F (25°C)	
to nitrogen	1×10^{-8}
to methane	2×10^{-8}
to oxygen	3×10^{-8}
to helium	10×10^{-8}
to carbon dioxide	19×10^{-8}
Tensile strength, psi	1000–2500
Elongation, % at break	200–600
Hardness, Shore A	40–95
Abrasion resistance	Excellent
Maximum temperature, continuous use	180–200°F/82–93°C
Impact resistance	Excellent
Compression set, %	15–35
Machining qualities	Can be ground
Resistance to sunlight	Excellent
Effect of aging	Little effect
Resistance to heat	Good

[a] These are representative values since they may be altered by compounding.

Chlorinated and aromatic hydrocarbons, organic esters, aromatic hydroxy compounds, and certain ketones have an adverse effect on neoprene and consequently only limited serviceability can be expected with them. Highly oxidizing acid and salt solutions also cause surface deterioration and loss of strength. Included in this category are nitric acid and concentrated sulfuric acid. Table 4.11 provides the compatibility of neoprene with selected corrodents. Reference 1 provides a more detailed listing.

The advantages of neoprene include its good chemical resistance, heat resistance to 200°F/93°C, good oil resistance, better resistance to ozone, sunlight, and weather than natural rubber, its excellent abrasion resistance, and the fact

TABLE **4.11** Compatibility of Neoprene with Selected Corrodents[a]

Chemical	Maximum temp.		Chemical	Maximum temp.	
	°F	°C		°F	°C
Acetaldehyde	200	93	Ammonium nitrate	200	93
Acetamide	200	93	Ammonium persulfate	200	93
Acetic acid 10%	160	71	Ammonium phosphate	150	66
Acetic acid 50%	160	71	Ammonium sulfate 10–	150	66
Acetic acid 80%	160	71	40%		
Acetic acid, glacial	x	x	Ammonium sulfide	160	71
Acetic anhydride	x	x	Ammonium sulfite		
Acetone	x	x	Amyl acetate	x	x
Acetyl chloride	x	x	Amyl alcohol	200	93
Acrylic acid	x	x	Amyl chloride	x	x
Acrylonitrile	140	60	Aniline	x	x
Adipic acid	160	71	Antimony trichloride	140	60
Allyl alcohol	120	49	Aqua regia 3:1	x	x
Allyl chloride	x	x	Barium carbonate	150	66
Alum	200	93	Barium chloride	150	66
Aluminum acetate			Barium hydroxide	230	110
Aluminum chloride,	150	66	Barium sulfate	200	93
aqueous			Barium sulfide	200	93
Aluminum chloride, dry			Benzaldehyde	x	x
Aluminum fluoride	200	93	Benzene	x	x
Aluminum hydroxide	180	82	Benzene sulfonic acid	100	38
Aluminum nitrate	200	93	10%		
Aluminum oxychloride			Benzoic acid	150	66
Aluminum sulfate	200	93	Benzyl alcohol	x	x
Ammonia gas	140	60	Benzyl chloride	x	x
Ammonium bifluoride	x	x	Borax	200	93
Ammonium carbonate	200	93	Boric acid	150	66
Ammonium chloride	150	66	Bromine gas, dry	x	x
10%			Bromine gas, moist	x	x
Ammonium chloride	150	66	Bromine liquid	x	x
50%			Butadiene	140	60
Ammonium chloride,	150	66	Butyl acetate	60	16
sat.			Butyl alcohol	200	93
Ammonium fluoride 10%	200	93	n-Butylamine		
Ammonium fluoride	200	93	Butyl phthalate		
25%			Butyric acid	x	x
Ammonium hydroxide	200	93	Calcium bisulfide		
25%			Calcium bisulfite	x	x
Ammonium hydroxide,	200	93	Calcium carbonate	200	93
sat.			Calcium chlorate	200	93

TABLE 4.11   Continued

Chemical	Maximum temp. °F	Maximum temp. °C	Chemical	Maximum temp. °F	Maximum temp. °C
Calcium chloride	150	66	Cyclohexanol	x	x
Calcium hydroxide 10%	230	110	Dichloroacetic acid	x	x
Calcium hydroxide, sat.	230	110	Dichloroethane (ethyl-	x	x
Calcium hypochlorite	x	x	ene dichloride)		
Calcium nitrate	150	66	Ethylene glycol	100	38
Calcium oxide	200	93	Ferric chloride	160	71
Calcium sulfate	150	66	Ferric chloride 50% in	160	71
Caprylic acid			water		
Carbon bisulfide	x	x	Ferric nitrate 10–50%	200	93
Carbon dioxide, dry	200	93	Ferrous chloride	90	32
Carbon dioxide, wet	200	93	Ferrous nitrate	200	93
Carbon disulfide	x	x	Fluorine gas, dry	x	x
Carbon monoxide	x	x	Fluorine gas, moist	x	x
Carbon tetrachloride	x	x	Hydrobromic acid,	x	x
Carbonic acid	150	66	dilute		
Cellosolve	x	x	Hydrobromic acid 20%	x	x
Chloracetic acid, 50%	x	x	Hydrobromic acid 50%	x	x
water			Hydrochloric acid 20%	x	x
Chloracetic acid	x	x	Hydrochloric acid 38%	x	x
Chlorine gas, dry	x	x	Hydrocyanic acid 10%	x	x
Chlorine gas, wet	x	x	Hydrofluoric acid 30%	x	x
Chlorine, liquid	x	x	Hydrofluoric acid 70%	x	x
Chlorobenzene	x	x	Hydrofluoric acid 100%	x	x
Chloroform	x	x	Hypochlorous acid	x	x
Chlorosulfonic acid	x	x	Iodine solution 10%	80	27
Chromic acid 10%	140	60	Ketones, general	x	x
Chromic acid 50%	100	38	Lactic acid 25%	140	60
Chromyl chloride			Lactic acid, concen-	90	32
Citric acid 15%	150	66	trated		
Citric acid, concen-	150	66	Magnesium chloride	200	93
trated			Malic acid		
Copper acetate	160	71	Manganese chloride	200	93
Copper carbonate			Methyl chloride	x	x
Copper chloride	200	93	Methyl ethyl ketone	x	x
Copper cyanide	160	71	Methyl isobutyl ketone	x	x
Copper sulfate	200	93	Muriatic acid	x	x
Cresol	x	x	Nitric acid 5%	x	x
Cupric chloride 5%	200	93	Nitric acid 20%	x	x
Cupric chloride 50%	160	71	Nitric acid 70%	x	x
Cyclohexane	x	x	Nitric acid, anhydrous	x	x

TABLE **4.11** Continued

Chemical	Maximum temp.		Chemical	Maximum temp.	
	°F	°C		°F	°C
Nitrous acid, concentrated	x	x	Sodium hypochlorite, concentrated	x	x
Oleum	x	x	Sodium sulfide to 50%	200	93
Perchloric acid 10%			Stannic chloride	200	93
Perchloric acid 70%	x	x	Stannous chloride	x	x
Phenol	x	x	Sulfuric acid 10%	150	66
Phosphoric acid 50–80%	150	66	Sulfuric acid 50%	100	38
Picric acid	200	93	Sulfuric acid 70%	x	x
Potassium bromide 30%	160	71	Sulfuric acid 90%	x	x
Salicylic acid			Sulfuric acid 98%	x	x
Silver bromide 10%			Sulfuric acid 100%	x	x
Sodium carbonate	200	93	Sulfuric acid, fuming	x	x
Sodium chloride	200	93	Sulfurous acid	100	38
Sodium hydroxide 10%	230	110	Thionyl chloride	x	x
Sodium hydroxide 50%	230	110	Toluene	x	x
Sodium hydroxide, concentrated	230	110	Trichloroacetic acid	x	x
			White liquor	140	60
Sodium hypochlorite 20%	x	x	Zinc chloride	160	71

[a] The chemicals listed are in the pure state or in a saturated solution unless otherwise indicated. Compatibility is shown to the maximum allowable temperature for which data are available. Incompatibility is shown by an x. A blank space indicates that data are unavailable.
Source: Ref. 1.

that it is unaffected by cold weather, rapid temperature changes, or mechanical stresses.

On the negative side is the fact that it is more expensive than natural rubber and more difficult to apply.

Typical applications include caustic storage and transportation tanks, sulfuric acid pickle tanks, chemical process and storage tanks, and miscellaneous process equipment accessories.

HYPALON (CHLOROSULFONATED POLYETHYLENE) CSM

Chlorosulfonated polyethylene rubber is manufactured by DuPont under the tradename Hypalon. In many respects it is similar to neoprene, but it does possess some advantages over neoprene in certain types of service. It has better ozone resistance and chemical resistance.

Hypalon will burn in an actual fire situation but is classified as self-extinguishing. If the flame is removed, the elastomer will stop burning. This phenomenon is due to its chlorine content which makes it more resistant to burning than exclusively hydrocarbon polymers.

Hypalon's resistance to abrasion is superior to that of natural rubber and many other elastomers by as much as 2 to 1. Good resistance to impact, crushing, cutting, and gouging, and other types of physical abuse are also present in this elastomer.

Hypalon can be used continuously in the temperature range of $-20°F/$ $-28°C$ to $250°F/121°C$. The physical and mechanical properties are shown in Table 4.12.

Hypalon is one of the most weather resistant elastomers available. Oxidation takes place at a very slow rate. Sunlight and ultraviolet light have little if any adverse effect on its physical properties. It is also inherently resistant to ozone attack without the need for the addition of special antioxidants or antiozonates to the formulation.

When properly compounded, Hypalon is highly resistant to attack by hydrocarbon oils and fuels, even at elevated temperatures. It is also resistant to such

TABLE 4.12 Physical and Mechanical Properties of Chlorosulfonated Polyethylene (Hypalon; CSM)[a]

Specific gravity	1.08–1.28
Brittle point	-40 to $-80°F/-40$ to $-62°C$
Dielectric strength, V/mil	500
Dielectric constant at 1000 Hz	8^{-10}
Dissipation factor at 1000 Hz	0.05–0.07
Tensile strength, psi	2500
Elongation, % at break	430–540
Hardness, Shore A	60
Abrasion resistance	Excellent
Maximum temperature, continuous use	$250°F/121°C$
Impact resistance	Good
Compression set, %	
at $158°F/70°C$	16
at $212°F/100°C$	25
at $250°F/121°C$	44
Resistance to sunlight	Excellent
Effect of aging	None
Resistance to heat	Good

[a] These are representative values since they may be altered by compounding.

oxidizing chemicals as sodium hypochlorite, sodium peroxide, ferric chloride, and sulfuric, chromic, and hydrofluoric acids. Concentrated hydrochloric acid (37%) at elevated temperatures above 158°F/70°C will attack Hypalon, but it can be handled without adverse effect at all concentrations below this temperature. Nitric acid at room temperature up to 60% concentration can also be handled without adverse effects.

Hypalon is also resistant to salt solutions, alcohols, and both weak and concentrated alkalies, and is generally unaffected by soil chemicals, moisture, and other deteriorating factors associated with burial in the earth. Long term contact with water has little or no effect on Hypalon.

Hypalon has poor resistance to aliphatic, aromatic, and chlorinated hydrocarbons. Refer to Table 4.13 for the compatibility of Hypalon with selected corrodents. Reference 1 provides a more detailed listing.

Hypalon's advantages include some oil resistance, excellent resistance to ozone, sunlight, weathering, chemicals, as well as resistance to permeation and abrasion. It is unaffected by cold weather or rapid temperature changes. Hypalon is difficult to apply, is more costly than natural rubber, is not resistant to oxygenated solvents, and requires a pressure cure.

Applications include linings of tanks and tank cars containing chromic and high concentrations of sulfuric and nitric acids, and other oxidizing chemicals. It is also used for pond liners.

EPDM

Ethylene-propylene rubber is a synthetic hydrocarbon based rubber made either from ethylene-propylene diene monomer (EPDM) or from ethylene-propylene terpolymer (EPT).

EPDM has exceptional heat resistance, being able to operate at temperatures of 300–350°F/148–176°C, while also finding applications as low as −70°F/−56°C.

As with other elastomeric materials EPDM can be compounded to improve specific properties. EPDM remains flexible at low temperatures. Standard compounds have brittle points of −90°F/−68°C or below.

This elastomer also exhibits good resistance to impact, abrasion, tearing, and cut growth over a wide temperature range. The physical and mechanical properties of ethylene-propylene rubber can be found in Table 4.14. Ethylene propylene terpolymer has similar physical properties.

Ethylene propylene rubber (EPDM) resists attack from oxygenated solvents (such as acetone, methyl ethyl ketone, ethyl acetate), weak acids and alkalies, detergents, phosphate esters, alcohols, and glycols. It exhibits exceptional resistance to hot water and high pressure steam. The elastomer, being hydrocar-

TABLE 4.13 Compatibility of Hypalon with Selected Corrodents[a]

Chemical	Maximum temp. °F	Maximum temp. °C	Chemical	Maximum temp. °F	Maximum temp. °C
Acetaldehyde	60	16	Barium hydroxide	200	93
Acetamide	x	x	Barium sulfate	200	93
Acetic acid 10%	200	93	Barium sulfide	200	93
Acetic acid 50%	200	93	Benzaldehyde	x	x
Acetic acid 80%	200	93	Benzene	x	x
Acetic acid, glacial	x	x	Benzene sulfonic acid 10%	x	x
Acetic anhydride	200	93			
Acetone	x	x	Benzoic acid	200	93
Acetyl chloride	x	x	Benzyl alcohol	140	60
Acrylonitrile	140	60	Benzyl chloride	x	x
Adipic acid	140	60	Borax	200	93
Allyl alcohol	200	93	Boric acid	200	93
Aluminum fluoride	200	93	Bromine gas, dry	60	16
Aluminum hydroxide	200	93	Bromine gas, moist	60	16
Aluminum nitrate	200	93	Bromine liquid	60	16
Aluminum sulfate	180	82	Butadiene	x	x
Ammonia gas	90	32	Butyl acetate	60	16
Ammonia carbonate	140	60	Butyl alcohol	200	93
Ammonium chloride 10%	190	88	Butyric acid	x	x
			Calcium bisulfite	200	93
Ammonium chloride 50%	190	88	Calcium carbonate	90	32
			Calcium chlorate	90	32
Ammonium chloride, sat.	190	88	Calcium chloride	200	93
Ammonium fluoride 10%	200	93	Calcium hydroxide 10%	200	93
Ammonium hydroxide 25%	200	93	Calcium hydroxide, sat.	200	93
			Calcium hypochlorite	200	93
Ammonium hydroxide, sat.	200	93	Calcium nitrate	100	38
			Calcium oxide	200	93
Ammonium nitrate	200	93	Calcium sulfate	200	93
Ammonium persulfate	80	27	Caprylic acid	x	x
Ammonium phosphate	140	60	Carbon dioxide, dry	200	93
Ammonium sulfate 10–40%	200	93	Carbon dioxide, wet	200	93
			Carbon disulfide	200	93
Ammonium sulfide	200	93	Carbon monoxide	x	x
Amyl acetate	60	16	Carbon tetrachloride	200	93
Amyl alcohol	200	93	Carbonic acid	x	x
Amyl chloride	x	x	Chloracetic acid	x	x
Aniline	140	60	Chlorine gas, dry	x	x
Antimony trichloride	140	60	Chlorine gas, wet	90	32
Barium carbonate	200	93	Chlorobenzene	x	x
Barium chloride	200	93	Chloroform	x	x

TABLE 4.13 Continued

Chemical	°F	°C	Chemical	°F	°C
Chlorosulfonic acid	x	x	Magnesium chloride	200	93
Chromic acid 10%	150	66	Manganese chloride	180	82
Chromic acid 50%	150	66	Methyl chloride	x	x
Chromyl chloride			Methyl ethyl ketone	x	x
Citric acid 15%	200	93	Methyl isobutyl ketone	x	x
Citric acid, concentrated	200	93	Muriatic acid	140	60
Copper acetate	x	x	Nitric acid 5%	100	38
Copper chloride	200	93	Nitric acid 20%	100	38
Copper cyanide	200	93	Nitric acid 70%	x	x
Copper sulfate	200	93	Nitric acid, anhydrous	x	x
Cresol	x	x	Oleum	x	x
Cupric chloride 5%	200	93	Perchloric acid 10%	100	38
Cupric chloride 50%	200	93	Perchloric acid 70%	90	32
Cyclohexane	x	x	Phenol	x	x
Cyclohexanol	x	x	Phosphoric acid 50–80%	200	93
Dichloroethane (ethylene dichloride)	x	x	Picric acid	80	27
			Potassium bromide 30%	200	93
Ethylene glycol	200	93	Sodium carbonate	200	93
Ferric chloride	200	93	Sodium chloride	200	93
Ferric chloride 50% in water	200	93	Sodium hydroxide 10%	200	93
			Sodium hydroxide 50%	200	93
Ferric nitrate 10–50%	200	93	Sodium hydroxide, con-centrated	200	93
Ferrous chloride	200	93			
Fluorine gas, dry	140	60	Sodium hypochlorite 20%	200	93
Hydrobromic acid, dilute	90	32	Sodium hypochlorite, concentrated		
Hydrobromic acid 20%	100	38			
Hydrobromic acid 50%	100	38	Sodium sulfide to 50%	200	93
Hydrochloric acid 20%	160	71	Stannic chloride	90	32
Hydrochloric acid 38%	140	60	Stannous chloride	200	93
Hydrocyanic acid 10%	90	32	Sulfuric acid 10%	200	93
Hydrofluoric acid 30%	90	32	Sulfuric acid 50%	200	93
Hydrofluoric acid 70%	90	32	Sulfuric acid 70%	160	71
Hydrofluoric acid 100%	90	32	Sulfuric acid 90%	x	x
Hypochlorous acid	x	x	Sulfuric acid 98%	x	x
Ketones, general	x	x	Sulfuric acid 100%	x	x
Lactic acid 25%	140	60	Sulfurous acid	160	71
Lactic acid, concen-trated	80	27	Toluene	x	x
			Zinc chloride	200	93

[a] The chemicals listed are in the pure state or in a saturated solution unless otherwise indicated. Compatibility is shown to the maximum allowable temperature for which data are available. Incompatibility is shown by an x.
Source: Ref. 1.

TABLE **4.14** Physical and Mechanical Properties of Ethylene-Propylene Rubber (EDPM)[a]

Specific gravity	0.85
Specific heat, cal/g	0.56
Brittle point	−90°F/−68°C
Resilience, %	
at 212°F/100°C	78
at 75°F/24°C	77
at 14°F/−10°C	63
Dielectric strength, V/mil	800
Insulation resistance, megohms/1000 ft	25,500
Insulation resistance, constant K, megohms/1000 ft	76,400
Permeability to air at 86°F (30°C), cm^3/cm^2-cm-s-atm	8.5×10^{-8}
Tensile strength, psi	To 3500
Elongation, % at break	560
Hardness, Shore A	30–90
Abrasion resistance	Good
Maximum temperature, continuous use	300°F/148°C
Impact resistance	Good
Compression set, %	
at 158°F/70°C	8–10
at 212°F/100°C	12–26
Resistance to sunlight	Excellent
Effect of aging	Nil
Resistance to heat	Excellent
Tear resistance	Good

[a]These are representative values since they may be altered by compounding.

bon based, is not resistant to hydrocarbon solvents or oils, chlorinated hydrocarbons, or turpentine.

Ethylene propylene terpolymer rubbers are in general resistant to most of the same corrodents as EPDM but do not have as broad a resistance to mineral acids and some organics.

Ethylene propylene rubbers are particularly resistant to sun, weather, and ozone attack. Ozone resistance is inherent in the polymer, and for all practical purposes it can be considered immune to ozone attack.

The compatibility of EPDM rubber with selected corrodents will be found in Table 4.15 with additional listings in Ref. 1.

The advantages of EPDM as a lining material are many:

It has a high heat resistance.
It is unaffected by cold or rapid temperature changes.

It has excellent resistance to ozone, sunlight, and weather.
It has good abrasion and chemical resistance.

On the negative side is the fact that it is neither oil nor flame resistant and it is difficult to apply.

EPDM has been used to line sodium hypochlorite storage and make-up tanks and has been used as a lining for other bleach solutions. Pumps have also been lined with EPDM.

URETHANE RUBBERS

The urethane rubbers are produced from a number of polyurethane polymers. The properties exhibited are dependent upon the specific polymer and the compounding. Urethane rubbers operate through the range of 50–212°F/10–100°C. Standard compounds exhibit good low temperature impact resistance and low brittle points. At very low temperature the material remains flexible and has good resistance to thermal shock. The standard compositions can be used at temperature as low as −80°F/−18°C.

Urethane rubbers have little or no resistance to burning. Some slight improvement can be made in this area by special compounding, but the material will still ignite in an actual fire situation.

Urethane has a low coefficient of friction, good abrasion resistance and good impact resistance. The physical and mechanical properties of urethane are shown in Table 4.16.

The urethane rubbers are resistant to most mineral and vegetable oils, greases and fuels, and to aliphatic and chlorinated hydrocarbons.

Aromatic hydrocarbons, polar solvents, esters, and ketones will attack urethane. Alcohols will soften and swell the urethane rubbers. These rubbers have limited service in weak acid solutions and cannot be used with concentrated acids. Nor are they resistant to steam or caustic, but they are resistant to the swelling and deteriorating effects of being immersed in water.

The urethane rubbers exhibit excellent resistance to ozone attack and have good resistance to weathering. However, extended exposure to ultraviolet light will reduce their physical properties and will cause the rubber to darken. Table 4.17 lists the compatibility of urethane with selected corrodents.

Urethane is used to line tanks and piping. It is also applied as a coating to protect concrete from chemical attack and to prevent seepage of the corrodent into the concrete and attacking the reinforcing steel. Protection from abrasion and erosion is also provided.

TABLE 4.15 Compatibility of EDPM Rubber with Selected Corrodents[a]

Chemical	°F	°C	Chemical	°F	°C
Acetaldehyde	200	93	Barium chloride	200	93
Acetamide	200	93	Barium hydroxide	200	93
Acetic acid 10%	140	60	Barium sulfate	200	93
Acetic acid 50%	140	60	Barium sulfide	140	60
Acetic acid 80%	140	60	Benzaldehyde	150	66
Acetic acid, glacial	140	60	Benzene	x	x
Acetic anhydride	x	x	Benzene sulfonic acid 10%	x	x
Acetone	200	93	Benzoic acid	x	x
Acetyl chloride	x	x	Benzyl alcohol	x	x
Acrylonitrile	140	60	Benzyl chloride	x	x
Adipic acid	200	93	Borax	200	93
Allyl alcohol	200	93	Boric acid	190	88
Allyl chloride	x	x	Bromine gas, dry	x	x
Alum	200	93	Bromine gas, moist	x	x
Aluminum fluoride	190	88	Bromine liquid	x	x
Aluminum hydroxide	200	93	Butadiene	x	x
Aluminum nitrate	200	93	Butyl acetate	140	60
Aluminum sulfate	190	88	Butyl alcohol	200	93
Ammonia gas	200	93	Butyric acid	140	60
Ammonium bifluoride	200	93	Calcium bisulfite	x	x
Ammonium carbonate	200	93	Calcium carbonate	200	93
Ammonium chloride 10%	200	93	Calcium chlorate	140	60
Ammonium chloride 50%	200	93	Calcium chloride	200	93
Ammonium chloride, sat.	200	93	Calcium hydroxide 10%	200	93
Ammonium fluoride 10%	200	93	Calcium hydroxide, sat.	200	93
Ammonium fluoride 25%	200	93	Calcium hypochlorite	200	93
Ammonium hydroxide 25%	100	38	Calcium nitrate	200	93
Ammonium hydroxide, sat.	100	38	Calcium oxide	200	93
Ammonium nitrate	200	93	Calcium sulfate	200	93
Ammonium persulfate	200	93	Carbon bisulfide	x	x
Ammonium phosphate	200	93	Carbon dioxide, dry	200	93
Ammonium sulfate 10–40%	200	93	Carbon dioxide, wet	200	93
			Carbon disulfide	200	93
Ammonium sulfide	200	93	Carbon monoxide	x	x
Amyl acetate	200	93	Carbon tetrachloride	200	93
Amyl alcohol	200	93	Carbonic acid	x	x
Amyl chloride	x	x	Cellosolve	200	93
Aniline	140	60	Chloracetic acid	160	71
Antimony trichloride	200	93	Chlorine gas, dry	x	x
Aqua regia 3:1	x	x	Chlorine gas, wet	x	x
Barium carbonate	200	93	Chlorine, liquid	x	x

TABLE 4.15 Continued

Chemical	Maximum temp. °F	Maximum temp. °C	Chemical	Maximum temp. °F	Maximum temp. °C
Chlorobenzene	x	x	Lactic acid, concentrated		
Chloroform	x	x	Magnesium chloride	200	93
Chlorosulfonic acid	x	x	Malic acid	x	x
Chromic acid 50%	x	x	Methyl chloride	x	x
Citric acid 15%	200	93	Methyl ethyl ketone	80	27
Citric acid, concentrated	200	93	Methyl isobutyl ketone	60	16
Copper acetate	100	38	Nitric acid 5%	60	16
Copper carbonate	200	93	Nitric acid 20%	60	16
Copper chloride	200	93	Nitric acid 70%	x	x
Copper cyanide	200	93	Nitric acid, anhydrous	x	x
Copper sulfate	200	93	Oleum	x	x
Cresol	x	x	Perchloric acid 10%	140	60
Cupric chloride 5%	200	93	Phosphoric acid 50–80%	140	60
Cupric chloride 50%	200	93	Picric acid	200	93
Cyclohexane	x	x	Potassium bromide 30%	200	93
Cyclohexanol	x	x	Salicylic acid	200	93
Dichloroethane (ethylene dichloride)	x	x	Sodium carbonate	200	93
			Sodium chloride	140	60
Ethylene glycol	200	93	Sodium hydroxide 10%	200	93
Ferric chloride	200	93	Sodium hydroxide 50%	180	82
Ferric chloride 50% in water	200	93	Sodium hydroxide, concentrated	180	82
Ferric nitrate 10–50%	200	93	Sodium hypochlorite 20%	200	93
Ferrous chloride	200	93	Sodium hypochlorite, concentrated	200	93
Ferrous nitrate	200	93			
Fluorine gas, moist	60	16	Sodium sulfide to 50%	200	93
Hydrobromic acid, dilute	90	32	Stannic chloride	200	93
Hydrobromic acid 20%	140	60	Stannous chloride	200	93
Hydrobromic acid 50%	140	60	Sulfuric acid 10%	150	66
Hydrochloric acid 20%	100	38	Sulfuric acid 50%	150	66
Hydrochloric acid 38%	90	32	Sulfuric acid 70%	140	60
Hydrocyanic acid 10%	200	93	Sulfuric acid 90%	x	x
Hydrofluoric acid 30%	60	16	Sulfuric acid 98%	x	x
Hydrofluoric acid 70%	x	x	Sulfuric acid 100%	x	x
Hydrofluoric acid 100%	x	x	Sulfuric acid, fuming	x	x
Hypochlorous acid	200	93	Toluene	x	x
Iodine solution 10%	140	60	Trichloroacetic acid	80	27
Ketones, general	x	x	White liquor	200	93
Lactic acid 25%	140	60	Zinc chloride	200	93

[a] The chemicals listed are in the pure state or in a saturated solution unless otherwise indicated. Compatibility is shown to the maximum allowable temperature for which data are available. Incompatibility is shown by an x. A blank space indicates that the data are unavailable.
Source: Ref. 1.

TABLE 4.16 Physical and Mechanical Properties of Urethane Rubbers[a]

Specific gravity	1.02–1.20
Brittle point	−85 to −100°F/−65 to −73°C
Dielectric strength, V/mil	450–500
Permeability to air	Good
Dielectric constant	
at 0.1 kHz	9.4
at 100 kHz	7.8
Tensile strength, psi	3000–8000
Elongation, % at break	270–400
Hardness, Shore A	30–80
Abrasion resistance	Excellent
Compression set, %	30–45
Machining qualities	Readily machined
Resistance to sunlight	Good
Effect of aging	Little effect
Resistance to heat	Good
Maximum temperature, continuous use	250°F/121°C

[a]These are representative values since they may be altered by compounding.

TABLE 4.17A Compatibility of Urethane Rubbers with Selected Corrodents[a]

Chemical	Maximum temp.		Chemical	Maximum temp.	
	°F	°C		°F	°C
Acetamide	x	x	Amyl alcohol	x	x
Acetic acid 10%	x	x	Aniline	x	x
Acetic acid 50%	x	x	Benzene	x	x
Acetic acid 80%	x	x	Benzoic acid	x	x
Acetic acid, glacial	x	x	Bromine water, dil	x	x
Acetone	x	x	Bromine water, sat.	x	x
Ammonia gas			Butane	100	38
Ammonium chloride 10%	90	32	Butyl acetate	x	x
Ammonium chloride 28%	90	32	Butyl alcohol	x	x
Ammonium chloride 50%	90	32	Calcium chloride, dil	x	x
Ammonium chloride, sat.	90	32	Calcium chloride, sat.	x	x
Ammonium hydroxide 10%	90	32	Calcium hydroxide 10%	x	x
			Calcium hydroxide 20%	90	32
Ammonium hydroxide 25%	90	32	Calcium hydroxide 30%	90	32
			Calcium hydroxide, sat.	90	32
Ammonium hydroxide, sat.	80	27	Cane sugar liquors		
Ammonium sulfate 10–40%			Carbon bisulfide		
			Carbon tetrachloride	x	x

TABLE 4.17A Continued

Chemical	Maximum temp. °F	°C	Chemical	Maximum temp. °F	°C
Carbonic acid	90	32	Jet fuel JP 5	x	x
Castor oil	90	32	Kerosene	90	32
Cellosolve	x	x	Lard oil	90	32
Chlorine water, sat.	x	x	Linseed oil	90	32
Chlorobenzene	x	x	Magnesium hydroxide	90	32
Chloroform	x	x	Mercury	90	32
Chromic acid 10%	x	x	Methyl alcohol	90	32
Chromic acid 30%	x	x	Methyl ethyl ketone	x	x
Chromic acid 40%	x	x	Methyl isobutyl ketone	x	x
Chromic acid 50%	x	x	Mineral oil	90	32
Citric acid 5%	x	x	Muriatic acid	x	x
Citric acid 10%	x	x	Nitric acid	x	x
Citric acid 15%	x	x	Oils and fats	80	27
Citric acid, concd			Phenol	x	x
Copper sulfate	90	32	Potassium bromide 30%	90	32
Corn oil			Potassium chloride 30%	90	32
Cottonseed oil	90	32	Potassium hydroxide to 50%	90	32
Cresol	x	x			
Dextrose	x	x	Potassium sulfate 10%	90	32
Dibutyl phthalate	x	x	Silver nitrate	90	32
Ethers, general	x	x	Soaps	x	x
Ethyl acetate	x	x	Sodium carbonate	x	x
Ethyl alcohol	x	x	Sodium chloride	80	27
Ethylene chloride			Sodium hydroxide to 50%	90	32
Ethylene glycol	90	32			
Fomaldehyde, dil	x	x	Sodium hypochlorite, all concn	x	x
Formaldehyde 37%	x	x			
Formaldehyde 50%	x	x	Sulfuric acid, all concn	x	x
Glycerine	90	32	Toluene	x	x
Hydrochloric acid, dil	x	x	Trichloroethylene	x	x
Hydrochloric acid 20%	x	x	Trisodium phosphate	90	32
Hydrochloric acid 35%	x	x	Turpentine	x	x
Hydrochloric acid 38%	x	x	Water, salt	x	x
Hydrochloric acid 50%	x	x	Water, sea	x	x
Isopropyl alcohol	x	x	Whiskey	x	x
Jet fuel JP 4	x	x	Xylene	x	x

[a] The chemicals listed are in the pure state or in a saturated solution unless otherwise indicated. Compatibility is shown to the maximum allowable temperature for which data are available. Incompatibility is shown by an x. A blank space indicates that data are unavailable.
Source: Ref. 1.

TABLE 4.17B Compatibility of Polyether Urethane Rubbers with Selected Corrodents[a]

Chemical	Maximum temp.		Chemical	Maximum temp.	
	°F	°C		°F	°C
Acetamide	x	x	Carbon bisulfide	x	x
Acetic acid 10%	x	x	Carbon tetrachloride	x	x
Acetic acid 50%	x	x	Carbonic acid	80	27
Acetic acid 80%	x	x	Castor oil	130	54
Acetic acid, glacial	x	x	Caustic potash	130	54
Acetone	x	x	Chloroacetic acid	x	x
Acetyl chloride	x	x	Chlorobenzene	x	x
Ammonium chloride 10%	130	54	Chloroform	x	x
			Chromic acid 10%	x	x
Ammonium chloride 28%	130	54	Chromic acid 30%	x	x
			Chromic acid 40%	x	x
Ammonium chloride 50%	130	54	Chromic acid 50%	x	x
			Citric acid 5%	90	32
Ammonium chloride, sat.	130	54	Citric acid 10%	90	32
Ammonium hydroxide 10%	110	43	Citric acid 15%	90	32
			Cottonseed Oil	130	38
Ammonium hydroxide 25%	110	43	Cresylic acid	x	x
			Cyclohexane	80	27
Ammonium hydroxide, sat.	100	38	Dioxane	x	x
			Diphenyl	x	x
Ammonium sulfate 10– 40%	120	49	Ethyl acetate	x	x
			Ethyl benzene	x	x
Amyl alcohol	x	x	Ethyl chloride	x	x
Aniline	x	x	Ethylene oxide	x	x
Benzene	x	x	Ferric chloride 75%	100	38
Benzoic acid	x	x	Fluorosilicic acid	100	38
Benzyl alcohol	x	x	Formic acid	x	x
Borax	90	32	Freon F-11	x	x
Butane	80	27	Freon F-12	x	x
Butyl acetate	x	x	Glycerine	130	38
Butyl alcohol	x	x	Hydrochloric acid, dil	x	x
Calcium chloride, dil	130	54	Hydrochloric acid 20%	x	x
Calcium chloride, sat.	130	54	Hydrochloric acid 35%	x	x
Calcium hydroxide 10%	130	54	Hydrochloric acid 38%	x	x
Calcium hydroxide 20%	130	54	Hydrochloric acid 50%	x	x
Calcium hydroxide 30%	130	54	Hydrofluoric acid, dil	x	x
Calcium hydroxide, sat.	130	54	Hydrofluoric acid 30%	x	x
Calcium hypochlorite 5%	130	54	Hydrofluoric acid 40%	x	x

TABLE 4.17B   Continued

Chemical	°F	°C	Chemical	°F	°C
Hydrofluoric acid 50%	x	x	Propane	x	x
Hydrofluoric acid 70%	x	x	Propyl acetate	x	x
Hydrofluoric acid 100%	x	x	Propyl alcohol	x	x
Isobutyl alcohol	x	x	Sodium chloride	130	54
Isopropyl acetate	x	x	Sodium hydroxide 30%	140	60
Lactic acid, all conc.	x	x	Sodium hypochlorite	x	x
Methyl ethyl ketone	x	x	Sodium peroxide	x	x
Methyl isobutyl ketone	x	x	Sodium phosphate, acid	130	54
Monochlorobenzene	x	x	Sodium phosphate, alk.	130	54
Muriatic acid	x	x	Sodium phosphate, neut.	130	54
Nitrobenzene	x	x	Toluene	x	x
Olive oil	120	49	Trichloroethylene	x	x
Phenol	x	x	Water, demineralized	130	54
Potassium acetate	x	x	Water, distilled	130	54
Potassium chloride 30%	130	54	Water, salt	130	54
Potassium hydroxide	130	54	Water, sea	130	54
Potassium sulfate 10%	130	54	Xylene	x	x

[a] The chemicals listed are in the pure state or in a saturated solution unless otherwise indicated. Compatibility is shown to the maximum allowable temperature for which data are available. Incompatibility is shown by an x. A blank space indicates that data are unavailable. *Source*: Ref. 1.

NITRILE RUBBER (NBR, BUNA-N)

Nitrile rubber is a copolymer of butadiene and acrylonitrile. The main advantages of the nitrile rubbers are their low cost, good oil and abrasion resistance, and good low temperature and swell characteristics.

The physical and mechanical properties of the nitrile rubbers are very similar to those of natural rubber. Buna-N does not have exceptional heat resistance. It has a maximum operating temperature of 200°F/93°C and has a tendency to harden at elevated temperatures. The nitrile rubbers will support combustion and burn. NBR has good abrasion resistance and tensile strength.

Table 4.18 provides the physical and mechanical properties of nitrile rubber.

The nitrile rubbers exhibit good resistance to solvents, oil, water, and hydraulic fluids. A very slight swelling occurs in the presence of aliphatic hydrocar-

TABLE 4.18 Physical and Mechanical Properties of Nitrile Rubber (NBR, Buna-N)[a]

Specific gravity	0.99
Refractive index	1.54
Brittle point	-32 to $-40°F/-1$ to $-40°C$
Swelling, % by volume	
in kerosene at 77°F/25°C	9–10
in benzene at 77°F/25°C	120
in acetone at 77°F/25°C	60–50
in mineral oil at 100°F/70°C	2–10
in air at 77°F/25°C	30–50
Tensile strength, psi	500–4000
Elongation, % at break	400
Hardness, Shore A	40–95
Abrasion resistance	Excellent
Maximum temperature, continuous use	250°F/120°C
Compression set	Good
Tear resistance	Excellent
Resilience, %	63–74
Machining qualities	Can be ground
Resistance to sunlight	Fair
Effect of aging	Highly resistant
Resistance to heat	Softens

[a]These are representative values since they may be altered by compounding.

bons, fatty acids, alcohols, and glycols. The deterioration of physical properties as a result of this swelling is small, making NBR suitable for gasoline and oil resistant applications.

The use of highly polar solvents such as acetone and methyl ethyl ketone, chlorinated hydrocarbons, ether, or esters should be avoided, since these materials will attack the nitrile rubbers. Table 4.19 shows the compatibility of the nitrile rubbers with selected corrodents.

Nitrile rubbers are unaffected by cold weather or rapid temperature change and provide excellent resistance to aliphatic hydrocarbons. They also have good abrasion resistance. However, this material is more costly than natural rubber and is more difficult to apply. Buna-N is used to line tanks and piping.

TABLE **4.19** Compatibility of Nitrile Rubber with Selected Corrodents[a]

Chemical	Maximum temp. °F	Maximum temp. °C	Chemical	Maximum temp. °F	Maximum temp. °C
Acetaldehyde	x	x	Ammonium phosphate	150	66
Acetamide	180	82	Ammonium sulfate 10–		
Acetic acid 10%	x	x	40%		
Acetic acid 50%	x	x	Ammonium sulfide		
Acetic acid 80%	x	x	Ammonium sulfite		
Acetic acid, glacial	x	x	Amyl acetate		
Acetic anhydride	x	x	Amyl alcohol	150	66
Acetone	x	x	Amyl chloride		
Acetyl chloride	x	x	Aniline	x	x
Acrylic acid	x	x	Antimony trichloride		
Acrylonitrile	x	x	Aqua regia 3:1		
Adipic acid	180	82	Barium carbonate		
Allyl alcohol	180	82	Barium chloride	125	52
Allyl chloride	x	x	Barium hydroxide		
Alum	150	66	Barium sulfate		
Aluminum acetate			Barium sulfide		
Aluminum chloride, aqueous	150	66	Benzaldehyde		
Aluminum chloride, dry			Benzene	150	66
Aluminum fluoride			Benzene sulfonic acid 10%		
Aluminum hydroxide	180	82	Benzoic acid	150	66
Aluminum nitrate	190	88	Benzyl alcohol		
Aluminum oxychloride			Benzyl chloride		
Aluminum sulfate	200	93	Borax		
Ammonia gas	190	88	Boric acid	150	66
Ammonium bifluoride			Bromine gas, dry		
Ammonium carbonate	x	x	Bromine gas, moist		
Ammonium chloride 10%			Bromine liquid		
Ammonium chloride 50%			Butadiene		
Ammonium chloride, sat.			Butyl acetate		
Ammonium fluoride 10%			Butyl alcohol		
Ammonium fluoride 25%			n-Butylamine		
Ammonium hydroxide 25%			Butyl phthalate		
Ammonium hydroxide, sat.			Butyric acid		
			Calcium bisulfide		
			Calcium bisulfite		
			Calcium carbonate		
			Calcium chlorate		
Ammonium nitrate	150	66	Calcium chloride		
Ammonium persulfate			Calcium hydroxide 10%		

TABLE 4.19 Continued

Chemical	Maximum temp. °F	Maximum temp. °C	Chemical	Maximum temp. °F	Maximum temp. °C
Calcium hydroxide, sat.			Dichloroethane (ethylene dichloride)		
Calcium hypochlorite	x	x	Ethylene glycol	100	38
Calcium nitrate			Ferric chloride	150	66
Calcium oxide			Ferric chloride 50% in water		
Calcium sulfate			Ferric nitrate 10–50%	150	66
Caprylic acid			Ferrous chloride		
Carbon bisulfide			Ferrous nitrate		
Carbon dioxide, dry			Fluorine gas, dry		
Carbon dioxide, wet			Fluorine gas, moist		
Carbon disulfide			Hydrobromic acid, dilute		
Carbon monoxide			Hydrobromic acid 20%		
Carbon tetrachloride			Hydrobromic acid 50%		
Carbonic acid	100	38	Hydrochloric acid 20%		
Cellosolve			Hydrochloric acid 38%		
Chloracetic acid, 50% water			Hydrocyanic acid 10%		
Chloracetic acid			Hydrofluoric acid 30%		
Chlorine gas, dry			Hydrofluoric acid 70%	x	x
Chlorine gas, wet			Hydrofluoric acid 100%	x	x
Chlorine, liquid			Hypochlorous acid		
Chlorobenzene			Iodine solution 10%		
Chloroform			Ketones, general		
Chlorosulfonic acid			Lactic acid 25%		
Chromic acid 10%			Lactic acid, concentrated		
Chromic acid 50%			Magnesium chloride		
Chromyl chloride			Malic acid		
Citric acid 15%			Manganese chloride		
Citric acid, concentrated			Methyl chloride		
Copper acetate			Methyl ethyl ketone		
Copper carbonate			Methyl isobutyl ketone		
Copper chloride			Muriatic acid		
Copper cyanide			Nitric acid 5%		
Copper sulfate			Nitric acid 20%	x	x
Cresol			Nitric acid 70%	x	x
Cupric chloride 5%			Nitric acid, anhydrous	x	x
Cupric chloride 50%			Nitrous acid, concentrated		
Cyclohexane			Oleum		
Cyclohexanol			Perchloric acid 10%		
Dichloroacetic acid					

TABLE 4.19 Continued

Chemical	Maximum temp. °F	Maximum temp. °C	Chemical	Maximum temp. °F	Maximum temp. °C
Perchloric acid 70%			Sodium sulfide to 50%		
Phenol	x	x	Stannic chloride	150	66
Phosphoric acid 50–80%	150	66	Stannous chloride		
Picric acid			Sulfuric acid 10%	150	66
Potassium bromide 30%			Sulfuric acid 50%	150	66
Salicylic acid			Sulfuric acid 70%	x	x
Silver bromide 10%			Sulfuric acid 90%	x	x
Sodium carbonate	125	52	Sulfuric acid 98%	x	x
Sodium chloride	200	93	Sulfuric acid 100%	x	x
Sodium hydroxide 10%	150	66	Sulfuric acid, fuming	x	x
Sodium hydroxide 50%			Sulfurous acid		
Sodium hydroxide, concentrated			Thionyl chloride		
			Toluene		
Sodium hypochlorite 20%	x	x	Trichloroacetic acid		
Sodium hypochlorite, concentrated	x	x	White liquor		
			Zinc chloride	150	66

[a] The chemicals listed are in the pure state or in a saturated solution unless otherwise indicated. Compatibility is shown to the maximum allowable temperature for which data are available. Incompatibility is shown by an x. A blank space indicates that data are unavailable.
Source: Ref. 1.

POLYESTER ELASTOMER (PE)

Polyester elastomer is strong, resilient, has good abrasion and good retention of properties at elevated temperatures. The mechanical properties of these elastomers are maintained up to 302°F/150°C, better than many rubbers. Above 240°F/120°C their tensile strengths far exceed those of other rubbers. Their hot strength and good resistance to hot fluids make PE elastomers suitable for many applications involving fluid containment.

All of the PE formulations have brittle points below −94°F/−70°C and exhibit resistance to stiffening at temperatures down to −40°F/−40°C.

The physical and mechanical properties of these elastomers are shown in Table 4.20.

In general, the fluid resistance of these polyester rubbers increases with increasing hardness. Since these rubbers contain no plasticizers, they are not susceptible to the solvent extraction or heat volatilization of such additives. Many fluids and chemicals will extract plasticizers from elastomers, causing a significant increase in stiffness (modulus) and volume shrinkage.

TABLE 4.20 Physical and Mechanical Properties of Polyester
Elastomer[a]

Specific gravity	1.17–1.25
Brittle point	−94°F/−70°C
Resilience, %	42–62
Coefficient of linear expansion, in./in.-°C	2×10^{-5}–21×10^{-5}
Dielectric strength, V/mil	645–900
Dielectric constant at 1 kHz	4.16–6
Permeability to air	Low
Tear strength	
lb/in.	631–1055
kn/m	110–185
Water absorption, %/24 h	0.6–.03
Tensile strength, psi	3700–5700
Elongation, % at break	350–450
Hardness, Shore D	40–72
Abrasion resistance	Good
Maximum temperature, continuous use	302°F/150°C
Impact resistance	Good
Compression set, %	
at 73°F/23°C	1–11
at 158°F/70°C	2–27
at 212°F/100°C	4–44
Resistance to sunlight	Good
Effect of aging	Nil
Resistance to heat	Excellent
Flexural modulus, psi	7,000–75,000

[a] These are representative values since they may be altered by compounding.

Overall, PE elastomers are resistant to the same class of chemicals and fluids that the polyurethanes are. However, PE has better high temperature properties than the polyurethanes and can be used satisfactorily at higher temperatures in the same fluids.

Polyester elastomers have excellent resistance to nonpolar materials such as oils and hydraulic fluids, even at elevated temperatures. At room temperature, PE elastomers are resistant to most polar fluids, such as acids, bases, amines, and glycols. Resistance is very poor at temperatures of 158°F/70°C or above. These rubbers should not be in applications requiring continuous exposure to polar fluids at elevated temperatures.

The compatibility of PE elastomers with selected corrodents is shown in Table 4.21.

PE elastomers are used to line tanks, ponds, swimming pools, and drums.

TABLE 4.21 Compatibility of Polyester Elastomers with Selected Corrodents[a]

Chemical	Maximum temp. °F	°C	Chemical	Maximum temp. °F	°C
Acetic acid 10%	80	27	Ethyl alcohol	80	27
Acetic acid 50%	80	27	Ethylene chloride	x	x
Acetic acid 80%	80	27	Ethylene glycol	80	27
Acetic acid, glacial	100	38	Formaldehyde, dil	80	27
Acetic acid, vapor	90	32	Formaldehyde 37%	80	27
Ammonium chloride 10%	90	32	Formic acid 10 to 85%	80	27
Ammonium chloride 28%	90	32	Glycerine	80	27
Ammonium chloride 50%	90	32	Hydrochloric acid, dil.	80	27
Ammonium chloride, sat.	90	32	Hydrochloric acid 20%	x	x
Ammonium sulfate 10–40%	80	27	Hydrochloric acid 35%	x	x
Amyl acetate	80	27	Hydrochloric acid 38%	x	x
Aniline	x	x	Hydrochloric acid 50%	x	x
Beer	80	27	Hydrofluoric acid, dil.	x	x
Benzene	80	27	Hydrofluoric acid 30%	x	x
Borax	80	27	Hydrofluoric acid 40%	x	x
Boric acid	80	27	Hydrofluoric acid 50%	x	x
Bromine, liquid	x	x	Hydrofluoric acid 70%	x	x
Butane	80	27	Hydrofluoric acid 100%	x	x
Butyl acetate	80	27	Hydrogen	80	27
Calcium chloride, dil	80	27	Hydrogen sulfide, dry	80	27
Calcium chloride, sat.	80	27	Lactic acid	80	27
Calcium hydroxide 10%	80	27	Methyl ethyl ketone	80	27
Calcium hydroxide 20%	80	27	Methylene chloride	x	x
Calcium hydroxide 30%	80	27	Mineral oil	80	27
Calcium hydroxide, sat.	80	27	Muriatic acid	x	x
Calcium hypochlorite 5%	80	27	Nitric acid	x	x
Carbon bisulfide	80	27	Nitrobenzene	x	x
Carbon tetrachloride	x	x	Ozone	80	27
Castor oil	80	27	Phenol	x	x
Chlorine gas, dry	x	x	Potassium dichromate 30%	80	27
Chlorine gas, wet	x	x	Potassium hydroxide	80	27
Chlorobenzene	x	x	Pyridine	80	27
Chloroform	x	x	Soap solutions	80	27
Chlorosulfonic acid	x	x	Sodium chloride	80	27
Citric acid 5%	80	27	Sodium hydroxide, all conc.	80	27
Citric acid 10%	80	27	Stannous chloride 15%	80	27
Citric acid 15%	80	27	Stearic acid	80	27
Citric acid, conc.	80	27	Sulfuric acid, all conc.	x	x
Copper sulfate	80	27	Toluene	80	27
Cottonseed oil	80	27	Water, demineralized	160	71
Cupric chloride to 50%	80	27	Water, distilled	160	71
Cyclohexane	80	27	Water, salt	160	71
Dibutyl phthalate	80	27	Water, sea	160	71
Dioctyl phthalate	80	27	Xylene	80	27
Ethyl acetate	80	27	Zinc chloride	80	27

[a]The chemicals listed are in the pure state or in a saturated solution unless otherwise indicated. Compatibility is shown to the maximum allowable temperature for which data are available. Incompatibility is shown by an x. A blank space indicates that data are unavailable.
Source: Ref. 1.

PERFLUOROELASTOMERS (FPM)

Perfluoroelastomers provide the elastomeric properties of fluoroelastomers and an exceptionally wide range of chemical resistance. Compared with other elastomeric compounds, they are more resistant to swelling and embrittlement and retain their elastomeric properties over the long term. In difficult environments there are no other elastomers that can outperform the perfluoroelastomers.

There are two categories of the perfluoroelastomers—fully fluorinated and partially fluorinated. Fully fluorinated materials include PTFE, FEP and PFA. These materials have the higher temperature limit (>300°F/150°C) and chemical resistances.

Partially fluorinated materials are ECTFE, ETFE, and PVDF. These materials have higher mechanical properties but lower temperature (<300°F/150°C) and chemical resistance. Fluoropolymer linings can cost 5–20 times more than thermoset linings so they are typically restricted to elevated temperature environment.

Fluoropolymer linings are installed using adhesive bonded fabric backed sheets, sprayed dispersions, rotolining, electrostatic sprayed powders, and by mechanically attaching loose sheet linings. General physical and mechanical properties of perfluoroelastomeric linings can be found in Table 4.22.

The corrosion resistance of the various perfluoroelastomers can be found in tables noted:

Perfluoroelastomer	Table
PVDF	Table 3.16
PTFE	Table 3.18
PFA	Table 3.22
FEP	Table 3.25
ECTFE	Table 3.27
ETFE	Table 3.29

FLUOROELASTOMERS (FKM)

Fluoroelastomers are fluorine containing hydrocarbon polymers with a saturated structure obtained by polymerizing fluorinated monomers such as vinylidene fluoride, hexafluoroprene, and tetrafluoroethylene. The result is a high performance rubber with exceptional resistance to oils and chemicals at elevated temperatures.

As with other rubbers, fluoropolymers are capable of being compounded with various additives to enhance specific properties for particular applications.

TABLE 4.22 Physical and Mechanical Properties of Perfluoroelastomers[a]

Specific gravity	1.9–2.0
Specific heat at 122–302°F (50–150°C), cal/g	0.226–0.250
Brittle point	−9 to −58°F/−23 to −50°C
Coefficient of friction (to steel)	0.25–0.60
Tear strength, psi	1.75–27
Coefficient of linear expansion	
in./°F	1.3×10^{-4}
in./°C	2.3×10^{-4}
Thermal conductivity, BTU-in./hr-°F-ft^2	
at 122°F (50°C)	1.3
at 212°F (100°C)	1.27
at 392°F (200°C)	1.19
at 572°F (300°C)	1.10
Dielectric constant, kV/mm	17.7
Dielectric constant at 1000 Hz	4.9
Dissipation factor at 1000 Hz	5×10^{-3}
Permeability ($\times 10^{-9}$ cm^3-cm/S-cm^2-cmHg P)	
to nitrogen, at room temperature	0.05
to oxygen, at room temperature	0.09
to helium, at room temperature	2.5
to hydrogen, at 199°F (93°C)	113
Tensile strength, psi	1850–3800
Elongation, % at break	20–190
Hardness, Shore A	65–95
Maximum temperature, continuous use	600°F/316°C
Abrasion resistance, NBS	121
Compression set, %	
at room temperature	15–40
at 212°F (100°C)	32–54
at 400°F (204°C)	63–82
at 500°F (260°C)	63–79
Resistance to sunlight	Excellent
Effect of aging	Nil
Resistance to heat	Excellent

[a] These are representative values since they may be altered by compounding.

Fluoropolymers are manufactured under various trade names by different manufacturers. Three typical materials are listed below:

Trade Name	Manufacturer
Viton	DuPont
Technoflon	Ausimont
Fluorel	3M

The general physical and mechanical properties of fluoroelastomers are similar to those of other synthetic rubbers. Fluoroelastomer compounds have good tensile strengths, ranging from 188 to 2900 psi. In general, the tensile strength of any elastomer tends to decrease at elevated temperatures; however, loss in tensile strength is much less with the fluoroelastomers.

The temperature resistance of the fluoroelastomers is exceptionally good over a wide temperature range. At high temperatures their mechanical properties are retained better than those of any other elastomer. Compounds remain usefully elastic indefinitely when exposed to aging up to 400°F/204°C. Continuous service limits are generally considered to be as follows:

3000 h at 450°F/232°C
1000 h at 500°F/260°C
240 h at 550°F/288°C
48 h at 600°F/313°C

These compounds are suitable for continuous use at 400°F/205°C depending upon the corrodent.

On the low temperature side these rubbers are generally serviceable in dynamic applications down to −10°F/−23°C. Flexibility at low temperatures is a function of the material thickness. The thinner the cross section, the less stiff the material is at every temperature. The brittle point at a thickness of 0.075 in. (1.9 mm) is in the neighborhood of −50°F/−45°C.

Fluoroelastomers are relatively impermeable to air and gases, ranking about midway between the best and the poorest elastomers in this respect. In all cases permeability increases rapidly with increasing temperature. This permeability can be modified considerably by the way they are compounded. Table 4.23 provides some data on the permeability as well as the physical and mechanical properties of the fluoroelastomers.

TABLE 4.23 Physical and Mechanical Properties of Fluoroelastomers[a]

Specific gravity	1.8
Specific heat	0.395
Brittle point	-25 to $-75°F/-32$ to $-59°C$
Coefficient of linear expansion	$88 \times 10^{-6}/°F$, $16 \times 10^{-5}/°C$
Thermal conductivity	
BTU-in./h-ft²°F at 100°F	1.58
kg-cal/cm-cm²-°C-h at 38°C	1.96
Electrical properties	
Dielectric constant at 1000 Hz	
at 75°F/24°C	10.5
at 300°F/149°C	7.1
at 390°F/199°C	9.1
Dissipation factor at 1000 Hz	
at 75°F/24°C	0.034
at 300°F/149°C	0.273
at 390°F/199°C	0.39–1.19
Permeability, cm³/cm²-cm-sec-atm at 75°F/24°C	
to air	0.0099×10^{-7}
to helium	0.892×10^{-7}
to nitrogen	0.0054×10^{-7}
at 86°F/30°C	
to carbon dioxide	0.59×10^{-7}
to oxygen	0.11×10^{-7}
Tensile strength, psi	1800–2900
Elongation, % at break	400
Hardness, Shore A	45–95
Abrasion resistance	Good
Maximum temperature, continuous use	400°F/205°C
Compression set, %	
at 70°F/21°C	21
at 300°F/149°C	32
at 392°F/200°C	98
Tear resistance	Good
Resistance to sunlight	Excellent
Effect of aging	Nil
Resistance to heat	Excellent

[a] These are representative values since they may be altered by compounding.

Being halogen containing polymers, these elastomers are more resistant to burning than are exclusively hydrocarbon rubbers. Normally compounded material will burn when directly exposed to flame, but will stop burning when the flame is removed. However, it must be remembered that under an actual fire condition the fluoroelastomers will burn. During combustion, fluorinated products such as hydrofluoric acid can be given off. Special compounding can improve the flame resistance. One such formulation has been developed for the space program that will not ignite under conditions of the NASA test, which specifies 100% oxygen at 6.2 psi absolute.

Fluoroelastomers have been approved by the U.S. Food and Drug Administration for use in repeated contact with food products. More details are available in the Federal Register Vol. 33 No. 5, Tuesday January 9, 1968 Part 121 Food Additive, Subpart F—Food Additives Resulting from Contact with Containers or Equipment and Food Additives Otherwise Affecting Food-Rubber Articles Intended for Repeated Use.

The fluoropolymers provide excellent resistance to oils, fuels, lubricants, most mineral acids, many aliphatic and aromatic hydrocarbons (carbon tetrachloride, benzene, toluene, xylene) that act as solvents for other rubbers, gasoline, naphtha, chlorinated solvents, and pesticides. Special formulations can be produced to obtain resistance to hot mineral acids.

These elastomers are not suitable for use with low molecular weight esters and ethers, ketones, certain amines, or hot anhydrous hydrofluoric or chlorosulfonic acids. Table 4.24 provides the compatibility of fluoroelastomers with selected corrodents.

TABLE 4.24 Compatibility of Fluoroelastomers with Selected Corrodents[a]

Chemical	Maximum temp. °F	°C	Chemical	Maximum temp. °F	°C
Acetaldehyde	x	x	Allyl alcohol	190	88
Acetamide	210	199	Allyl chloride	100	38
Acetic acid 10%	190	88	Alum	190	88
Acetic acid 50%	180	82	Aluminum acetate	180	82
Acetic acid 80%	180	82	Aluminum chloride,	400	204
Acetic acid, glacial	x	x	aqueous		
Acetic anhydride	x	x	Aluminum fluoride	400	204
Acetone	x	x	Aluminum hydroxide	190	88
Acetyl chloride	400	204	Aluminum nitrate	400	204
Acrylic acid	x	x	Aluminum oxychloride	x	x
Acrylonitrile	x	x	Aluminum sulfate	390	199
Adipic acid	190	82	Ammonia gas	x	x

TABLE 4.24 Continued

Chemical	Maximum temp. °F	°C	Chemical	Maximum temp. °F	°C
Ammonium bifluoride	140	60	Butyl alcohol	400	204
Ammonium carbonate	190	88	n-Butylamine	x	x
Ammonium chloride 10%	400	204	Butyl phthalate	80	27
Ammonium chloride 50%	300	149	Butyric acid	120	49
Ammonium chloride, sat.	300	149	Calcium bisulfide	400	204
Ammonium fluoride 10%	140	60	Calcium bisulfite	400	204
Ammonium fluoride 25%	140	60	Calcium carbonate	190	88
Ammonium hydroxide 25%	190	88	Calcium chlorate	190	88
Ammonium hydroxide, sat.	190	88	Calcium chloride	300	149
Ammonium nitrate	x	x	Calcium hydroxide 10%	300	149
Ammonium persulfate	140	60	Calcium hydroxide, sat.	400	204
Ammonium phosphate	180	82	Calcium hypochlorite	400	204
Ammonium sulfate 10–40%	180	82	Calcium nitrate	400	204
			Calcium sulfate	200	93
Ammonium sulfide	x	x	Carbon bisulfide	400	204
Amyl acetate	x	x	Carbon dioxide, dry	80	27
Amyl alcohol	200	93	Carbon dioxide, wet	x	x
Amyl chloride	190	88	Carbon disulfide	400	204
Aniline	230	110	Carbon monoxide	400	204
Antimony trichloride	190	88	Carbon tetrachloride	350	177
Aqua regia 3:1	190	88	Carbonic acid	400	204
Barium carbonate	250	121	Cellosolve	x	x
Barium chloride	400	204	Chloraceticacid, 50% water	x	x
Barium hydroxide	400	204			
Barium sulfate	400	204	Chloracetic acid	x	x
Barium sulfide	400	204	Chlorine gas, dry	190	88
Benzaldehyde	x	x	Chlorine gas, wet	190	88
Benzene	400	204	Chlorine, liquid	190	88
Benzene sulfonic acid 10%	190	88	Chlorobenzene	400	204
			Chloroform	400	204
Benzoic acid	400	204	Chlorosulfonic acid	x	x
Benzyl alcohol	400	204	Chromic acid 10%	350	177
Benzyl chloride	400	204	Chromic acid 50%	350	177
Borax	190	88	Citric acid 15%	300	149
Boric acid	400	204	Citric acid, conc.	400	204
Bromine gas, dry, 25%	180	82	Copper acetate	x	x
Bromine gas, moist, 25%	180	82	Copper carbonate	190	88
Bromine liquid	350	177	Copper chloride	400	204
Butadiene	400	204	Copper cyanide	400	204
Butyl acetate	x	x	Copper sulfate	400	204

TABLE 4.24 Continued

Chemical	Maximum temp. °F	Maximum temp. °C	Chemical	Maximum temp. °F	Maximum temp. °C
Cresol	x	x	Nitric acid 20%	400	204
Cupric chloride 5%	180	82	Nitric acid 70%	190	88
Cupric chloride 50%	180	82	Nitric acid, anhydrous	190	88
Cyclohexane	400	204	Nitrous acid, concen-	90	32
Cyclohexanol	400	204	trated		
Dichloroethane (ethylene	190	88	Oleum	190	88
dichloride)			Perchloric acid 10%	400	204
Ethylene glycol	400	204	Perchloric acid 70%	400	204
Ferric chloride	400	204	Phenol	210	99
Ferric chloride 50% in	400	204	Phosphoric acid 50–80%	300	149
water			Picric acid	400	204
Ferric nitrate 10–50%	400	204	Potassium bromide 30%	190	88
Ferrous chloride	180	82	Salicylic acid	300	149
Ferrous nitrate	210	99	Sodium carbonate	190	88
Fluorine gas, dry	x	x	Sodium chloride	400	204
Fluorine gas, moist	x	x	Sodium hydroxide 10%	x	x
Hydrobromic acid, dilute	400	204	Sodium hydroxide 50%	x	x
Hydrobromic acid 20%	400	204	Sodium hydroxide, con-	x	x
Hydrobromic acid 50%	400	204	centrated		
Hydrochloric acid 20%	350	177	Sodium hypochlorite 20%	400	204
Hydrochloric acid 38%	350	177	Sodium hypochlorite, con-	400	204
Hydrocyanic acid 10%	400	204	centrated		
Hydrofluoric acid 30%	210	99	Sodium sulfide to 50%	190	88
Hydrofluoric acid 70%	350	177	Stannic chloride	400	204
Hydrofluoric acid 100%	x	x	Stannous chloride	400	204
Hypochlorous acid	400	204	Sulfuric acid 10%	350	149
Iodine solution 10%	190	88	Sulfuric acid 50%	350	149
Ketones, general	x	x	Sulfuric acid 70%	350	149
Lactic acid 25%	300	149	Sulfuric acid 90%	350	149
Lactic acid, concentrated	400	204	Sulfuric acid 98%	350	149
Magnesium chloride	390	199	Sulfuric acid 100%	180	82
Malic acid	390	199	Sulfuric acid, fuming	200	93
Manganese chloride	180	82	Sulfurous acid	400	204
Methyl chloride	190	88	Thionyl chloride	x	x
Methyl ethyl ketone	x	x	Toluene	400	204
Methyl isobutyl ketone	x	x	Trichloroacetic acid	190	88
Muriatic acid	350	149	White liquor	190	88
Nitric acid 5%	400	204	Zinc chloride	400	204

[a] The chemicals listed are in the pure state or in a saturated solution unless otherwise indicated. Compatibility is shown to the maximum allowable temperature for which data are available. Incompatibility is shown by an x. A blank space indicates that data are unavailable.
Source: Ref. 1.

The main applications for the fluoroelastomers are in those applications requiring resistance to high operating temperatures together with high chemical resistance to aggressive media, and to those characterized by severe operating conditions that no other elastomer can withstand. Linings and coatings are such applications. Their chemical stability solves the problem of chemical corrosion by making it possible to use them for such purposes as:

A protective lining for power station stacks operated with high sulfur fuels. Tank linings for the chemical industry.

The advantages of the fluoroelastomers lies in their high degree of resistance to aggressive media. Their cost is less than the perfluoroelastomers but greater than other elastomeric materials used for the lining of vessels.

REFERENCES

1. Schweitzer, Philip A. *Corrosion Resistance Tables*, 4th ed., vols. 1–3, Marcel Dekker, New York, 1995.
2. Schweitzer, Philip A. *Corrosion Resistant Piping Systems*, Marcel Dekker, New York, 1994.

5

Liquid Applied Linings

On many occasions linings and coatings terminologies are used interchangeably. In this book *linings* refer to coatings that are subject to immersion, such as the interior of a vessel, and *coatings* alone refer to applications of protective materials used to protect against atmospheric corrosion.

Of the various coating applications the most critical is that of a tank lining application. Liquid applied linings are coatings that may be spray applied or troweled. The material must be resistant to the corrodent and be free of pinholes through which the corrosive could penetrate and reach the substrate. The severe attack that many corrosives have on the bare tank emphasizes the importance of using the correct material and the correct procedure in lining a tank to obtain a perfect coating. It is also essential that the tank be designed and constructed in the proper manner to permit a perfect lining to be applied.

In a tank lining, there are usually four areas of contact with the stored product that may lead to different types of corrosion. These areas are the vapor phase (the area above the liquid level), the interphase (the area where the vapor phase meets the liquid phase), the liquid phase (the area always immersed), and the bottom of the tank (where moisture and other contaminants of greater density may settle). Each of these areas can, at one time or another, be more severely attacked than the rest, depending on the type of material contained, the impurities present, and the amounts of oxygen and water present. In view of this it is necessary to understand the corrosion resistance of the lining material under each condition, not only the immersed condition.

TABLE 5.1 Criteria for Tank Linings

1. Design of the vessel
2. Lining selection
3. Shell construction
4. Shell preparation
5. Lining application
6. Cure of the lining material
7. Inspection of the lining
8. Safety
9. Causes of failure
10. Operating instructions

Other factors that have an effect on the performance of the lining material are vessel design, vessel preparation prior to lining, application techniques of the lining material, curing of the lining, inspection, operating instructions, and temperature limitations. In general, the criteria for tank linings are given in Table 5.1.

DESIGN OF THE VESSEL

All vessels to be lined should be of welded construction. Riveted tanks will expand or contract, thus damaging the lining and causing leakage. Butt welding is preferred, but lap welding may be used providing a fillet weld is used and all sharp edges are ground smooth. Refer to Figure 5.1. Butt welds need not be

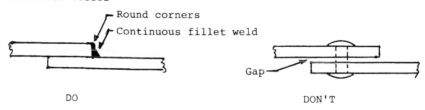

FIGURE 5.1 Butt welding preferred rather than lap welding or riveted construction.

Clean

Grind smooth

DO

Weld spatter

Weld flux

DON'T

FIGURE 5.2 Remove all weld spatter and grind smooth.

ground flush but they must be ground to a smooth rounded contour. A good way to judge a weld is to run your finger over it. Sharp edges can be detected easily. All weld spatter must be removed. Refer to Figure 5.2.

Any sharp prominence may result in a spot where the film thickness will be inadequate and noncontinuous, thus causing premature failure.

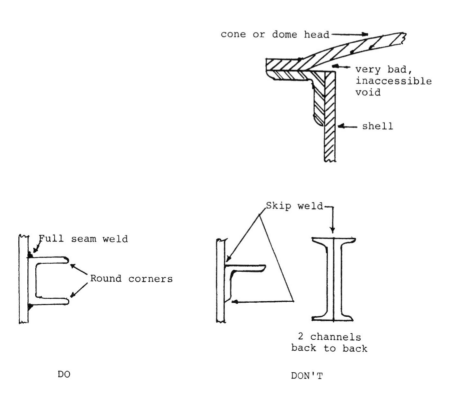

cone or dome head

very bad, inaccessible void

shell

Full seam weld

Round corners

DO

Skip weld

2 channels back to back

DON'T

FIGURE 5.3 Avoid all pockets or crevices that cannot be properly sandblasted and lined.

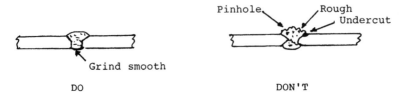

FIGURE 5.4 All joints must be continuous solid welded and ground smooth.

If possible, avoid the use of bolted joints. Should it be necessary to use a bolted joint, it should be made of corrosion resistant materials and sealed shut. The mating surface of steel surfaces should be gasketted. The lining material should be applied prior to bolting.

Do not use construction that will result in the creation of pockets or crevices that will not drain or that cannot be properly sandblasted and lined. Refer to Figure 5.3.

All joints must be continuous and solid welded. All welds must be smooth with no porosity, holes, high spots, lumps, or pockets. Refer to Figure 5.4.

All sharp edges must be ground to a minimum of $\frac{1}{8}$ in. radius. Refer to Figure 5.5.

Outlets must be of flanged or pad type rather than threaded. If pressure requirements permit, use slip-on flanges since the inside diameter of the attaching weld is readily available for radiusing and grinding. If pressure dictates the use of weld neck flanges, the inside diameter of the attaching weld is in the throat of the nozzle. It is, therefore, more difficult to repair surface irregularities such as weld undercutting by grinding. See Figure 5.6.

Stiffening members should be placed on the outside of the vessel rather than on the inside. Refer to Figure 5.7.

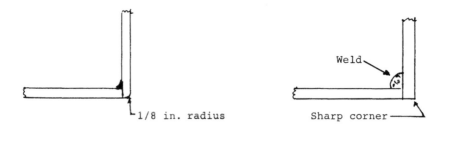

FIGURE 5.5 Grind all sharp edges to a minimum $\frac{1}{8}$ in. radius.

Flanged outlet

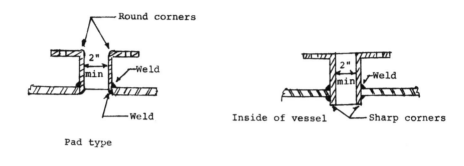

Pad type

Nozzle lining detail

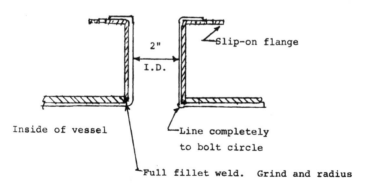

FIGURE 5.6 Typical vessel outlets.

Tanks larger than 25 ft in diameter may require three manways for working entrances. Usually two are located at the bottom (180° apart) and one at the top. The minimum opening is 20 in., but 30 in. openings are preferred.

On occasion an alloy is used to replace the steel bottom of a vessel. Under these conditions galvanic corrosion will take place. If a lining is applied to the steel and for several inches (usually 6–8 in.) onto the alloy, any discontinuity in

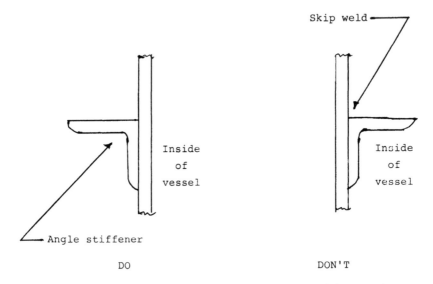

Figure 5.7 Stiffening members should be on the outside of the vessel.

the lining will become anodic. Once corrosion starts, it progresses rapidly because of the bare exposed alloy cathodic area. Without the lining, galvanic corrosion would cause the steel to corrode at the weld area, but at a much lower rate. The recommended practice therefore is to line the alloy completely as well as the steel, thereby eliminating the possible occurrence of a large cathode-to-small anode area. Refer to Figure 5.8.

It is important that processing liquor is not directed against the side of the tank but rather toward the center. Other appurtenances inside the tank must be located for accessibility of lining. Heating elements should be placed with a mini-

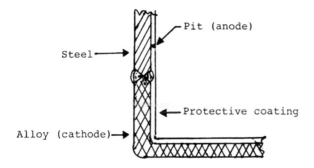

Figure 5.8 Potential galvanic action.

mum clearance of 6 in. Baffles, agitator base plates, pipe, ladders, and other items can either be lined in place, or detached and lined before installation. The use of complex shapes such as angles, channels, and I beams should be avoided. Sharp edges should be ground smooth and they should be fully welded. Spot welding or intermittent welding should not be permitted. Gouges, hackles, deep scratches, slivered steel, or other surface flaws should be ground smooth.

Concrete tanks should be located above the water table. They require special lining systems. Unless absolutely necessary, expansion joints should be avoided. Small tanks do not normally require expansion joints. Larger tanks can make use of a chemical resistant joint such as polyvinyl chloride (PVC). Any concrete curing compound used must be compatible with the lining material or removed before application. Form joints must be made as smooth as possible. Adequate steel reinforcement must be used in a strong, dense, concrete mix to reduce movement and cracking. The lining manufacturer should be consulted for special instructions. Concrete tanks should be lined only by a licensed applicator.

LINING SELECTION

The primary function of a lining system is to protect the substrate. An equally important consideration is product purity protection. The purity of the liquid must not be contaminated with byproducts of corrosion or leachate from the lining system itself. Selection of a lining material for a stationary storage tank dedicated to holding one product at more or less constant times and temperature conditions is relatively easy since such tanks present predictable conditions for lining selection. Conversely, tanks that see intermittent storage of a variety of chemicals or solvents, such as product carriers, present a more difficult problem since the parameters of operation vary. Consideration must be given to the effect of cummulative cargoes. In addition abrasion resistance must be considered if the product in the tank is changed regularly with complete cleaning required between loadings. Workers and equipment will abrade the lining.

When selecting a lining system, it is necessary to determine all of the conditions to which the lining will be subjected. The conditions to be considered fall into two general categories:

1. Chemicals to which the lining will be exposed
2. Conditions of operation

As a result it is necessary to verify that the chemical resistance and physical/ mechanical properties of the lining material are suitable for the application.

In order to specify a lining material, it is necessary to know specifically what is being handled and under what conditions. The following information must be known about the material being handled.

1. What are the primary chemicals being handled and at what concentrations?
2. Are there any secondary chemicals and if so at what concentrations?
3. Are there any trace impurities or chemicals present? This is extremely important since concentrations in the ppm range can cause problems.
4. Are there any solids present, and if so what are the particle sizes and concentrations?
5. Will there be any agitation?
6. What are the fluid purity requirements?
7. What will be present in the vapor phase above the liquid?

The answers to these questions will narrow the selection to those coatings that are compatible. Table 5.2 provides a list of typical lining materials and their general area of application.

Answers to the next set of questions will narrow the selection down to those materials that are compatible as well as to those lining systems that have the required mechanical and/or physical properties.

TABLE 5.2   Typical Lining Materials

Lining materials	Applications
High-bake phenolic	Excellent resistance to acids, solvents, food products, beverages and water. Most widely used lining material, but has poor flexibility compared with other lining materials.
Modified air-dry phenolics (catalyst required)	Nearly equal in resistance to high-bake phenolics. May be formulated for excellent resistance to alkalies, solvents, salt water, deionized water, freshwater, and mild acids. Excellent for dry products.
Modified PVC polyvinyl chloroacetals air-cured	Excellent resistance to strong mineral acids and water. Most popular lining for water storage tanks; used in water immersion service (potable and marine) and beverage processing.
PVC Plastisols	Acid resistant, must be heat cured.
Hypalon	Chemical salts.
Epoxy (amine catalyst)	Good alkali resistance. Fair to good resistance to solvents, mild acids, and dry food products. Finds application in covered hopper-car linings and nuclear containment facilities.
Epoxy polyamide	Good resistance to water and brines. Used in storage tanks and nuclear containment facilities. Poor acid resistance and fair alkali resistance.

TABLE 5.2 Continued

Lining materials	Applications
Epoxy polyester	Good abrasion resistance. Used for covered hopper-car linings. Poor solvent resistance.
Epoxy coal tar	Excellent resistance to mild acids, mild alkalies, salt water and freshwater. Poor solvent resistance. Used for crude oil storage tanks, sewage disposal plants, and water works.
Coal tar	Excellent water resistance. Used for water tanks.
Asphalts	Good acid and water resistance.
Neoprene	Good acid and flame resistance. Used for chemical processing equipment.
Polysulfide	Good water and solvent resistance.
Butyl rubber	Good water resistance.
Styrene-butadiene polymers	Finds application in food and beverage processing and in the lining of concrete tanks.
Rubber latex	Excellent alkali resistance. Finds application in caustic tanks (50–73%) at 180°F/82°C to 250°F/121°C.
Urethanes	Superior abrasion resistance. Excellent resistance to strong mineral acids and alkalies. Fair solvent resistance. Used to line dishwashers and washing machines.
Vinyl ester	Excellent resistance to strong acids and better resistance up to 350°F/193°C to 400°F/204°C, depending upon thickness
Vinyl urethanes	Finds application in food processing, hopper cars and wood tanks.
Fluoropolymers	High chemical resistance and fire resistance. Used in SO_2 scrubber service.
Vinylidene chloride latex	Excellent fuel oil resistance.
Alkyds, epoxy esters, oleoresinous primers	Water immersion applications and as primers for other top coats.
Inorganic zinc, water based postcure, and water based self-cure	Jet fuel storage tanks and petroleum products.
Inorganic zinc solvent based self curing	Excellent resistance to most organic solvents (aromatics, ketones, and hydrocarbons), excellent water resistance Difficult to clean. May be sensitive to decomposition products of materials stored in tanks.
Furan	Most acid resistant organic polymer. Used for stack linings and chemical treatment tanks.

1. What is the normal operating temperature and temperature range?
2. What peak temperatures can be reached during shutdown, start-up, process upset, etc.?
3. Will any mixing areas exist where exothermic heat of mixing temperatures may develop?
4. What is the normal operating pressure?
5. What vacuum conditions and range are possible during operation, start-up, shutdown, and upset conditions?
6. Is grounding necessary?

Manufacturers and/or other corrosion engineers should be consulted for case histories of identical applications. Included in the case history should be the name of the applicator who applied the lining, application conditions, type of equipment used, degree of application difficulty, and other special procedures required. A lining with superior chemical resistance will fail rapidly if it cannot be properly applied, so it is advantageous to learn from the experience of others.

In order to maximize sales, lining manufacturers formulate their products to meet as broad a range as possible of chemical and solvent environments. Consequently, a tank lining may be listed as suitable for in excess of 100 products with varying degrees of compatibility. However, there is a potential for failure if the list is viewed only from the standpoint of the products approved for service.

If more than one of these materials listed as being compatible with the lining is to be used, consideration must be given as to the sequence of use in which the chemicals or solvents will be stored or carried in the tank. This is particularly critical when the cargo is water miscible, for example, methanol or cellosolve, and is followed by water ballast. A sequence such as this creates excessive softening of the film and makes recovery of the lining film more difficult, thus prone to early failure.

Certain tank lining systems may have excellent resistance to specific chemicals for a given period of time, after which they must be cleaned and allowed to recover for a designated period of time in order to return to their original resistance level. Thirty days is a common period of time for this recovery process between chemicals such as acrylonitrile and solvents such as methanol.

In some cases the density of cure can be increased by loading a hot mildly aggressive alkaline material on to a newly applied lining. This permits the subsequent loading of an aggressive solvent at a later date. Ketimine epoxy is such an example. There have been cases where three or four consecutive hot mild cargoes have increased the density of the lining to such an extent that the ketimine epoxy lining was resistant to methanol. Under normal circumstances ketimine epoxy is not compatible with methanol.

When case histories are unavailable or manufacturers are unable to make a recommendation it will be necessary to conduct tests. This can occur in the case of a proprietary material being handled or if a solution may contain unknown chemicals. Sample panels of several lining systems should be tested for a minimum of 90 days, with a 6 month test being preferable. Because of normal time requirements, 90 days is a standard.

The test must be conducted at the maximum operating temperature to which the lining will be subjected and should simulate actual operating conditions, including washing cycles, cold wall, and effects of insulation.

Other factors to be considered in the lining selection include service life, maintenance cycles, operating cycles, and reliability of the lining. Different protective coatings provide different degrees of protection for different periods of time at a variety of costs. Allowable downtime of the facility for inspection and maintenance must also be considered, in terms of frequency and length of time.

Once the lining system has been selected, recommendations from the manufacturer as to a competent applicator should be requested, and contact should be made with previous customers.

SHELL CONSTRUCTION

In the design section several features of construction were discussed. It is important that the finished vessel shell be thoroughly inspected to ensure that the vessel has been fabricated and finished in accordance with the specifications. Such items as sharp corners and rough welds may have been overlooked by the fabricator.

On occasion it may be necessary that certain parts of the tank, such as the bottom plate for a center post, need to be dismantled and lined separately. This particular section would then be reassembled after the tank is blast cleaned and lined.

SHELL PREPARATION

In order for the lining material to obtain maximum adhesion to the substrate surface, it is essential that the surface be absolutely clean. All steel surfaces to be lined must be abrasive blasted to a white metal in accordance with SSPC Specification SPS-63 or NACE Specification 1. A white metal blast is defined as removing all rust, scales, paints, and so on to a clean white metal that has a uniform gray-white appearance. No streaks or stains of rust or any other contaminants are permitted on the surface. At times a near-white blast cleaned surface equal to SSPC-SP10 may be used. Since this is less expensive, it should be used providing the lining manufacturer permits it.

All dust and spent abrasive must be removed from the surface by vacuum cleaning or brushing. After blasting all workers coming into contact with the clean surface should wear clean, protective gloves and clothing to prevent contamination of the cleaned surface. Any contamination may cause premature failure by osmotic blistering or adhesion loss. The first coat must be applied before the surface starts to rust. If the blasted surface changes color, or rust bloom begins to form, the surface must be reblasted. Dehumidifiers and temperature controls are helpful.

It is important that no moisture or oil passes through the compressor and on to the blasted surface. Use a white rag to determine the quality. A black light may also be used to determine oil contamination. One hundred percent oil-free air can be supplied by rotary screw, two stage lubrication-oil-free compressors.

Concrete surfaces must be clean, dry, and properly cured before applying the lining. All protrusions and form joints must be removed. All surfaces must be roughened by sandblasting to remove all loose, weak, or protruding concrete to open all voids and provide the necessary profile for mechanical adhesion of the lining. All dust must be removed by brushing or vacuuming. The lining manufacturer should be contacted for special priming and caulking methods.

LINING APPLICATION

The primary concern in applying a lining to a vessel is to deposit a void-free film of the specified thickness on the surface. Any area that is considerably less than the specified thickness may have a noncontinuous film. Additionally, pinholes in the lining may cause premature failure.

Films that exceed the specified thickness always pose the danger of entrapping solvents which can lead to poor adhesion, excessive brittleness, improper cure, and subsequent poor performance. Avoid dry spraying of the lining material since this causes the lining to be porous. If thinners other than those recommended by the manufacturer are used, poor film formation may result. Do not permit application to take place below the temperature recommended by the manufacturer.

When selecting an applicator for the lining, it is important that the applicator selected be very familiar and experienced in applying the lining to be used. Too often the lowest bidder is selected without adequately considering the quality of workmanship, with the result of a tank lining failure. In order for a tank lining to be effective a nearly perfect application is required. In view of this a conscientious and knowledgeable applicator is needed. Before awarding the lining contract, evaluate the applicator to assure that the tank lining contractor is experienced in applying the recommended lining.

When evaluating the qualifications of the tank lining contractor, ask what jobs he has done using the specified lining material and check out his references. If possible visit his facilities and inspect his workmanship before placing him on the bidders' list. Precautions taken at this point will be repaid by assuring total performance.

CURE OF THE APPLIED LINING

Proper curing is essential if the lining is to provide the corrosion protection for which it was selected. Each coat must be cured using proper air circulation techniques. In order to obtain proper air circulation, it is necessary that the tank has at least two openings, one at the top and one at the bottom.

Since most solvents used in lining materials are heavier than air, the fresh air intake will be at the top of the vessel and the exhaust at the bottom. The temperature of the fresh air intake should be over 50°F/10°C with a relative humidity of less than 89%. If possible the fresh air intake should be fed by forced air fans.

A faster and more positive cure will be accomplished by using a warm forced air cure between coats and as a final cure. This will produce a dense film and tighter cross-linking, which provides superior resistance to solvents and moisture permeability.

Before placing the vessel in service the lining should be washed down with water to remove any loose overspray. For linings in contact with food products, a final warm, forced air cure and water wash is essential.

It is important that sufficient time is allowed to permit the lining to obtain a full cure. This usually requires 3–7 days. So not skimp on this time.

When the tank is placed in service, operating instructions should be prepared and should include the maximum temperature to be used. The outside of the tank should be labeled. Do not exceed X°F/X°C. This tank has been lined with X. It is to be used only for X service.

INSPECTION OF THE LINING

Having a qualified inspector available throughout the project is highly recommended to guarantee a satisfactory lining application. The inspector should be involved with the job from the beginning. He should have an understanding of the design criteria of the vessel and the reasons for the specific design configuration. The inspector should participate in the following functions:

1. Pre-work meetings
2. Selection of the contractor for fabrication and lining application
3. Surface preparation inspection
4. Coating application inspection
5. Daily inspection reports
6. Final acceptance report

The inspector should be involved with the selection of the fabricator and applicator of the lining. A knowledgeable and conscientious fabricator and applicator is necessary if a nearly perfect lining is to be achieved. Again the lowest bidder is not necessarily the best choice. The inspector should evaluate the fabricator and applicator before awarding the contract. In general it is better to have the vessel fabricated and lined by the same contractor if possible. Evaluation should be made as to the experience the contractor has in applying the lining to be used. His facility should be visited and qualifications verified by checking on previous lining jobs he has done with your specified lining material.

Before application the surface must be properly prepared to receive the material as detailed in the sections on shell construction and shell preparation. In addition the surface profile should be checked. An adequate ''key'' must be provided to furnish a proper anchor pattern. Too little anchor pattern will result in too smooth a surface and therefore poor adhesion. Too deep a profile will require additional lining material. In general the profile depth should be approximately 25% of the total lining thickness. On this basis if a 6 mil lining is specified the profile depth should be 1.5 mil.

The type of abrasive employed will determine the profile depth. Profile depth in the field can be determined by one of many ways.

The Keane-Tator Profile Comparator contains a metal disc with nominal surface profiles of 0.5, 1, 2, 3, and 4 mil. This flashlight magnifier is used as a handy pocket type comparator to check on sandblast cleaned surfaces. Metal discs for comparisons of anchor patterns prepared with grit or shot blast are also available.

A Testex tape has been developed that is pressed into the profile, then removed, and this profile that remains on the tape is measured with a micrometer, subtracting the thickness of the tape.

Clemtex offers a series of four steel coupons with profile gauges ranging from 1 to 4 mil.

Once the surface has been prepared, the inspector must work quickly so that the application of the lining to the surface is not delayed. Normally the steel may stand unprotected for a few hours before beginning the application without any detrimental effects.

Inspections should be made during and after each lining application. The lining should be checked for porosities and pinholing on the first visual inspection. After repairs of the visible defects, inspection may be done by using low voltage (75 V or less). Holiday detectors that ring, buzz, or light up to show electrical contact through a porosity within the lining to the metal or concrete surfaces. By checking the lining in this manner during the first and second coats, such defects can be touched up and made free of porosities before the final top coat.

Visual inspections are performed with either the unaided eye or by the use of a magnifying glass. In some cases the use of telescopic observation or the use of low power magnification may be required. A Pike magnifier is one of several types that may be used.

The inspector is able to identify areas that have been missed, damaged areas, or thin areas by employing these visual techniques. Using instruments provides the inspector with a means to make an accurate appraisal of what dimensional requirements have been met by the applicator.

During and after each coating the inspector should prepare an inspection report on the applicable items shown in Table 5.3.

After the final coat has been applied and the lining has been properly cured the following tests should be conducted to verify that the lining has been properly cured.

Sandpaper Test

When not properly cured, some lining materials will remain slightly tacky. When abraded with fine sandpaper, no material should be seen on the face of the sandpaper. It should be removed as a fine powdery residue.

Hardness Test

Using your fingernail is a satisfactory way of determing hardness. If desired a Barcol impressor, or a pencil hardness test may be employed.

TABLE 5.3 Inspection Report for Liquid Applied Linings

Item	Tank linings	Concrete surfacer	Concrete top coats
Pinholes	x	x	x
Blisters	x	x	x
Color and gloss uniformity			x
Bubbling	x	x	x
Fish eyes	x	x	x
Orange peel	x	x	x
Mud cracking	x	x	
Curing properties	x	x	x
Runs and sags	x	x	x
Film thickness, dry	x		
Film thickness, wet		x	x
Holidays, missed areas	x	x	x
Dry spray	x	x	x
Foreign contaminents	x	x	x
Mechanical damage	x	x	x
Uniformity	x	x	x
Adhesion	x	x	x

Adhesion

A pocket knife is the best instrument to use to test adhesion. Cut a "V" in the film and pick off the coating at the vertex. The lining should be very difficult to remove, indicating good adhesion.

Film Thickness

Dry film thicknesses on steel surfaces are determined by magnetic and eddy current nondestructive test instruments. The most popular instruments employ the magnetic principle measuring magnetic attraction, which is inversely proportional to the lining thickness. Examples are the BSA-Tinsley thickness gauge and the Elcometer 157 pulloff gauge.

The pulloff gauge is a rough guide to determine if the protective lining is within the thickness specification. The manufacturer's stated accuracy is 15%, provided that the gauge is used within a true vertical plane. The accuracy is reduced when the gauge is used in a horizontal or overhead position.

The pulloff gauge has other limitations besides accuracy:

1.   The eye must record the coating thickness as the magnet breaks away from the lining.
2.   Erroneous readings will result if the magnet is allowed to slide over the lining before breakaway or pulloff.

The type 7000 Tinsley gauge contains a dial like scale with a balanced pointer that is not affected by angular positions. A direct readout from a locked-in zero reset is given with an accuracy of $\pm 10\%$.

The "banana" type thickness gauge is a more sophisticated version of the magnetic pulloff principle. A permanent magnet is mounted at one end of a balanced, pivoted arm assembly, and a coil spring is attached to the pivot and to a calibrated rotatable dial. The rotatable dial is moved forward until the magnet sticks to the lining. This unknown force is determined by rotating the dial backward, applying tension to the spring. When the spring tension exceeds the unknown magnetic attractive force, the magnet breaks contact with the coated surface. An audible click will be heard and the lining thickness will be shown on the calibrated dial.

There are several gauges available which make use of the guided or controlled magnetic pulloff principle. These include Inspector thickness gauge, Model 111/1E manufactured by Elcometer Instruments Ltd., the Mikrotest thickness gauge Model 102/FIM and Mikrotest 11 FM manufactured by Elktro Physik, Cologne, West Germany.

These gauges will measure lining thickness in any position without recalibrating, because the pivot arm is balanced. The Inspector gauge has an external calibration adjustment making use of a screwdriver slot located below the name-plate.

A magnetic reluctance technique is made use of in the Elcometric thickness gauge. Reluctance is the characteristic of a material to resist the creation of a magnetic flux in that material (e.g., iron has less reluctance than air).

In this gauge a permanent magnet is located between two soft iron poles resembling a horseshoe magnet. The magnet is adjustable to produce an air gap. In the center of this horseshoe is a meter-pointer assembly with a soft iron vane, which creates a magnetic circuit with an indicating device that requires no power supply or battery.

When the Elcometer gauge is placed onto a dry lining applied to steel, the magnetic flux will change in strength across the air gap in the magnetic circuit, causing the meter pointer to move across a calibrated scale, indicating the lining thickness in either micrometers or mils.

When using the Elcometer, it is essential that the gauge be held at a right angle to the surface being measured since tilting it will give erroneous measurements. The gauge must be recalculated when changing from the vertical to the horizontal or to an overhead position.

A small zero knob is located on the side of the instrument to permit external calibration. To take readings, it is necessary to depress a small button. This gauge is very sensitive to residual magnetism in the substrate, surface roughness, edge effect, and tilt of head. Blind hole measurements cannot be made with this gauge.

Film thickness measurements should not be made close to the edge of a steel surface since the magnetic properties of the steel influence the reading, causing distorted results. It is recommended that measurements be made at least 1 in. away from the edge. Distorted readings can also result from angles, corners, welds, crevices, and joints. Always measure a clean surface, never an oily or dirty one.

Electronic gauges are also available that are more accurate than the mechanical units previously discussed. These include the Model 158 Minitector thickness gauge and Model 102/F100 Minitest thickness gauge. The Minitectic gauge is portable and uses standard transistor radio batteries. It is manufactured by Elcometer Instruments Ltd. The Minitest gauge is battery operated and comes with an automatic battery power off switch to extend battery life. It is manufactured by Elektic Physik.

General Electric produces a Model B thickness gauge which requires a 115 V ac source and is not portable for field use.

SAFETY DURING APPLICATION

Many of the lining materials contain solvents, making it necessary to take certain safety precautions. It is necessary that all lining materials and thinners be kept away from any source of open flame. This means that welding in adjacent areas must be discontinued during application and "no smoking" must be the rule during application.

A proper air supply and ventilation must be provided during the tank lining work. Vapor concentration inside of the vessel should be checked on a regular basis to ensure that the maximum allowable vapor concentration is not reached. For most solvents a vapor concentration of between 2 and 12% in the air is sufficient to cause an explosion. As long as the concentration is kept below the lower level, no explosion will take place. All electrical equipment must be grounded.

Since flammable solvents are being exhausted from the tank, precautions must be taken on the exterior of the tank. These flammable vapors will travel a considerable distance along the ground. No flames, sparks, or ungrounded equipment can be nearby.

Those applying the lining should wear fresh airline respirators and protective cream on exposed parts of the body. Water should be available for flushing accidental spills from the skin. Never allow one worker in the tank alone.

OSHA issues a form called the Material Safety Data Sheet. The manufacturers supply this form by listing all toxicants or hazardous materials and provide a list of the solvents used. Also included are the threshold limit values (TLV) for each substance. Explosive hazards, flash points, and temperature limits are established for safe application of each lining material. These Material Safety Data Sheets should be kept on file in the job superintendent's office and at the first-aid station.

CAUSES OF FAILURE

Most types of failure are the result of misuse of the tank lining which may result in blistering, cracking, hardening or softening, peeling, staining, burning, and undercutting. A frequent cause of failure is overheating during operation. When a very heavily pigmented surface or thick film begins to shrink, stresses are formed on the surface resulting in cracks. These cracks do not always expose the substrate and may not penetrate. Under these conditions the best practice is to remove those areas and recoat according to standard repair procedures.

Aging or poor resistance to the corrosive can result in hardening or softening. As the lining ages, particularly epoxy and phenolic amines, it becomes brittle and may chip from the surface. Peeling can result from improperly cured surfaces, poor surface preparation, or a wet or dirty surface. Staining results when there is a reaction of the corrosive on the surface of the lining or slight staining from impurities in the corrosive. The true cause must be determined by scraping or detergent-washing the lining. If the stain is removed and softening of the film is not apparent, failure has not occurred.

Any of the above defects can result in undercutting. After the corrosive penetrates to the substrate, corrosion will proceed to extend under the film areas that have not been penetrated or failed. Some linings are more resistant to undercutting or underfilm corrosion than others. Usually if the lining exhibits good adhesive properties, and if the prime coat is chemically resistant to the corrosive environment, underfilm corrosion will be greatly retarded.

In addition, a tank lining must not impart any impurities to the material contained. The application is a failure if any color, taste, smell, or other contamination is imparted to the product, even though the lining is intact. Such contamination can be caused by the extraction of impurities from the lining leading to blistering between coats or to metal.

If the lining is unsuited for the service, complete failure may occur by softening, dissolution, and finally complete disintegration of the lining. This type of problem is prevalent between the interphase and bottom of the tank. At the bottom of the tank and throughout the liquid phase, penetration is of great concern.

The vapor phase of the tank is subject to corrosion from concentrated vapors mixed with oxygen present and can cause extensive corrosion.

OPERATING INSTRUCTIONS

When the tank is placed into service, specific instructions should be given as to what the tank is to be used for, temperature limitations, cleaning instructions, and information on the lining material.

The outside of the tank should be labeled. Do not exceed X°F/X°C. This tank has been lined with X. It is to be used only for X service.

6

Specific Liquid Applied Lining Materials

Most liquid applied lining materials are capable of being formulated to meet requirements for specific applications. Corrosion data referring to the suitability of a lining material for a specific corrodent indicates that a formulation is available to meet these conditions. Since all formulations available may not be suitable, the manufacturer must be checked as to the suitability of his formulation.

PHENOLICS

Phenolics were the first commercial synthetic resins introduced in the early 1900s. They are available as 100% phenolic baking resins, oil modified and phenolic dispersions.

Baked Phenolics

This series of linings has been an industry standard for many years. It is most commonly used for immersion service in acids, ammonias, petroleum products, dairy and food products/components, and industrial chemicals. After baking, these linings become nontoxic, odorless, and tasteless, meeting the requirements of the U.S. Federal Register, Food and Drug Regulations, Title 21, Chapter 1, Paragraph 175, 300 and M.I.D. of the U.S.D.A.

TABLE 6.1 Compatibility of Phenolics with Selected Corrodents[a]

Chemical	Maximum temp. °F	Maximum temp. °C	Chemical	Maximum temp. °F	Maximum temp. °C
Acetic acid 10%	212	100	Cresol	300	149
Acetic acid, glacial	70	21	Ethylene glycol	70	21
Acetic anhydride	70	21	Ferric chloride	300	149
Acetone	x	x	50% in water	300	149
Aluminum chloride,	90	32	Hydrobromic acid, dilute	200	93
aqueous			Hydrobromic acid 20%	200	93
Aluminum sulfate	300	149	Hydrobromic acid 50%	200	93
Ammonia gas	90	32	Hydrochloric acid 20%	300	149
Ammonium carbonate	90	32	Hydrochloric acid 38%	300	149
Ammonium chloride 10%	80	27	Hydrofluoric acid 30%	x	x
Ammonium chloride 50%	80	27	Hydrofluoric acid 60%	x	x
Ammonium chloride, sat.	80	27	Hydrofluoric acid 100%	x	x
Ammonium hydroxide 25%	x	x	Lactic acid 25%	160	71
Ammonium hydroxide, sat.	x	x	Methyl ethyl ketone	160	71
Ammonium nitrate	160	71	Methyl isobutyl ketone	x	x
Ammonium sulfate 10–40%	300	149	Muriatic acid	160	71
Aniline	x	x	Nitric acid 5%	300	149
Benzaldehyde	70	21	Nitric acid 20%	x	x
Benzene	160	71	Nitric acid 70%	x	x
Benzenesulfonic acid 10%	70	21	Nitric acid, anhydrous	x	x
Benzoic acid			Nitrous acid, conc.	x	x
Benzyl alcohol	70	21	Phosphoric acid 50–80%	x	x
Butadiene	x	x	Picric acid	212	100
Butyl phthalate	160	71	Sodium hydroxide 10%	300	149
Calcium chlorate	300	149	Sodium hydroxide 50%	x	x
Calcium hypochlorite 10%	x	x	Sodium hydroxide, concd	x	x
Carbon dioxide, dry	300	149	Sodium hypochlorite 15%	x	x
Carbon dioxide, wet	300	149	Sodium hypochlorite,	x	x
Carbon tetrachloride	200	93	concd	x	x
Carbonic acid	200	93	Sulfuric acid 10%	250	121
Chlorine gas, wet	x	x	Sulfuric acid 50%	250	121
Chlorine, liquid	x	x	Sulfuric acid 70%	200	93
Chlorobenzene	260	127	Sulfuric acid 90%	70	21
Chloroform	160	71	Sulfuric acid 98%	x	x
Chromic acid 50%	x	x	Sulfuric acid 100%	x	x
Chromyl chloride	x	x	Sulfurous acid	80	27
Citric acid, conc.	160	71	Thionyl chloride	200	93
Copper acetate	160	71	Zinc chloride	300	149

[a]The chemicals listed are in the pure state or in a saturated solution unless otherwise indicated. Compatibility is shown to the maximum allowable temperature for which data are available. Incompatibility is shown by an x. A blank space indicates that the data are unavailable.
Source: Ref. 1.

In general, baked phenolic linings have excellent resistance to acids, solvents, food products, beverages, and water. It has poor flexibility compared with other lining materials, with a dry heat resistance of 400°F/204°C.

Refer to Table 6.1 for the compatibility of phenolics with selected corrodents. Reference 1 provides a more comprehensive listing.

Modified Air Dried Phenolics

These linings are nearly equivalent to high bake phenolics with a dry heat resistance of 150°F/65°C. They may be formulated for excellent resistance to alkalies, solvents, fresh water, salt, deionized water, and mild acid resistance.

EPOXY

Epoxy resins can be formulated with a wide range of properties. These medium to high priced resins are noted for their adhesion. Epoxy linings provide excellent chemical and corrosion resistance. They exhibit good resistance to alkalies, non-oxidizing acids, and many solvents. Typically, epoxies are compatible with the following materials at 200°F/93°C unless otherwise noted:

a. Acids

Acetic acid 10% to 150°F/66°C

Benzoic acid

Butyric acid

Fatty acids

Hydrochloric acid 10%

Sulfuric acid 20% to 180°F/82°C

Rayon spin bath

Oxalic acid

b. Bases

Sodium hydroxide 50% to 180°F/82°C

Sodium sulfide 10%

Trisodium phosphate

Magnesium hydroxide

c. Salts, metallic salts

Aluminum

Calcium

Iron

Magnesium

Potassium

Sodium

Most ammonium salts

d. Alcohols, solvents

Methyl

Ethyl

Isopropyl to 150°F/66°C

Benzene to 150°F/66°C

Ethylacetate to 150°F/66°C

Naphtha

Toluene

Xylene

TABLE 6.2 Compatibility of Epoxy with Selected Corrodents[a]

Chemical	Maximum temp. °F	Maximum temp. °C	Chemical	Maximum temp. °F	Maximum temp. °C
Acetaldehyde	150	66	Barium sulfide	300	149
Acetamide	90	32	Benzaldehyde	x	x
Acetic acid 10%	190	88	Benzene	160	71
Acetic acid 50%	110	43	Benzenesulfonic acid 10%	160	71
Acetic acid 80%	110	43	Benzoic acid	200	93
Acetic anhydride	x	x	Benzyl alcohol	x	x
Acetone	110	43	Benzyl chloride	60	16
Acetyl chloride	x	x	Borax	250	121
Acrylic acid	x	x	Boric acid 4%	200	93
Acrylonitrile	90	32	Bromine gas, dry	x	x
Adipic acid	250	121	Bromine gas, moist	x	x
Allyl alcohol	x	x	Bromine, liquid	x	x
Allyl chloride	140	60	Butadiene	100	38
Alum	300	149	Butyl acetate	170	77
Aluminum chloride, aqueous 1%	300	149	Butyl alcohol	140	60
			n-Butylamine	x	x
Aluminum chloride, dry	90	32	Butyric acid	210	99
Aluminum fluoride	180	82	Calcium bisulfide		
Aluminum hydroxide	180	82	Calcium bisulfite	200	93
Aluminum nitrate	250	121	Calcium carbonate	300	149
Aluminum sulfate	300	149	Calcium chlorate	200	93
Ammonia gas, dry	210	99	Calcium chloride 37.5%	190	88
Ammonium bifluoride	90	32	Calcium hydroxide, sat.	180	82
Ammonium carbonate	140	60	Calcium hypochlorite 70%	150	66
Ammonium chloride, sat.	180	82	Calcium nitrate	250	121
Ammonium fluoride 25%	150	66	Calcium sulfate	250	121
Ammonium hydroxide 25%	140	60	Caprylic acid	x	x
Ammonium hydroxide, sat.	150	66	Carbon bisulfide	100	38
Ammonium nitrate 25%	250	121	Carbon dioxide, dry	200	93
Ammonium persulfate	250	121	Carbon disulfide	100	38
Ammonium phosphate	140	60	Carbon monoxide	80	27
Ammonium sulfate 10–40%	300	149	Carbon tetrachloride	170	77
Ammonium sulfite	100	38	Carbonic acid	200	93
Amyl acetate	80	27	Cellosolve	140	60
Amyl alcohol	140	60	Chloroacetic acid, 92% water	150	66
Amyl chloride	80	27			
Aniline	150	66	Chloroacetic acid	x	x
Antimony trichloride	180	82	Chlorine gas, dry	150	66
Aqua regia 3:1	x	x	Chlorine gas, wet	x	x
Barium carbonate	240	116	Chlorobenzene	150	66
Barium chloride	250	121	Chloroform	110	43
Barium hydroxide 10%	200	93	Chlorosulfonic acid	x	x
Barium sulfate	250	121	Chromic acid 10%	110	43

TABLE **6.2** Continued

Chemical	Maximum temp. °F	°C	Chemical	Maximum temp. °F	°C
Chromic acid 50%	x	x	Methyl ethyl ketone	90	32
Citric acid 15%	190	88	Methyl isobutyl ketone	140	60
Citric acid 32%	190	88	Muriatic acid	140	60
Copper acetate	200	93	Nitric acid 5%	160	71
Copper carbonate	150	66	Nitric acid 20%	100	38
Copper chloride	250	121	Nitric acid 70%	x	x
Copper cyanide	150	66	Nitric acid, anhydrous	x	x
Copper sulfate 17%	210	99	Nitrous acid, conc.	x	x
Cresol	100	38	Oleum	x	x
Cupric chloride 5%	80	27	Perchloric acid 10%	90	32
Cupric chloride 50%	80	27	Perchloric acid 70%	80	27
Cyclohexane	90	32	Phenol	x	x
Cyclohexanol	80	27	Phosphoric acid 50–80%	110	43
Dichloroacetic acid	x	x	Picric acid	80	27
Dichloroethane (ethylene dichloride)	x	x	Potassium bromide 30%	200	93
			Salicylic acid	140	60
Ethylene glycol	300	149	Sodium carbonate	300	149
Ferric chloride	300	149	Sodium chloride	210	99
Ferric chloride 50% in water	250	121	Sodium hydroxide 10%	190	88
			Sodium hydroxide 50%	200	93
Ferric nitrate 10–50%	250	121	Sodium hypochlorite 20%	x	x
Ferrous chloride	250	121	Sodium hypochlorite, concd	x	x
Ferrous nitrate					
Fluorine gas, dry	90	32	Sodium sulfide to 10%	250	121
Hydrobromic acid, dilute	180	82	Stannic chloride	200	93
Hydrobromic acid 20%	180	82	Stannous chloride	160	71
Hydrobromic acid 50%	110	43	Sulfuric acid 10%	140	60
Hydrochloric acid 20%	200	93	Sulfuric acid 50%	110	43
Hydrochloric acid 38%	140	60	Sulfuric acid 70%	110	43
Hydrocyanic acid 10%	160	71	Sulfuric acid 90%	x	x
Hydrofluoric acid 30%	x	x	Sulfuric acid 98%	x	x
Hydrofluoric acid 70%	x	x	Sulfuric acid 100%	x	x
Hydrofluoric acid 100%	x	x	Sulfuric acid, fuming	x	x
Hypochlorous acid	200	93	Sulfurous acid 20%	240	116
Ketones, general	x	x	Thionyl chloride	x	x
Lactic acid 25%	220	104	Toluene	150	66
Lactic acid, conc.	200	93	Trichloroacetic acid	x	x
Magnesium chloride	190	88	White liquor	90	32
Methyl chloride	x	x	Zinc chloride	250	121

[a] The chemicals listed are in the pure state or in a saturated solution unless otherwise indicated. Compatibility is shown to the maximum allowable temperature for which data are available. Incompatibility is shown by an x. A blank space indicates that the data are unavailable. *Source*: Ref. 1.

e. Miscellaneous

Distilled water	Jet fuel
Seawater	Gasoline
White liquor	Diesel fuel
Sour crude oil	Black liquor

Epoxies are not satisfactory for use with:

Bromine water	Hydrogen peroxide
Chromic acid	Sulfuric acid above 70%
Bleaches	Wet chlorine gas
Fluorine	Wet sulfur dioxide
Methylene chloride	

Refer to Table 6.2 for the compatibilities of epoxy with selected corrodents and Ref. 1 for a more comprehensive listing.

Epoxy resins must be cured with cross-linking agents (hardeners) or catalysts to develop desired properties. Cross-linking takes place at the epoxy and hydroxyl groups, which are the reaction sites. The primary types of curing agents used for linings are aliphatic amines and catalytic curing agents.

Epoxy (Amine Catalyst)

These are widely used since the curing of the epoxies takes place at room temperature. High exothermic temperatures develop during the curing reaction which limits the mass of material that can be cured. Amine cured linings exhibit good alkali resistance, fair to good resistance to mild acids, solvents, and dry food products. They are widely used for hopper car linings and nuclear containment facilities. The maximum allowable temperature is 275°F/135°C.

Baked Epoxy

Catalytic curing agents require a temperature of 200°F/93°C or higher to react. They exhibit excellent resistance to acids, alkalies, solvents, inorganic salts, and water. The maximum operating temperature is 325°F/163°C, somewhat higher than that of the amine cured epoxies.

FURANS

Furan polymers have excellent resistance to strong concentrated mineral acids, caustics, and combinations of solvents with acids and bases. These furans are subject to many different formulations, making them suitable for specific applications.

In general, furan formulations are compatible with the following:

a. Solvents

Acetone	Ethanol
Benzene	Ethyl acetate
Carbon disulfide	Methanol
Chlorobenzene	Methyl ethyl ketone
Perchlorethylene	Trichlorethylene
Styrene	Xylene
Toluene	

b. Acids

Acetic acid	Phosphoric acid
Hydrochloric acid	Sulfuric acid 60% to 150°F/66°C
Nitric acid 5%	

c. Bases

Diethylamine	Sodium sulfide
Sodium carbonate	Sodium hydroxide 50%

d. Water

Demineralized	Distilled

e. Others

Pulp mill liquor

The furan resins are not satisfactory for use with oxidizing media, such as chromic or nitric acids, peroxides, hypochlorites, chlorine, phenol, and concentrated sulfuric acid. Refer to Table 6.3 for the compatibility of furan resins with selected corrodents and to Ref. 1 for a more comprehensive listing.

VINYL ESTER

Vinyl ester resins are available in various formulations. Halogenated modifications are available where fire resistance and ignition resistance are major concerns. They are resistant up to 400°F/204°C.

Vinyl esters can be used to handle most hot, highly chlorinated and acid mixtures at elevated temperatures. They also provide excellent resistance to strong mineral acids and bleaching solutions. Vinyl esters excel in alkaline and bleach environments and are used extensively in the very corrosive conditions found in the pulp and paper industry. Refer to Table 6.4 for the compatibility of vinyl ester with selected corrodents and to Ref. 1 for a more comprehensive listing.

TABLE 6.3 Compatibility of Furan Resins with Selected Corrodents[a]

Chemical	Maximum temp. °F	Maximum temp. °C	Chemical	Maximum temp. °F	Maximum temp. °C
Acetaldehyde	x	x	Benzyl chloride	140	60
Acetic acid 10%	212	100	Borax	140	60
Acetic acid 50%	160	71	Boric acid	300	149
Acetic acid	80	27	Bromine gas, dry	x	x
Acetic acid, glacial	80	27	Bromine gas, moist	x	x
Acetic anhydride	80	27	Bromine, liquid 3% max.	300	149
Acetone	80	27	Butadiene		
Acetyl chloride	200	93	Butyl acetate	260	127
Acrylic acid	80	27	Butyl alcohol	212	100
Acrylonitrile	80	27	n-Butylamine	x	x
Adipic acid 25%	280	138	Butyric acid	260	127
Allyl alcohol	300	149	Calcium bisulfite	260	127
Allyl chloride	300	149	Calcium chloride	160	71
Alum 5%	140	60	Calcium hydroxide, sat.	260	127
Aluminum chloride, aqueous	300	149	Calcium hypochlorite	x	x
Aluminum chloride, dry	300	149	Calcium nitrate	260	127
Aluminum fluoride	280	138	Calcium oxide		
Aluminum hydroxide	260	127	Calcium sulfate	260	127
Aluminum sulfate	160	71	Caprylic acid	250	121
Ammonium carbonate	240	116	Carbon bisulfide	160	71
Ammonium hydroxide 25%	250	121	Carbon dioxide, dry	90	32
Ammonium hydroxide, sat.	200	93	Carbon dioxide, wet	80	27
Ammonium nitrate	250	121	Carbon disulfide	260	127
Ammonium persulfate	260	127	Carbon tetrachloride	212	100
Ammonium phosphate	260	127	Cellosolve	240	116
Ammonium sulfate 10–40%	260	127	Chloroacetic acid, 50% water	100	38
Ammonium sulfide	260	127	Chloroacetic acid	240	116
Ammonium sulfite	240	116	Chlorine gas, dry	260	127
Amyl acetate	260	127	Chlorine gas, wet	260	127
Amyl alcohol	278	137	Chlorine, liquid	x	x
Amyl chloride	x	x	Chlorobenzene	260	127
Aniline	80	27	Chloroform	x	x
Antimony trichloride	250	121	Chlorosulfonic acid	260	127
Aqua regia 3:1	x	x	Chromic acid 10%	x	x
Barium carbonate	240	116	Chromic acid 50%	x	x
Barium chloride	260	127	Chromyl chloride	250	121
Barium hydroxide	260	127	Citric acid 15%	250	121
Barium sulfide	260	127	Citric acid, conc.	250	121
Benzaldehyde	80	27	Copper acetate	260	127
Benzene	160	71	Copper carbonate		
Benzenesulfonic acid 10%	160	71	Copper chloride	260	127
Benzoic acid	260	127	Copper cyanide	240	116
Benzyl alcohol	80	27	Copper sulfate	300	149

TABLE **6.3** Continued

Chemical	Maximum temp. °F	°C	Chemical	Maximum temp. °F	°C
Cresol	260	127	Nitric acid 5%	x	x
Cupric chloride 5%	300	149	Nitric acid 20%	x	x
Cupric chloride 50%	300	149	Nitric acid 70%	x	x
Cyclohexane	140	60	Nitric acid, anhydrous	x	x
Cyclohexanol			Nitrous acid, conc.	x	x
Dichloroacetic acid	x	x	Oleum	190	88
Dichloroethane (ethylene di-chloride)	250	121	Perchloric acid 10%	x	x
			Perchloric acid 70%	260	127
Ethylene glycol	160	71	Phenol	x	x
Ferric chloride	260	127	Phosphoric acid 50%	212	100
Ferric chloride 50% in water	160	71	Picric acid		
Ferric nitrate 10–50%	160	71	Potassium bromide 30%	260	127
Ferrous chloride	160	71	Salicylic acid	260	127
Ferrous nitrate			Silver bromide 10%		
Fluorine gas, dry	x	x	Sodium carbonate	212	100
Fluorine gas, moist	x	x	Sodium chloride	260	127
Hydrocyanic acid 10%	160	71	Sodium hydroxide 10%	x	x
Hydrobromic acid, dilute	212	100	Sodium hydroxide 50%	x	x
Hydrobromic acid 20%	212	100	Sodium hydroxide, conc.	x	x
Hydrobromic acid 50%	212	100	Sodium hypochlorite 15%	x	x
Hydrochloric acid 20%	212	100	Sodium hypochlorite, conc.	x	x
Hydrochloric acid 38%	80	27	Sodium sulfide to 10%	260	127
Hydrofluoric acid 30%	230	110	Stannic chloride	260	127
Hydrofluoric acid 70%	140	60	Stannous chloride	250	121
Hydrofluoric acid 100%	140	60	Sulfuric acid 10%	160	71
Hypochlorous acid	x	x	Sulfuric acid 50%	80	27
Iodine solution 10%	x	x	Sulfuric acid 70%	80	27
Ketones, general	100	38	Sulfuric acid 90%	x	x
Lactic acid 25%	212	100	Sulfuric acid 98%	x	x
Lactic acid, conc.	160	71	Sulfuric acid 100%	x	x
Magnesium chloride	260	127	Sulfuric acid, fuming	x	x
Malic acid 10%	260	127	Sulfurous acid	160	71
Manganese chloride	200	93	Thionyl chloride	x	x
Methyl chloride	120	49	Toluene	212	100
Methyl ethyl ketone	80	27	Trichloroacetic acid 30%	80	27
Methyl isobutyl ketone	160	71	White liquor	140	60
Muriatic acid	80	27	Zinc chloride	160	71

[a] The chemicals listed are in the pure state or in a saturated solution unless otherwise indicated. Compatibility is shown to the maximum allowable temperature for which data are available. Incompatibility is shown by an x. A blank space indicates that data are unavailable.
Source: Ref. 1.

TABLE 6.4   Compatibility of Vinyl Ester with Selected Corrodents[a]

Chemical	Maximum temp. °F	Maximum temp. °C	Chemical	Maximum temp. °F	Maximum temp. °C
Acetaldehyde	x	x	Ammonium sulfate 10–	220	104
Acetamide			40%		
Acetic acid 10%	200	93	Ammonium sulfide	120	49
Acetic acid 50%	180	82	Ammonium sulfite	220	104
Acetic acid 80%	150	66	Amyl acetate	110	38
Acetic acid, glacial	150	66	Amyl alcohol	210	99
Acetic anhydride	100	38	Amyl chloride	120	49
Acetone	x	x	Aniline	x	x
Acetyl chloride	x	x	Antimony trichloride	160	71
Acrylic acid	100	38	Aqua regia 3:1	x	x
Acrylonitrile	x	x	Barium carbonate	260	127
Adipic acid	180	82	Barium chloride	200	93
Allyl alcohol	90	32	Barium hydroxide	150	66
Allyl chloride	90	32	Barium sulfate	200	93
Alum	240	116	Barium sulfide	180	82
Aluminum acetate	210	99	Benzaldehyde	x	x
Aluminum chloride,	260	127	Benzene	x	x
aqueous			Benzenesulfonic acid	200	93
Aluminum chloride, dry	140	60	10%		
Aluminum fluoride	100	38	Benzoic acid	180	82
Aluminum hydroxide	200	93	Benzyl alcohol	100	38
Aluminum nitrate	200	93	Benzyl chloride	90	32
Aluminum oxychloride			Borax	210	99
Aluminum sulfate	250	121	Boric acid	200	93
Ammonia gas	100	38	Bromine gas, dry	100	38
Ammonium bifluoride	150	66	Bromine gas, moist	100	38
Ammonium carbonate	150	66	Bromine, liquid	x	x
Ammonium chloride 10%	200	93	Butadiene		
Ammonium chloride 50%	200	93	Butyl acetate	80	27
Ammonium chloride, sat.	200	93	Butyl alcohol	120	49
Ammonium fluoride 10%	140	60	n-Butylamine	x	x
Ammonium fluoride 25%	140	60	Butyric acid	130	54
Ammonium hydroxide	100	38	Calcium bisulfide		
25%			Calcium bisulfite	180	82
Ammonium hydroxide,	130	54	Calcium carbonate	180	82
sat.			Calcium chlorate	260	127
Ammonium nitrate	250	121	Calcium chloride	180	82
Ammonium persulfate	180	82	Calcium hydroxide 10%	180	82
Ammonium phosphate	200	93	Calcium hydroxide, sat.	180	82

TABLE 6.4 Continued

Chemical	Maximum temp. °F	°C	Chemical	Maximum temp. °F	°C
Calcium hypochlorite	180	82	Dichloroethane (ethylene	110	43
Calcium nitrate	210	99	dichloride)		
Calcium oxide	160	71	Ethylene glycol	210	99
Calcium sulfate	250	116	Ferric chloride	210	99
Caprylic acid	220	104	Ferric chloride 50% in	210	99
Carbon bisulfide	x	x	water		
Carbon dioxide, dry	200	93	Ferric nitrate 10–50%	200	93
Carbon dioxide, wet	220	104	Ferrous chloride	200	93
Carbon disulfide	x	x	Ferrous nitrate	200	93
Carbon monoxide	350	177	Fluorine gas, dry	x	x
Carbon tetrachloride	180	82	Fluorine gas, moist	x	x
Carbonic acid	120	49	Hydrobromic acid, dil	180	82
Cellosolve	140	60	Hydrobromic acid 20%	180	82
Chloroacetic acid, 50%	150	66	Hydrobromic acid 50%	200	93
water			Hydrochloric acid 20%	220	104
Chloroacetic acid	200	93	Hydrochloric acid 38%	180	82
Chlorine gas, dry	250	121	Hydrocyanic acid 10%	160	71
Chlorine gas, wet	250	121	Hydrofluoric acid 30%	x	x
Chlorine, liquid	x	x	Hydrofluoric acid 70%	x	x
Chlorobenzene	110	43	Hydrofluoric acid 100%	x	x
Chloroform	x	x	Hypochlorous acid	150	66
Chlorosulfonic acid	x	x	Iodine solution 10%	150	66
Chromic acid 10%	150	66	Ketones, general	x	x
Chromic acid 50%	x	x	Lactic acid 25%	210	99
Chromyl chloride	210	99	Lactic acid, conc.	200	93
Citric acid 15%	210	99	Magnesium chloride	260	127
Citric acid, conc.	210	99	Malic acid 10%	140	60
Copper acetate	210	99	Manganese chloride	210	99
Copper carbonate			Methyl chloride		
Copper chloride	220	104	Methyl ethyl ketone	x	x
Copper cyanide	210	99	Methyl isobutyl ketone	x	x
Copper sulfate	240	116	Muriatic acid	180	82
Cresol	x	x	Nitric acid 5%	180	82
Cupric chloride 5%	260	127	Nitric acid 20%	150	66
Cupric chloride 50%	220	104	Nitric acid 70%	x	x
Cyclohexane	150	66	Nitric acid, anhydrous	x	x
Cyclohexanol	150	66	Nitrous acid 10%	150	66
Dibutyl phthalate	200	93	Oleum	x	x
Dichloroacetic acid	100	38	Perchloric acid 10%	150	66

TABLE 6.4 Continued

Chemical	Maximum temp. °F	°C	Chemical	Maximum temp. °F	°C
Perchloric acid 70%	x	x	Sodium sulfide to 50%	220	104
Phenol	x	x	Stannic chloride	210	99
Phosphoric acid 50–80%	210	99	Stannous chloride	200	93
Picric acid	200	93	Sulfuric acid 10%	200	93
Potassium bromide 30%	160	71	Sulfuric acid 50%	210	99
Salicylic acid	150	66	Sulfuric acid 70%	180	82
Silver bromide 10%			Sulfuric acid 90%	x	x
Sodium carbonate	180	82	Sulfuric acid 98%	x	x
Sodium chloride	180	82	Sulfuric acid 100%	x	x
Sodium hydroxide 10%	170	77	Sulfuric acid, fuming	x	x
Sodium hydroxide 50%	220	104	Sulfurous acid 10%	120	49
Sodium hydroxide, conc.			Thionyl chloride	x	x
			Toluene	120	49
Sodium hypochlorite 20%	180	82	Trichloroacetic acid 50%	210	99
Sodium hypochlorite, conc.	100	38	White liquor	180	82
			Zinc chloride	180	82

[a] The chemicals listed are in the pure state or in a saturated solution unless otherwise indicated. Compatibility is shown to the maximum allowable temperature for which data are available. Incompatibility is shown by an x. A blank space indicates that data are unavailable.
Source: Ref. 1.

EPOXY POLYAMIDE

Polyamide resins can be reacted with epoxies to form durable protective linings with a temperature resistance of 225°F/107°C dry and 150°F/60°C wet. The chemical resistance of epoxy polyamides is inferior to that of the amine cured epoxies. They are partially resistant to acids, acid salts, alkaline and organic solvents and resistant to moisture. Refer to Table 6.5 for the compatibility of epoxy polyamides with selected corrodents and to Ref. 1 for a more comprehensive listing.

COAL TAR EPOXY

Coal tar epoxy combines the moisture resistance of coal tar with the chemical resistance and abrasion resistance of epoxy. It possesses excellent resistance to saltwater, freshwater, mild acids, and mild alkalies. Coal tar epoxy has poor solvent resistance. The temperature resistance of coal tar epoxy is 225°F/107°C dry and 150°F/66°C wet.

TABLE 6.5 Compatibility of Epoxy Polyamide with Selected Corrodents[a]

Chemical	Maximum temp. °F	°C	Chemical	Maximum temp. °F	°C
Acetaldehyde	x	x	Lard oil	x	x
Acetic acid all conc.	x	x	Lauric acid	x	x
Acetic acid vapors	x	x	Linseed oil	100	38
Acetone	x	x	Magnesium chloride 50%	100	38
Aluminum chloride dry	100	38	Mercuric chloride	100	38
Aluminum fluoride	x	x	Mercuric nitrate	100	38
Ammonium chloride all	100	38	Methyl alcohol	100	38
Ammonium hydroxide 25%	100	38	Methyl sulfate	x	x
Aqua regia 3:1	x	x	Methylene chloride	x	x
Benzene	x	x	Mineral oil	100	38
Boric acid	140	60	Nitric acid	x	x
Bromine gas dry	x	x	Oil vegetable	100	38
Bromine gas moist	x	x	Oleum	x	x
Calcium chloride	110	43	Oxalic acid all conc.	100	38
Calcium hydroxide all	140	60	Perchloric acid	x	x
Citric acid all conc.	100	38	Petroleum oils, sour	100	38
Diesel fuels	100	38	Phenol	x	x
Ethanol	100	38	Phosphoric acid	x	x
Ferric chloride	100	38	Potassium chloride 30%	100	38
Formaldehyde to 50%	100	38	Potassium hydroxide 50%	100	38
Formic acid	x	x	Propylene glycol	100	38
Glucose	100	38	Sodium chloride	110	43
Green liquor	100	38	Sodium hydroxide to 50%	100	38
Hydrobromic acid	x	x	Sulfur dioxide wet	100	38
Hydrochloric acid dil	100	38	Sulfuric acid	x	x
Hydrochloric acid 20%	x	x	Water demineralized	110	43
Hydrofluoric acid dil	100	38	Water distilled	130	54
Hydrofluoric acid 30%	x	x	Water salt	130	54
Hydrofluoric acid vapors	100	38	Water sea	110	43
Hydrogen sulfide dry	100	38	Water sewage	100	38
Hydrogen sulfide wet	100	38	White liquor	150	66
Iodine	x	x	Wines	100	38
Lactic acid	x	x	Xylene	x	x

[a] The chemicals listed are in the pure state or in a saturated solution unless otherwise indicated. Compatibility is shown to the maximum allowable temperature for which data are available. Incompatibility is shown by an x. A blank space indicates that data are unavailable.
Source: Ref. 1.

TABLE 6.6 Compatibility of Coal Tar Epoxy with Selected Corrodents[a]

Chemical	°F	°C	Chemical	°F	°C
Acetaldehyde	x	x	Lard oil	x	x
Acetic acid to 20%	100	38	Lauric acid	x	x
Acetic acid vapors	100	38	Linseed oil	100	38
Acetone	x	x	Magnesium chloride 50%	90	32
Aluminum chloride dry	100	38	Mercuric chloride	100	38
Aluminum fluoride	120	49	Mercuric nitrate	100	38
Ammonium chloride dry	100	71	Methyl alcohol	100	38
Ammonium hydroxide 25%	110	43	Methyl sulfate	x	x
Aqua regia 3:1	x	x	Methylene chloride	x	x
Benzene	x	x	Mineral oil	100	38
Boric acid	100	38	Nitric acid	x	x
Bromine gas dry	100	38	Oil vegetable	100	38
Bromine gas moist	x	x	Oleum	x	x
Calcium chloride	100	38	Oxalic acid all conc.	100	38
Calcium hydroxide all	100	38	Perchloric acid	x	x
Citric acid all conc.	100	38	Petroleum oils, sour	100	38
Diesel fuels	100	38	Phenol	x	x
Ethanol	100	38	Phosphoric acid	x	x
Ferric chloride	100	38	Potassium chloride 30%	100	38
Formaldehyde to 50%	100	38	Potassium hydroxide 50%	100	38
Formic acid	x	x	Propylene glycol	100	38
Glucose	100	38	Sodium chloride	110	43
Green liquor	100	38	Sodium hydroxide to 50%	100	38
Hydrobromic acid	x	x	Sulfur dioxide wet	100	38
Hydrochloric acid dil	100	38	Sulfuric acid	x	x
Hydrochloric acid 20%	x	x	Water demineralized	100	38
Hydrofluoric acid dil	100	38	Water distilled	100	38
Hydrofluoric acid 30%	x	x	Water salt	130	54
Hydrofluoric acid vapors	110	43	Water sea	90	32
Hydrogen sulfide dry	100	38	Water sewage	100	38
Hydrogen sulfide wet	100	38	White liquor	100	38
Iodine	x	x	Wines	100	38
Lactic acid	x	x	Xylene	x	x

[a] The chemicals listed are in the pure state or in a saturated solution unless otherwise indicated. Compatibility is shown to the maximum allowable temperature for which data are available. Incompatibility is shown by an x. A blank space indicates that data are unavailable. *Source*: Ref. 1.

Refer to Table 6.6 for the compatibility of coal tar epoxy with selected corrodents. Reference 1 has a more comprehensive listing.

Coal tar epoxy finds application as a lining for crude oil storage tanks, in sewage disposal plants and water works.

COAL TAR

Coal tar linings exhibit excellent resistance to moisture and good resistance to weak acids and alkalies, petroleum oils, and salts. However, they have a low maximum operating temperature. At 100°F/38°C the coal tar lining will soften. Table 6.7 provides the compatibility of coal tar with selected corrodents, while Ref. 1 provides a more comprehensive listing.

URETHANES

The urethanes exhibit excellent resistance to strong mineral acids and alkalies with only fair solvent resistance. They are also resistant to greases, fuels, aliphatic and chlorinated hydrocarbons.

Aromatic hydrocarbons, polar solvents, esters, and ketones will attack urethane. They have a maximum operating temperature of 250°F/121°C dry and 150°F/66°C wet and a superior abrasion resistance. They are used to line dishwashers and washing machines.

Refer to Table 6.8 for the compatibility of urethanes with selected corrodents and to Reference 1 for a more comprehensive listing.

NEOPRENE

Neoprene solvent solutions are prepared by dissolving neoprene in standard rubber solvents. These solutions can be formulated in a range of viscosities suitable for brush, spray, or roller application. Areas of application include linings for storage tanks, industrial equipment, and chemical processing equipment.

Neoprene possesses excellent resistance to attack from solvents, waxes, fats, oils, greases, and many other pertroleum based products. It also exhibits excellent service when it is in contact with aliphatic compounds (methyl and ethyl alcohols, ethylene glycols, etc.), and aliphatic hydrocarbons. It is also resistant to dilute mineral acids, inorganic salt solutions, or alkalies.

Chlorinated and aromatic hydrocarbons, organic esters, aromatic hydroxy compounds, and certain ketones will attack neoprene. Refer to Table 4.11 for the compatibility of neoprene with selected corrodents and to Ref. 1 for a more comprehensive listing.

TABLE 6.7 Compatibility of Coal Tar with Selected Corrodents[a]

Chemical	°F	°C	Chemical	°F	°C
Acetaldehyde	x	x	Lard oil	x	x
Acetic acid all conc.			Lauric acid	x	x
Acetic acid vapors			Linseed oil		
Acetone	x	x	Magnesium chloride 50%		
Aluminum chloride dry			Mercuric chloride		
Aluminum fluoride			Mercuric nitrate		
Ammonium chloride all			Methyl alcohol		
Ammonium hydroxide 25%			Methyl sulfate	x	x
Aqua regia 3:1	x	x	Methylene chloride	x	x
Benzene	x	x	Mineral oil		
Boric acid			Nitric acid	x	x
Bromine gas dry	x	x	Oil vegetable		
Bromine gas moist	x	x	Oleum	x	x
Calcium chloride			Oxalic acid all conc.		
Calcium hydroxide all conc.			Perchloric acid	x	x
Citric acid all conc.			Petroleum oils, sour		
Diesel fuels			Phenol	x	x
Ethanol			Phosphoric acid	x	x
Ferric chloride			Potassium chloride 30%		
Formaldehyde to 50%	x	x	Potassium hydroxide 50%		
Formic acid	x	x	Propylene glycol		
Glucose			Sodium chloride		
Green liquor	x	x	Sodium hydroxide to 50%		
Hydrobromic acid	x	x	Sulfur dioxide wet		
Hydrochloric acid dil	x	x	Sulfuric acid	x	x
Hydrochloric acid 20%	x	x	Water demineralized	90	32
Hydrofluoric acid dil	x	x	Water distilled		
Hydrofluoric acid 30%	x	x	Water salt		
Hydrofluoric acid vapors			Water sea	90	32
Hydrogen sulfide dry			Water sewage	90	32
Hydrogen sulfide wet			White liquor		
Iodine	x	x	Wines		
Lactic acid	x	x	Xylene	x	x

The chemicals listed are in the pure state or in a saturated solution unless otherwise indicated. Compatibility is shown to the maximum allowable temperature for which data are available. Incompatibility is shown by an x. A blank space indicates that data are unavailable.
Source: Ref. 1.

TABLE 6.8 Compatibility of Urethranes with Selected Corrodents[a]

Chemical	Maximum temp. °F	°C	Chemical	Maximum temp. °F	°C
Acetaldehyde	x	x	Lard oil	90	32
Acetic acid to 20%	90	32	Lauric acid		
Acetic acid vapors	90	32	Linseed oil	90	32
Acetone	90	32	Magnesium chloride 50%	90	32
Aluminum chloride dry			Mercuric chloride		
Aluminum fluoride			Mercuric nitrate		
Ammonium chloride all	90	32	Methyl alcohol	90	32
Ammonium hydroxide 25%	90	32	Methyl sulfate		
Aqua regia 3:1	x	x	Methylene chloride	x	x
Benzene	x	x	Mineral oil	90	32
Boric acid	90	32	Nitric acid	x	x
Bromine gas dry			Oil vegetable		
Bromine gas moist			Oleum	x	x
Calcium chloride	80	27	Oxalic acid all conc.		
Calcium hydroxide all	90	32	Perchloric acid	x	x
Citric acid all conc.			Petroleum oils, sour		
Diesel fuels			Phenol	x	x
Ethanol	90	32	Phosphoric acid		
Ferric chloride	90	32	Potassium chloride 30%	90	32
Formaldehyde to 50%			Potassium hydroxide 50%	90	32
Formic acid	x	x	Propylene glycol		
Glucose	x	x	Sodium chloride	80	27
Green liquor			Sodium hydroxide to 50%	90	32
Hydrobromic acid			Sulfur dioxide wet		
Hydrochloric acid dil	x	x	Sulfuric acid 10%	90	32
Hydrochloric acid 20%	x	x	Water demineralized		
Hydrofluoric acid dil			Water distilled	90	32
Hydrofluoric acid 30%			Water salt	x	x
Hydrofluoric acid vapors			Water sea	80	27
Hydrogen sulfide dry			Water sewage		
Hydrogen sulfide wet			White liquor		
Iodine			Wines	x	x
Lactic acid			Xylene	x	x

The chemicals listed are in the pure state or in a saturated solution unless otherwise indicated. Compatibility is shown to the maximum allowable temperature for which data are available. Incompatibility is shown by an x. A blank space indicates that data are unavailable.
Source: Ref. 1.

TABLE 6.9 Compatibility of Polysulfides with Selected Corrodents[a]

Chemical	Maximum temp. °F	°C	Chemical	Maximum temp. °F	°C
Acetaldehyde			Lauric acid		
Acetic acid all conc.	80	27	Linseed oil	150	66
Acetic acid vapors	90	32	Magnesium chloride		
Acetone	80	27	50%		
Aluminum chloride dry			Mercuric chloride		
Aluminum fluoride			Mercuric nitrate		
Ammonium chloride all	140	60	Methyl alcohol	80	27
Ammonium hydroxide	x	x	Methyl sulfate		
25%			Methylene chloride		
Aqua regia 3:1			Mineral oil	80	27
Benzene	x	x	Nitric acid	x	x
Boric acid			Oil vegetable	x	x
Bromine gas dry			Oleum	x	x
Bromine gas moist			Oxalic acid all conc.	x	x
Calcium chloride	150	66	Perchloric acid		
Calcium hydroxide all	x	x	Petroleum oils, sour		
Citric acid all conc.	x	x	Phenol	x	x
Diesel fuels	80	27	Phosphoric acid	x	x
Ethanol	150	66	Potassium chloride 30%		
Ferric chloride			Potassium hydroxide	80	27
Formaldehyde to 50%	80	27	50%		
Formic acid			Propylene glycol		
Glucose			Sodium chloride	80	27
Green liquor			Sodium hydroxide to	x	x
Hydrobromic acid			50%		
Hydrochloric acid dil	x	x	Sulfur dioxide wet		
Hydrochloric acid 20%	x	x	Sulfuric acid	x	x
Hydrofluoric acid dil	x	x	Water demineralized	80	27
Hydrofluoric acid 30%	x	x	Water distilled	80	27
Hydrofluoric acid vapors			Water salt	80	27
Hydrogen sulfide dry			Water sea	80	27
Hydrogen sulfide wet			Water sewage	80	27
Iodine			White liquor		
Lactic acid	x	x	Wines	x	x
Lard oil			Xylene	80	27

[a] The chemicals listed are in the pure state or in a saturated solution unless otherwise indicated. Compatibility is shown to the maximum allowable temperature for which data are available. Incompatibility is shown by an x. A blank space indicates that data are unavailable.
Source: Ref. 1.

POLYSULFIDE RUBBERS

The polysulfides have an outstanding resistance to solvents, oils, gasoline, and aliphatic and aromatic hydrocarbon solvents, as well as very good water resistance.

Refer to Table 6.9 for the compatibility of polysulfides with selected corrodents and to Ref. 1 for a more comprehensive listing.

Polysulfides find application in the lining of jet fuel tanks.

POLYVINYL CHLORIDE (PVC) PLASTISOLS

A plastisol is a liquid dispersion of a fine-particle-size polyvinyl chloride in a plasticizer. PVC linings are commonly called vinyls. They are noted for their toughness, chemical resistance, and durability. They are available as solutions, dispersions, and lattices. PVC powders have essentially the same properties as liquids. Polyvinyl chloride plastisol and powder coatings have limited adhesion and require primers. These linings must be heat cured.

Vinyl plastisol linings are popular for use as an acid resisting lining. Refer to Table 3.5 for the compatibility of the plastisols with selected corrodents.

TEFZEL (ETFE)

Tefzel is the trademark of DuPont. It is inert to strong mineral acids, inorganic bases, halogens, and strong metal salt solutions. Even carboxylic acids, anhydrides, aromatic and aliphatic hydrocarbons, alcohols, ketones, aldehydes, ethers, chlorocarbons, and classic polymer solvents have little effect on Tefzel. Very strong oxidizing acids near their boiling points, such as nitric at high concentration will affect Tefzel in varying degrees. So will strong organic bases such as amines and sulfonic acids. Refer to Table 3.29 for the compatibility of ETFE with selected corrodents.

The typical Tefzel lining thickness is nominal 0.040 in. thick on all interior surfaces and flange faces. Coating thicknesses from 0.020 to 0.090 in. are available depending on application requirements and part geometry. For linings or coatings applied on carbon steel or stainless steel the continuous service temperature range is from $-25°F/-32°C$ to $220°F/104°C$.

PERFLUOROALKOXY (PFA)

PFA is manufactured by DuPont. It is not degraded by systems commonly encountered in chemical processes. It is inert to strong mineral acids, inorganic bases, inorganic oxidizing agents, salt solutions, and such organic compounds as organic acids, anhydrides, aromatics, aliphatic hydrocarbons, alcohols, aldehydes, ketones, esters, ethers, chlorocarbons, fluorocarbons, and mixtures of the above. Refer to Table 3.22 for the compatibility of PFA with selected corrodents.

PFA will be attacked by certain halogen complexes containing fluorine. These include chlorine trifluoride, bromine trifluoride, iodine pentafluoride, and fluorine itself. It is also attacked by such metals as sodium or potassium, particularly in their molten states.

Standard lining thickness is nominal 0.040 in. on interior and wetted surfaces. When abrasion is a problem, a thickness of 0.090 in. is available. Coatings applied on carbon or stainless steel have a continuous service temperature range of from $-60°F/-51°C$ to $400°F/204°C$.

ECTFE (HALAR)

ECTFE is manufactured under the tradename of Halar by Ausimont. Halar exhibits outstanding chemical resistance. It is virtually unaffected by all corrosive chemicals commonly encountered in industry including strong mineral and oxidizing acids, alkalies, metal etchants, liquid oxygen, and essentially all organic solvents except hot amines (e.g., aniline dimethylamine). As with other fluoropolymers, Halar will be attacked by metallic sodium and potassium. Refer to Table 3.27 for the compatibility of ECTFE with selected corrodents.

Coating thicknesses range from 0.010 to 0.040 in. with a temperature range from cryogenic to $320°F/160°C$. In addition to its corrosion resistance, the material has excellent impact strength and abrasion resistance. Previously glass lined tanks can be refurbished.

FLUOROELASTOMERS

Fluoroelastomers are fluorine containing hydrocarbon polymers manufactured under various tradenames such as Viton by DuPont, Technoflon by Ausimont, and Fluorel by 3M. The fluoroelastomers provide excellent resistance to oils, fuels, lubricants, most mineral acids, many aliphatic and aromatic hydrocarbons (carbon tetrachloride, benzene, toluene, xylene) that act as solvents for rubbers, gasoline, naphtha, chlorinated solvents, and pesticides.

These materials are not suitable for use with low molecular weight esters, ethers, ketones, and certain amines, nor with hot anhydrous hydrofluoric or chlorosulfonic acids. Their solubility in low molecular weight ketones is an advantage in producing solutions for lining and coating applications. Refer to Table 4.24 for the compatibility of fluoroelastomers with selected corrodents.

Fluoroelastomer linings have an allowable temperature of $-40°F/-40°C$ to $400°F/206°C$. They are available in various colors and can be custom formulated to meet specific applications.

FLUORINATED ETHYLENE PROPYLENE (FEP)

FEP, with a few exceptions, exhibits the same corrosion resistance as PTFE but at a lower temperature. It is resistant to practically all chemicals, the exception being extremely potent oxidizers such as chlorine trifluoride and related compounds. Some chemicals will attack FEP when at or near the service temperature limit in high concentrations. Refer to Table 3.25 for the compatibility of FEP with selected corrodents.

Coating thicknesses range from 0.010 to 0.060 in., with a maximum service temperature of 390°F/199°C. Previously glass lined tanks can be refurbished with this lining.

POLYVINYLIDENE FLUORIDE (PVDF)

Polyvinylidene fluoride is resistant to most acids, alkalies, aliphatic and aromatic hydrocarbons, alcohols, and strong oxidizing agents. Highly polar solvents such as acetone and ethyl acetate may cause swelling. When used with strong alkalies, stress cracking results. Refer to Table 3.16 for the compatibility of PVDF with selected corrodents.

PVDF linings have an operating temperature range of from −4°F/−20°C to 280°F/138°C. Lining thicknesses range from 0.010 to 0.040 in. Typical applications include vessels, agitators, pump housings, centrifuge housings, manway covers, and dust collectors. Previously glass lined tanks and accessories can be refurbished to like-new condition.

SILICONES

The silicone resins can be used in contact with dilute acids and alkalies, alcohols, animal and vegetable oils, and lubricating oils. They are also resistant to aliphatic hydrocarbons, but aromatic solvents such as benzene, toluene, gasoline, and chlorinated solvents will cause excessive swelling.

Typically, the silicon atoms will have one or more side groups attached to them, generally phenol (C_6H_5-), methyl (CH_3), or vinyl ($CH_2=CH-$) units. These groups impart properties such as solvent resistance, lubricity, and reactivity with organic chemicals and polymers. Since these side groups affect the corrosion resistance of the resin, it is necessary to check with the supplier as to the properties of the resin being supplied. Table 6.10 lists the corrosion resistance of methyl appended silicone with selected corrodents. Reference 1 provides additional listing.

The silicone resins are relatively high priced and are noted for their high temperature resistance and moisture resistance. Maximum allowable operating

TABLE 6.10   Compatibility of Methyl Appended Silicone with Selected
Corrodents[a]

Chemical	Maximum temp.		Chemical	Maximum temp.	
	°F	°C		°F	°C
Acetic acid 10%	90	32	Hydrobromic acid 50%	x	x
Acetic acid 50%	90	32	Hydrochloric acid 20%	90	32
Acetic acid 80%	90	32	Hydrochloric acid 38%	x	x
Acetic acid, glacial	90	32	Hydrofluoric acid 30%	x	x
Acetone	100	43	Lactic acid, all conc.	80	27
Acrylic acid 75%	80	27	Lactic acid, conc.	80	27
Acrylonitrile	x	x	Magnesium chloride	400	204
Alum	220	104	Methyl alcohol	410	210
Aluminum sulfate	410	210	Methyl ethyl ketone	x	x
Ammonium chloride 10%	x	x	Methyl isobutyl ketone	x	x
Ammonium chloride 50%	80	27	Nitric acid 5%	80	27
Ammonium chloride, sat.	80	27	Nitric acid 20%	x	x
Ammonium fluoride 25%	80	27	Nitric acid 70%	x	x
Ammonium hydroxide 25%	x	x	Nitric acid, anhydrous	x	x
Ammonium nitrate	210	99	Oleum	x	x
Amyl acetate	80	27	Phenol	x	x
Amyl alcohol	x	x	Phosphoric acid 50–80%	x	x
Amyl chloride	x	x	Propyl alcohol	400	204
Aniline	x	x	Sodium carbonate	300	149
Antimony trichloride	80	27	Sodium chloride 10%	400	204
Aqua regia 3:1	x	x	Sodium hydroxide 10%	90	27
Benzene	x	x	Sodium hydroxide 50%	90	27
Benzyl chloride	x	x	Sodium hydroxide, conc.	90	27
Boric acid	390	189	Sodium hypochlorite 20%	x	x
Butyl alcohol	80	27	Sodium sulfate	400	204
Calcium bisulfide	400	204	Stannic chloride	80	27
Calcium chloride	300	149	Sulfuric acid 10%	x	x
Calcium hydroxide 30%	200	99	Sulfuric acid 50%	x	x
Calcium hydroxide, sat.	400	201	Sulfuric acid 70%	x	x
Carbon bisulfide	x	x	Sulfuric acid 90%	x	x
Carbon disulfide	x	x	Sulfuric acid 98%	x	x
Carbon monoxide	400	204	Sulfuric acid 100%	x	x
Carbonic acid	400	204	Sulfuric acid, fuming	x	x
Chlorobenzene	x	x	Sulfurous acid	x	x
Chlorosulfonic acid	x	x	Tartaric acid	400	204
Ethylene glycol	400	204	Tetrahydrofuran	x	x
Ferric chloride	400	204	Toluene	x	x

TABLE 6.10 Continued

Chemical	Maximum temp. °F	°C	Chemical	Maximum temp. °F	°C
Tributyl phosphate	x	x	Water, distilled	210	99
Turpentine	x	x	Water, salt	210	99
Vinegar	400	204	Water, sea	210	99
Water, acid mine	210	99	Xylene	x	x
Water, demineralized	210	99	Zinc chloride	400	204

[a] The chemicals listed are in the pure state or in a saturated solution unless otherwise indicated. Compatibility is shown to the maximum allowable temperature. Incompatibility is shown by an x.
Source: Ref. 1.

temperature is 572°F/300°C. They are also suitable for operation at cryogenic temperatures.

A second high temperature formulation with aluminum can be operated up to 1200°F/649°C. This high temperature type requires baking for a good cure. It is also water repellent.

The silicones are used primarily as linings for high temperature exhaust stacks, ovens, and space heaters.

ASPHALT

Asphalt consists of solids from crude oil refining in aliphatic solvents and exhibits good moisture resistance and good resistance to weak acids, alkalies, and salts. It softens at 100°F/38°C. Applications include chemical fume atmospheres.

POLYESTERS, UNSATURATED

Unsaturated polyesters are produced via a condensation reaction between a dibasic organic acid or anhydride and a difunctional alcohol. The prepolymer is then dissolved in an unsaturated monomer, such as styrene. To initiate a cross-linking, a free radical source such as an organic peroxide is added to the liquid resin. The types and ratio of components used to manufacture the prepolymers, the manufacturing procedure, and the molecular weight of the prepolymer will determine the properties of the cured polyester.

Isophthalic Polyesters

The isophthalic polyesters are the most common type used for chemical service applications. They have a relatively wide range of corrosion resistance being

TABLE 6.11 Compatibility of Isophthalic Polyester with Selected Corrodents[a]

Chemical	Maximum temp. °F	°C	Chemical	Maximum temp. °F	°C
Acetaldehyde	x	x	Barium sulfate	160	71
Acetic acid 10%	180	82	Barium sulfide	90	32
Acetic acid 50%	110	43	Benzaldehyde	x	x
Acetic acid 80%	x	x	Benzene	x	x
Acetic acid, glacial	x	x	Benzenesulfonic acid 10%	180	82
Acetic anhydride	x	x	Benzoic acid	180	82
Acetone	x	x	Benzyl alcohol	x	x
Acetyl chloride	x	x	Benzyl chloride	x	x
Acrylic acid	x	x	Borax	140	60
Acrylonitrile	x	x	Boric acid	180	82
Adipic acid	220	104	Bromine gas, dry	x	x
Allyl alcohol	x	x	Bromine gas, moist	x	x
Allyl chloride	x	x	Bromine, liquid	x	x
Alum	250	121	Butyl acetate	x	x
Aluminum chloride, aqueous	180	82	Butyl alcohol	80	27
Aluminum chloride, dry	170	77	n-Butylamine	x	x
Aluminum fluoride 10%	140	60	Butyric acid 25%	129	49
Aluminum hydroxide	160	71	Calcium bisulfide	160	71
Aluminum nitrate	160	71	Calcium bisulfite	150	66
Aluminum sulfate	180	82	Calcium carbonate	160	71
Ammonia gas	90	32	Calcium chlorate	160	71
Ammonium carbonate	x	x	Calcium chloride	180	82
Ammonium chloride 10%	160	71	Calcium hydroxide 10%	160	71
Ammonium chloride 50%	160	71	Calcium hydroxide, sat.	160	71
Ammonium chloride, sat.	180	82	Calcium hypochlorite 10%	120	49
Ammonium fluoride 10%	90	32	Calcium nitrate	140	60
Ammonium fluoride 25%	90	32	Calcium oxide	160	71
Ammonium hydroxide 25%	x	x	Calcium sulfate	160	71
Ammonium hydroxide, sat.	x	x	Caprylic acid	160	71
Ammonium nitrate	160	71	Carbon bisulfide	x	x
Ammonium persulfate	160	71	Carbon dioxide, dry	160	71
Ammonium phosphate	160	71	Carbon dioxide, wet	160	71
Ammonium sulfate 10%	180	82	Carbon disulfide	x	x
Ammonium sulfide	x	x	Carbon monoxide	160	71
Ammonium sulfite	x	x	Carbon tetrachloride	x	x
Amyl acetate	x	x	Carbonic acid	160	71
Amyl alcohol	160	71	Cellosolve	x	x
Amyl chloride	x	x	Chloroacetic acid, 50% water	x	x
Aniline	x	x	Chloroacetic acid to 25%	150	66
Antimony trichloride	160	71	Chlorine gas, dry	160	71
Aqua regia 3:1	x	x	Chlorine gas, wet	160	71
Barium carbonate	190	88	Chlorine, liquid	x	x
Barium chloride	140	60	Chlorobenzene	x	x
Barium hydroxide	x	x	Chloroform	x	x

TABLE **6.11** Continued

Chemical	Maximum temp.		Chemical	Maximum temp.	
	°F	°C		°F	°C
Chlorosulfonic acid	x	x	Methyl ethyl ketone	x	x
Chromic acid 10%	x	x	Methyl isobutyl ketone	x	x
Chromic acid 50%	x	x	Muriatic acid	160	71
Chromyl chloride	140	60	Nitric acid 5%	120	49
Citric acid 15%	160	71	Nitric acid 20%	x	x
Citric acid, conc.	200	93	Nitric acid 70%	x	x
Copper acetate	160	71	Nitric acid, anhydrous	x	x
Copper chloride	180	82	Nitrous acid, conc.	120	49
Copper cyanide	160	71	Oleum	x	x
Copper sulfate	200	93	Perchloric acid 10%	x	x
Cresol	x	x	Perchloric acid 70%	x	x
Cupric chloride 5%	170	77	Phenol	x	x
Cupric chloride 50%	170	77	Phosphoric acid 50–80%	180	82
Cyclohexane	80	27	Picric acid	x	x
Dichloroacetic acid	x	x	Potassium bromide 30%	160	71
Dichloroethane (ethylene dichloride)	x	x	Salicylic acid	100	38
			Sodium carbonate 20%	90	32
Ethylene glycol	120	49	Sodium chloride	200	93
Ferric chloride	180	82	Sodium hydroxide 10%	x	x
Ferric chloride 50% in water	160	71	Sodium hydroxide 50%	x	x
Ferric nitrate 10–50%	180	82	Sodium hydroxide, conc.	x	x
Ferrous chloride	180	82	Sodium hypochlorite 20%	x	x
Ferrous nitrate	160	71	Sodium hypochlorite, conc.	x	x
Fluorine gas, dry	x	x	Sodium sulfide to 50%	x	x
Fluorine gas, moist	x	x	Stannic chloride	180	82
Hydrobromic acid, dilute	120	49	Stannous chloride	180	82
Hydrobromic acid 20%	140	60	Sulfuric acid 10%	160	71
Hydrobromic acid 50%	140	60	Sulfuric acid 50%	150	66
Hydrochloric acid 20%	160	71	Sulfuric acid 70%	x	x
Hydrochloric acid 38%	160	71	Sulfuric acid 90%	x	x
Hydrocyanic acid 10%	90	32	Sulfuric acid 98%	x	x
Hydrofluoric acid 30%	x	x	Sulfuric acid 100%	x	x
Hydrofluoric acid 70%	x	x	Sulfuric acid, fuming	x	x
Hydrofluoric acid 100%	x	x	Sulfurous acid	x	x
Hypochlorous acid	90	32	Thionyl chloride	x	x
Ketones, general	x	x	Toluene	110	43
Lactic acid 25%	160	71	Trichloroacetic acid 50%	170	77
Lactic acid, conc.	160	71	White liquor	x	x
Magnesium chloride	180	82	Zinc chloride	180	82
Malic acid	90	32			

[a] The chemicals listed are in the pure state or in a saturated solution unless otherwise indicated. Compatibility is shown to the maximum allowable temperature for which data are available. Incompatibility is shown by an x. A blank space indicates that data are unavailable.

satisfactory for use up to 125°F/52°C in such acids as 10% acetic, benzoic, citric, oleic, 25% phosphoric, tartaric, 10–25% sulfuric, and fatty acids. Most inorganic salts are also compatible with isophthalic polyesters. Solvents such as amyl alcohol, ethylene glycol, formaldehyde, gasoline, kerosine, and naphtha are also compatible.

The isophthalic polyesters are not resistant to acetone, amyl acetate, benzene, carbon disulfide, solutions of alkaline salts of potassium and sodium, hot distilled water, or higher concentrations of oxidizing acids. Refer to Table 6.11 for the compatibility of isophthalic polyesters with selected corrodents and to Ref. 1 for a more comprehensive listing.

Bisphenol-A Fumurate Polyesters

This is a premium grade corrosion resistant resin. It costs approximately one-third more than an isophthalic resin. The bisphenol polyesters are superior in their corrosion resistance to the isophthalic polyesters. They show good performance with moderate alkaline solutions and excellent resistance to the various categories of bleaching agents. The bisphenol polyesters will break down in highly concentrated acids or alkalies.

The resins can be used in the handling of the following materials:

a. Acids (to 200°F/93°C)

Acetic	Fatty acids	Stearic
Benzoic	Hydrochloric 10%	Sulfonic 30%
Boric	Lactic	Tannic
Butyric	Maleic	Tartaric
Chloroacetic 15%	Oleic	Trichloroacetic 50%
Chromic 5%	Oxalic	Rayon spin bath
Citric	Phosphoric 80%	

b. Salt (solutions to 200°F/93°C)

All aluminum salts	Copper salts	Zinc salts
Most ammonium salts	Iron salts	Most plating solutions
Calcium salts		

c. Alkalies

Ammonium hydroxide 5% to 160°F/71°C
Calcium hydroxide 25% to 160°F/71°C
Calcium hypochlorite 20% to 200°F/93°C
Chlorine dioxide 15% to 200°F/93°C
Potassium hydroxide 25% to 160°F/71°C
Sodium hydroxide 25% to 160°F/71°C
Sodium chlorite to 200°F/93°C
Sodium hydrosulfite to 200°F/93°C

d. Solvents
All solvents the isophthalics are resistant to, plus:

Sour crude oil	Alcohols at ambient temperature
Glycerine	Linseed oil

e. Gases to 200°F/93°C

Carbon dioxide	Sulfur dioxide, dry
Carbon monoxide	Sulfur dioxide, wet
Chlorine, dry	Sulfur trioxide
Chlorine, wet	Rayon waste gases 150°F/66°C

Solvents such as benzene, carbon disulfide, ether, methyl ethyl ketone, toluene, xylene, trichloroethylene, and trichloroethane will attack the resin. Sulfuric acid above 70%, sodium hydroxide, and 30% chromic acid will also attack the resin. Refer to Table 6.12 for the compatibility of bisphenol A–fummurate polyester and Table 6.13 for the compatibility of hydrogenated bisphenol A–bisphenol A polyester with selected corrodents. Reference 1 provides a more comprehensive listing.

Halogenated Polyesters

Halogenated resins consist of chlorinated or brominated polyesters. These are also known as chlorendic polyesters when chlorinated. These resins have the highest heat resistance of any chemically resistant polyester. They are also inherently fire retardant. A noncombustible rating of 20 can be achieved, making this the safest possible polyester for stacks, hoods, or wherever a fire hazard may exist.

Excellent resistance is exhibited in contact with oxidizing acids and solutions, such as 35% nitric acid at elevated temperatures and 70% nitric acid at room temperature, 40% chromic acid, chlorine water, wet chlorine, and 15% hypochlorites. They also resist neutral and acid salts, nonoxidizing acids, organic acids, mercaptans, ketones, aldehydes, alcohols, glycols, organic esters, and fats and oils. Table 6.14 is an application guide for chlorinated polyesters. Reference 1 provides a detailed listing.

These polyesters are not resistant to highly alkaline solutions of sodium hydroxide, concentrated sulfuric acid, alkaline solutions with pH greater than 10, aliphatic, primary and aromatic amines, amides, and other alkaline organics, phenol, and acid halides. Table 6.15 lists the compatibility of halogenated polyesters with selected corrodents.

HYPALON (CHLOROSULFONATED POLYETHYLENE)

Hypalon in many respects is similar to neoprene, but it does possess some advantages over neoprene in certain types of service. It has better heat resistance and

TABLE 6.12 Compatibility of Bisphenol A—Fumarate Polyester with Selected Corrodents[a]

Chemical	Maximum temp.		Chemical	Maximum temp.	
	°F	°C		°F	°C
Acetaldehyde	x	x	Barium carbonate	200	93
Acetic acid 10%	220	104	Barium chloride	220	104
Acetic acid 50%	160	171	Barium hydroxide	150	66
Acetic acid 80%	160	171	Barium sulfate	220	104
Acetic acid, glacial	x	x	Barium sulfide	140	60
Acetic anhydride	110	43	Benzaldehyde	x	x
Acetone	x	x	Benzene	x	x
Acetyl chloride	x	x	Benzenesulfonic acid 10%	200	93
Acrylic acid	100	38	Benzoic acid	180	82
Acrylonitrile	x	x	Benzyl alcohol	x	x
Adipic acid	220	104	Benzyl chloride	x	x
Allyl alcohol	x	x	Borax	220	104
Allyl chloride	x	x	Boric acid	220	104
Alum	220	104	Bromine gas, dry	90	32
Aluminum chloride, aqueous	200	93	Bromine gas, moist	100	38
			Bromine, liquid	x	x
Aluminum fluoride 10%	90	32	Butyl acetate	80	27
Aluminum hydroxide	160	71	Butyl alcohol	80	27
Aluminum nitrate	200	93	n-Butylamine	x	x
Aluminum sulfate	200	93	Butyric acid	220	93
Ammonia gas	200	93	Calcium bisulfite	180	82
Ammonium carbonate	90	32	Calcium carbonate	210	99
Ammonium chloride 10%	200	93	Calcium chlorate	200	93
Ammonium chloride 50%	220	104	Calcium chloride	220	104
Ammonium chloride, sat.	220	104	Calcium hydroxide 10%	180	82
Ammonium fluoride 10%	180	82	Calcium hydroxide, sat.	160	71
Ammonium fluoride 25%	120	49	Calcium hypochlorite 10%	80	27
Ammonium hydroxide 25%	100	38	Calcium nitrate	220	93
Ammonium hydroxide 20%	140	60	Calcium sulfate	220	93
Ammonium nitrate	220	104	Caprylic acid	160	71
Ammonium persulfate	180	82	Carbon bisulfide	x	x
Ammonium phosphate	80	27	Carbon dioxide, dry	350	177
Ammonium sulfate 10–40%	220	104	Carbon dioxide, wet	210	99
Ammonium sulfide	110	43	Carbon disulfide	x	x
Ammonium sulfite	80	27	Carbon monoxide	350	177
Amyl acetate	80	27	Carbon tetrachloride	110	43
Amyl alcohol	200	93	Carbonic acid	90	32
Amyl chloride	x	x	Cellosolve	140	60
Aniline	x	x	Chloroacetic acid, 50% water	140	60
Antimony trichloride	220	104			
Aqua regia 3:1	x	x	Chloroacetic acid to 25%	80	27

TABLE 6.12   Continued

Chemical	Maximum temp. °F	Maximum temp. °C	Chemical	Maximum temp. °F	Maximum temp. °C
Chlorine gas, dry	200	93	Lactic acid, conc.	220	104
Chlorine gas, wet	200	93	Magnesium chloride	220	104
Chlorine, liquid	x	x	Malic acid	160	71
Chlorobenzene	x	x	Methyl ethyl ketone	x	x
Chloroform	x	x	Methyl isobutyl ketone	x	x
Chlorosulfonic acid	x	x	Muriatic acid	130	54
Chromic acid 10%	x	x	Nitric acid 5%	160	71
Chromic acid 50%	x	x	Nitric acid 20%	100	38
Chromyl chloride	150	66	Nitric acid 70%	x	x
Citric acid 15%	220	104	Nitric acid, anhydrous	x	x
Citric acid, conc.	220	104	Oleum	x	x
Copper acetate	180	82	Phenol	x	x
Copper chloride	220	104	Phosphoric acid 50–80%	220	104
Copper cyanide	220	104	Picric acid	110	43
Copper sulfate	220	104	Potassium bromide 30%	200	93
Cresol	x	x	Salicylic acid	150	66
Cyclohexane	x	x	Sodium carbonate	160	71
Dichloroacetic acid	100	38	Sodium chloride	220	104
Dichloroethane	x	x	Sodium hydroxide 10%	130	54
(ethylene dichloride)	220	104	Sodium hydroxide 50%	220	104
Ethylene glycol	220	104	Sodium hydroxide, conc.	200	93
Ferric chloride	220	104	Sodium hypochlorite 20%	x	x
Ferric chloride 50% in	220	104	Sodium sulfide to 50%	210	99
water			Stannic chloride	200	93
Ferric nitrate 10–50%	220	104	Stannous chloride	220	104
Ferrous chloride	220	104	Sulfuric acid 10%	220	104
Ferrous nitrate	220	104	Sulfuric acid 50%	220	104
Fluorine gas, moist			Sulfuric acid 70%	160	71
Hydrobromic acid, dilute	220	104	Sulfuric acid 90%	x	x
Hydrobromic acid 20%	220	104	Sulfuric acid 98%	x	x
Hydrobromic acid 50%	160	71	Sulfuric acid 100%	x	x
Hydrochloric acid 20%	190	88	Sulfuric acid, fuming	x	x
Hydrochloric acid 38%	x	x	Sulfurous acid	110	43
Hydrocyanic acid 10%	200	93	Thionyl chloride	x	x
Hydrofluoric acid 30%	90	32	Toluene	x	x
Hypochlorous acid 20%	90	32	Trichloroacetic acid 50%	180	82
Iodine solution 10%	200	104	White liquor	180	82
Lactic acid 25%	210	99	Zinc chloride	250	121

[a] The chemical listed are in the pure state or in a saturated solution unless otherwise indicated. Compatibility is shown to the maximum allowable temperature for which data are available. Incompatibility is shown by an x. A blank space indicates that data are unavailable.
Source: Ref. 1.

TABLE 6.13 Compatibility of Hydrogenated Bisphenol A—Bisphenol A
Polyester with Selected Corrodents[a]

Chemical	Maximum temp.		Chemical	Maximum temp.	
	°F	°C		°F	°C
Acetic acid 10%	200	93	Chloroacetic acid, 50%	90	32
Acetic acid 50%	160	71	water		
Acetic anhydride	x	x	Chlorine gas, dry	210	99
Acetone	x	x	Chlorine gas, wet	210	99
Acetyl chloride	x	x	Chloroform	x	x
Acrylonitrile	x	x	Chromic acid 50%	x	x
Aluminum acetate			Citric acid 15%	200	93
Aluminum chloride,	200	93	Citric acid, conc.	210	99
aqueous			Copper acetate	210	99
Aluminum fluoride	x	x	Copper chloride	210	99
Aluminum sulfate	200	93	Copper cyanide	210	99
Ammonium chloride, sat.	200	93	Copper sulfate	210	99
Ammonium nitrate	200	93	Cresol	x	x
Ammonium persulfate	200	93	Cyclohexane	210	99
Ammonium sulfide	100	38	Dichloroethane (ethylene	x	x
Amyl acetate	x	x	dichloride)		
Amyl alcohol	200	93	Ferric chloride	210	99
Amyl chloride	90	32	Ferric chloride 50% in	200	93
Aniline	x	x	water		
Antimony trichloride	80	27	Ferric nitrate 10–50%	200	93
Aqua regia 3:1	x	x	Ferrous chloride	210	99
Barium carbonate	180	82	Ferrous nitrate	210	99
Barium chloride	200	93	Hydrobromic acid 20%	90	32
Benzaldehyde	x	x	Hydrobromic acid 50%	90	32
Benzene	x	x	Hydrochloric acid 20%	180	82
Benzoic acid	210	99	Hydrochloric acid 38%	190	88
Benzyl alcohol	x	x	Hydrocyanic acid 10%	x	x
Benzyl chloride	x	x	Hydrofluoric acid 30%	x	x
Boric acid	210	99	Hydrofluoric acid 70%	x	x
Bromine, liquid	x	x	Hydrofluoric acid 100%	x	x
Butyl acetate	x	x	Hypochlorous acid 50%	210	99
n-Butylamine	x	x	Lactic acid 25%	210	99
Butyric acid	x	x	Lactic acid, conc.	210	99
Calcium bisulfide	120	49	Magnesium chloride	210	99
Calcium chlorate	210	99	Methyl ethyl ketone	x	x
Calcium chloride	210	99	Methyl isobutyl ketone	x	x
Calcium hypochlorite 10%	180	82	Muriatic acid	190	88
Carbon bisulfide	x	x	Nitric acid 5%	90	32
Carbon disulfide	x	x	Oleum	x	x
Carbon tetrachloride	x	x	Perchloric acid 10%	x	x

TABLE 6.13 Continued

Chemical	Maximum temp.		Chemical	Maximum temp.	
	°F	°C		°F	°C
Perchloric acid 70%	x	x	Sulfuric acid 50%	210	99
Phenol	x	x	Sulfuric acid 70%	90	32
Phosphoric acid 50–80%	210	99	Sulfuric acid 90%	x	x
Sodium carbonate 10%	100	38	Sulfuric acid 98%	x	x
Sodium chloride	210	99	Sulfuric acid 100%	x	x
Sodium hydroxide 10%	100	38	Sulfuric acid, fuming	x	x
Sodium hydroxide 50%	x	x	Sulfurous acid 25%	210	99
Sodium hydroxide, conc.	x	x	Toluene	90	32
Sodium hypochlorite 10%	160	71	Trichloroacetic acid	90	32
Sulfuric acid 10%	210	99	Zinc chloride	200	93

[a]The chemicals listed are in the pure state or in a saturated solution unless otherwise indicated. Compatibility is shown to the maximum allowable temperature for which data are available. Incompatibility is shown by an x. A blank space indicates that data are unavailable.
Source: Ref. 1.

TABLE 6.14 General Application Guide for Chlorinated Polyester

Environment	Comments
Acid halides	Not recommended
Acids, mineral nonoxidizing	Resistant to 250°F/121°C
Acids, organic	Resistant to 250°F/121°C; glacial acetic acid to 120°F/49°C
Alcohols	Resistant to 180°F/82°C
Aldehydes	Resistant to 180°F/82°C
Alkaline solutions pH > 10	Not recommended for continuous exposure
Amines, aliphatic, primary aromatic	Can cause severe attack
Amides, other alkaline organics	Can cause severe attack
Esters, organic	Resistant to 180°F/82°C
Fats and oils	Resistant to 200°F/95°C
Glycols	Resistant to 180°F/82°C
Ketones	Resistant to 180°F/82°C
Mercaptans	Resistant to 180°F/82°C
Phenol	Not recommended
Salts, acid	Resistant to 250°F/121°C
Salts, neutral	Resistant to 250°F/121°C
Water, demineralized, distilled, deionized, steam and condensate	Resistant to 212°F/100°C; Lowest absorption of any polyester

TABLE 6.15 Compatibility of Halogenated Polyester with Selected Corrodents[a]

Chemical	Maximum temp. °F	Maximum temp. °C	Chemical	Maximum temp. °F	Maximum temp. °C
Acetaldehyde	x	x	Barium carbonate	250	121
Acetic acid 10%	140	60	Barium chloride	250	121
Acetic acid 50%	90	32	Barium hydroxide	x	x
Acetic acid, glacial	110	43	Barium sulfate	180	82
Acetic anhydride	100	38	Barium sulfide	x	x
Acetone	x	x	Benzaldehyde	x	x
Acetyl chloride	x	x	Benzene	90	32
Acrylic acid	x	x	Benzenesulfonic acid 10%	120	49
Acrylonitrile	x	x	Benzoic acid	250	121
Adipic acid	220	104	Benzyl alcohol	x	x
Allyl alcohol	x	x	Benzyl chloride	x	x
Allyl chloride	x	x	Borax	190	88
Alum 10%	200	93	Boric acid	180	82
Aluminum chloride, aqueous	120	49	Bromine gas, dry	100	38
			Bromine gas, moist	100	38
Aluminum fluoride 10%	90	32	Bromine, liquid	x	x
Aluminum hydroxide	170	77	Butyl acetate	80	27
Aluminum nitrate	160	71	Butyl alcohol	100	38
Aluminum oxychloride			n-Butylamine	x	x
Aluminum sulfate	250	121	Butyric acid 20%	200	93
Ammonia gas	150	66	Calcium bisulfide	x	x
Ammonium carbonate	140	60	Calcium bisulfite	150	66
Ammonium chloride 10%	200	93	Calcium carbonate	210	99
Ammonium chloride 50%	200	93	Calcium chlorate	250	121
Ammonium chloride, sat.	200	93	Calcium chloride	250	121
Ammonium fluoride 10%	140	60	Calcium hydroxide, sat.	x	x
Ammonium fluoride 25%	140	60	Calcium hypochlorite 20%	80	27
Ammonium hydroxide 25%	90	32	Calcium nitrate	220	104
			Calcium oxide	150	66
Ammonium hydroxide, sat.	90	32	Calcium sulfate	250	121
			Caprylic acid	140	60
Ammonium nitrate	200	93	Carbon bisulfide	x	x
Ammonium persulfate	140	60	Carbon dioxide, dry	250	121
Ammonium phosphate	150	66	Carbon dioxide, wet	250	121
Ammonium sulfate 10–40%	200	93	Carbon disulfide	x	x
			Carbon monoxide	170	77
Ammonium sulfide	120	49	Carbon tetrachloride	120	49
Ammonium sulfite	100	38	Carbonic acid	160	71
Amyl acetate	190	85	Cellosolve	80	27
Amyl alcohol	200	93	Chloroacetic acid, 50% water	100	38
Amyl chloride	x	x			
Aniline	120	49	Chloroacetic acid 25%	90	32
Antimony trichloride 50%	200	93	Chlorine gas, dry	200	93
Aqua regia 3:1	x	x	Chlorine gas, wet	220	104

TABLE 6.15 Continued

Chemical	Maximum temp. °F	Maximum temp. °C	Chemical	Maximum temp. °F	Maximum temp. °C
Chlorine, liquid	x	x	Methyl ethyl ketone	x	x
Chlorobenzene	x	x	Methyl isobutyl ketone	80	27
Chloroform	x	x	Muriatic acid	190	88
Chlorosulfonic acid	x	x	Nitric acid 5%	210	99
Chromic acid 10%	180	82	Nitric acid 20%	80	27
Chromic acid 50%	140	60	Nitric acid 70%	80	27
Chromyl chloride	210	99	Nitrous acid, conc.	90	32
Citric acid 15%	250	121	Oleum	x	x
Citric acid, conc.	250	121	Perchloric acid 10%	90	32
Copper acetate	210	99	Perchloric acid 70%	90	32
Copper chloride	250	121	Phenol 5%	90	32
Copper cyanide	250	121	Phosphoric acid 50–80%	250	121
Copper sulfate	250	121	Picric acid	100	38
Cresol	x	x	Potassium bromide 30%	230	110
Cyclohexane	140	60	Salicylic acid	130	54
Dibutyl phthalate	100	38	Sodium carbonate 10%	190	88
Dichloroacetic acid	100	38	Sodium chloride	250	121
Dichloroethane (ethylene dichloride)	x	x	Sodium hydroxide 10%	110	43
Ethylene glycol	250	121	Sodium hydroxide 50%	x	x
Ferric chloride	250	121	Sodium hydroxide, conc.	x	x
Ferric chloride 50% in water	250	121	Sodium hypochlorite 20%	x	x
			Sodium hypochlorite, conc.	x	x
Ferric nitrate 10–50%	250	121	Sodium sulfide to 50%	x	x
Ferrous chloride	250	121	Stannic chloride	80	27
Ferrous nitrate	160	71	Stannous chloride	250	121
Hydrobromic acid, dil	200	93	Sulfuric acid 10%	260	127
Hydrobromic acid 20%	160	71	Sulfuric acid 50%	200	93
Hydrobromic acid 50%	200	93	Sulfuric acid 70%	190	88
Hydrochloric acid 20%	230	110	Sulfuric acid 90%	x	x
Hydrochloric acid 38%	180	82	Sulfuric acid 98%	x	x
Hydrocyanic acid 10%	150	66	Sulfuric acid 100%	x	x
Hydrofluoric acid 30%	120	49	Sulfuric acid, fuming	x	x
Hypochlorous acid 10%	100	38	Sulfurous acid 10%	80	27
Lactic acid 25%	200	93	Thionyl chloride	x	x
Lactic acid, conc.	200	93	Toluene	110	43
Magnesium chloride	250	121	Trichloroacetic acid 50%	200	93
Malic acid 10%	90	32	White liquor	x	x
Methyl chloride	80	27	Zinc chloride	200	93

[a] The chemicals listed are in the pure state or in a saturated solution unless otherwise indicated. Compatibility is shown to the maximum allowable temperature for which data are available. Incompatibility is shown by an x. A blank space indicates that data are unavailable.
Source: Ref. 1.

better chemical resistance. Proper compounding will permit Hypalon to exhibit good resistance to wear and abrasion, and high impact resistance.

The chlorine content protects Hypalon against the attack of microorganisms, and it will not promote the growth of mold, mildew, fungus, or bacteria. To maintain this feature, it is important that proper compounding procedures be followed.

The maximum continuous operating temperature is 250°F/121°C. On the low temperature side conventional compounds can be used continuously down to 0 to −20°F/−18 to −28°C.

When properly compounded, Hypalon is resistant to attack by hydrocarbon oils and fuels, even at elevated temperatures. It is also resistant to such oxidizing chemicals as sodium hypochlorite, sodium peroxide, ferric chlorides, and sulfuric, chromic, and hydrofluoric acids. Concentrated hydrochloric acid (37%) at elevated temperatures above 158°F/70°C will attack Hypalon, but can be handled without adverse effect at all concentrations below this temperature. Nitric acid at room temperature and up to 60% concentration can also be handled without adverse effect.

Hypalon is also resistant to salt solutions, alcohols, and both weak and concentrated alkalies. Long term contact with water has little effect on Hypalon.

Hypalon has poor resistance to aliphatic, aromatic, and chlorinated hydrocarbons, aldehydes, and ketones.

Refer to Table 4.13 for the compatibility of Hypalon with selected corrodents.

Hypalon is used to line railroad tank cars and other tanks containing acids and other oxidizing chemicals.

REFERENCE

1. Schweitzer, Philip A. *Corrosion Resistance Tables*, 4th ed., vols. 1–3, Marcel Dekker, New York, 1995.

7

Masonry Linings

Masonry linings consist of three basic items: chemically inert masonry units such as brick, stone, or block; bonding mortar or mortars; and membrane materials. Such linings are used to protect vessels from mechanical damage, (such as blows or abrasion), thermal damage by providing sufficient insulation to reduce thermal exposure to a level the substrate can accept, or to limit the penetration of the chemical corrosive to the substrate to prevent corrosion of the substrate.

As with any lining system, the conditions under which it will operate must be investigated. It is important that the following conditions be known in order to make the proper selection of materials.

1. The anticipated chemical environment, including all chemicals and their concentrations, all cleaning materials that may be used, and whether or not by-products will be produced during the reaction.
2. Complete information on temperature ranges during processing as well as during any cleaning cycle, start-up and shutdown. If there is a possibility of any heat of reaction taking place, this must be considered.
3. Pressure and vacuum ranges must be known during the processing, as well as during start-up and shutdown.
4. With some masonry materials thermal shock could pose a problem. Therefore, it is necessary to know whether or not there will be any abrupt temperature changes.

SHELL DESIGN

Several factors must be taken into account when considering selection of the shell material. The shell must:

1. Contain the process pressure
2. Support the weight of the masonry lining via its foundation
3. Have an interior surface compatible with the membrane adhesion system
4. Be cost effective
5. Offer ease of fabrication

The ideal materials meeting these needs are carbon steel and concrete. All steel shells to be lined with masonry must meet the following requirements:

1. Steel vessels which are to receive a masonry lining must conform to ASME boiler and pressure vessel code, Section VIII Division 1, and in particular the design detailed in paragraph UG-22 item 4, which states that in lined vessels, loading due to the lining itself (such as weights and stresses) must be taken into account to determine wall thickness and supports.

   As a rough guide, an increase above normal design strengths of between 3600 and 4300 psi may be used when operating pressures do not exceed 3 atm and temperatures do not exceed 212°F/100°C, with a vessel diameter of less than 10 ft. If the pressure is increased to 5 atm and the temperature to 400°F/204°C, the increase in design strength would be in the 4300–7200 psi range.

2. The code, in paragraph UG-80, also stipulates an out of roundness maximum of 1% variation of all diameters from the normal diameter. This must not be exceeded in vessels to be masonry lined.

3. Nozzles of masonry lined vessels, which must be sleeved, cannot have an out of roundness exceeding 0.4%. Refer to Figure 7.1.

4. Supports must be located so as to support the vessel and its extra weight uniformly without distortion of the vessel.

   a. If the vessel is conical or dome headed and supported by a continuous skirt, the skirt should be centered directly under the masonry column and should be vented to provide adequate ventilation under the vessel.

   b. If the vessel is conical or dome headed and support legs are used, they should be centered under the masonry lining column tangentially to the vessel body.

   c. With a flat bottomed vessel, the bottom must be so constructed and supported as to be completely rigid, and well ventilated from the sides and underneath. This can usually be accomplished by means of cribbage or I beams underneath.

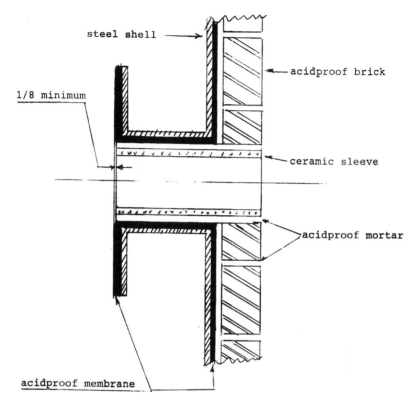

steel shell

acidproof brick

1/8 minimum

ceramic sleeve

acidproof mortar

acidproof membrane

FIGURE **7.1** Typical lined nozzle.

5. The interior of the vessel must be prepared as any vessel would that was to receive an elastomeric or sheet lining. Refer to pages 21–22 for details.
6. Nozzles must be reinforced by gusseting, straps, or strengtheners to prevent bending, or flexing in any way in relation to the body of the vessel.
7. The vessel should be located away from supports, walls, and adjacent equipment to permit air circulation and access by maintenance personnel.

Flat bottom vessels have several disadvantages:

1. They are difficult to drain and clean out.
2. They are hard to "stabilize." Stiffeners must be installed at frequent intervals in order to prevent the bottom from flopping up and down.
3. If a cylindrical tank is operating at an elevated temperature, the steel walls will expand and so will the flat bottom. The steel expands outward and the masonry also expands, but often not quite as much as

the steel. The weight of the column of masonry lining the walls rests on the bottom of the tank. The expansion of the masonry in the bottom is somewhat restrained by the friction created by the weight of this load pressing against the bottom and so creating a tendency for the masonry floor to relieve expansion pressure by heaving upwards. If a flexible membrane, such as rubber, is installed in the tank, it can flex with the movement of the bottom. However, the masonry lining is rigid and cannot flex. Consequently it will be disrupted by such flexing. In view of this, it is essential that any such vessel with a masonry lining have the flat bottom rigidly reinforced with channels or I beams at frequent intervals to prevent any movement. Flat bottom tanks should always be installed on dunnage, preferably 4 in. I beams, to permit circulation of air, inspection, and maintenance. Refer to Figure 7.2.

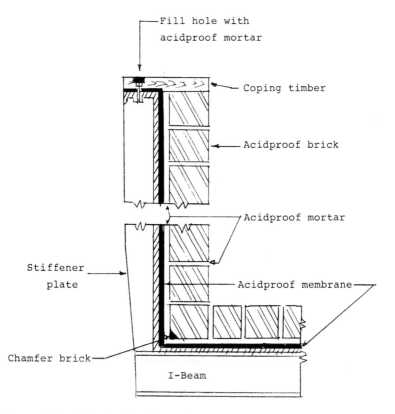

FIGURE 7.2 Typical reinforcing and support.

CONCRETE VESSELS

Concrete is ideally suited to being used as a supporting substrate for acid resistant masonry. It has a coefficient of thermal expansion closest to that of brickwork (5.3×10^{-6} as against $3.8 - 4.2 \times 10^{-6}$ for most shale and fireclay brick—while steel falls into the range of 6.8 to 7×10^{-6}). Concrete also has the ability to absorb blows and vibration without transmitting them to the masonry lining.

1. The vessel should be supported on a slab, not just support pillows or foundations. The slab should have sufficient reinforcing and strength to "float" the vessel without permitting or causing any deflection regardless of how unstable the ground may be below the slab.
2. The walls and bottom should be designed with sufficient reinforcing to prevent any deflection of the walls, taking into account the weight of the masonry lining, liquid of a specific gravity of 0.5 above that of the contained liquid at a height of not less than 6 in. from the top of the vessel. If the vessel is designed to empty through a weir or overflow, the design level of the liquid should be 1 in. above the level of the weir or overflow.
3. Forms used to cast the concrete in should be continuously, rigidly supported steel or plywood providing a continuous smooth surface. Any form marks must be removed and the surface ground smooth.
4. Reinforcing must be continuous from the bottom through the sides to the top. Rebars must be welded, not lapped and tied. Refer to Figure 7.3.
5. If oils or other lubricants are used as release agents on the forms, any residue remaining on the concrete must be removed by sandblasting before the membrane is installed.
6. It is preferable to maintain continuous pouring of the concrete, wet to wet. If not possible, continuous water stops must be provided when pouring wet to dry. Concrete adhesives applied to the dry face before pouring are beneficial.
7. All metal (e.g., tie wires) reaching through the concrete must be cut back at least $\frac{1}{2}$ in. below the surface. The voids created should be completely filled with a rich Portland cement-sand mix tamped into place and finished flush with the concrete surface.
8. Before installing the membrane, the vessel must be tested for watertightness. The vessel should not be lined if leaking.

In general, construction design, formulation of mix, and installation should conform to the best practice as recommended by the American Concrete Institute and/or Portland Cement Association.

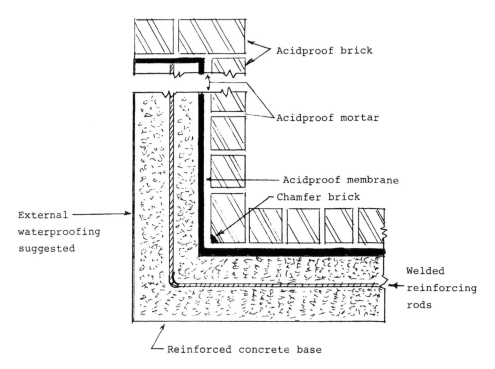

FIGURE 7.3 Typical acid proof construction for protection of concrete tank.

MEMBRANES

Membranes are installed between the substrate and masonry lining to act as a barrier to prevent the corrodent that passes through the masonry lining from reaching the substrate. These membranes can be classified as rigid or nonrigid, and as true or semimembranes.

Semimembranes permit limited amounts of corrodent to penetrate to the substrate, the amount of which is acceptable in terms of limited or very slow damage to the substrate. In some instances the formation of an inhibiting layer of a reaction product is formed, which prevents further attack. These membranes also may be rigid or nonrigid.

Chapters 4 to 6 provide details of various lining materials which may be used as membranes.

MASONRY MATERIALS

Masonry linings are installed to protect the substrate from mechanical damage such as blows or abrasion and/or to prevent thermal damage to the membrane and/or substrate by providing sufficient insulation to reduce thermal exposure to an acceptable level and/or to limit the penetration of the corrodent to the substrate to an acceptable level.

In most instances it is necessary to provide thermal insulation to protect the membrane. If too high a temperature reaches the membrane, particularly rigid materials with high coeficients of thermal expansion, they may pull loose from the substrate, bulge behind the brickwork, and in many cases push it out or cause joints to crack. Even if the brickwork remains tight, such membranes may "crimp," craze, and crack, allowing the corrodent to penetrate to the substrate. The designer must calculate the thermal drop through the full lining system to ensure that the above problems will be avoided. In order to do that, it is necessary to obtain the K factor for the masonry material, membrane, and substrate. Material manufacturers have this data.

TYPES OF BRICK

There are four major categories of corrosion resistant brick: red shale, fireclay, refractory, and carbon. Each type has somewhat different properties and compositions. However, all resist attack by most acids and are fired at 2000°F/1093°C; therefore, they are capable of handling process temperature at least that high. Their application temperatures, however, are determined by their resistance to thermal shock. In general, red shale is for ambient temperatures, fireclay for temperatures up to 1000°F/538°C, and refractory brick up to 3000°F/1649°C.

Most bricks contain silica and therefore cannot handle fluorides or caustics. The exceptions are carbon bricks which do not contain silica and therefore can be used to handle fluorides and caustics. Refer to Table 7.1 for relative properties of corrosion resistant brick.

Since bricks are porous, corrodents are not prevented from contacting the membrane. However, this porosity prevents rapid renewal of corrodents on the membrane surface. Consequently, if a chemical reaction occurs between the corrodent and the membrane, its rate will be slowed considerably.

One of the main purposes of the brick lining is to provide thermal protection to the membrane. The protection afforded depends not only on the brick's thickness but also on its thermal conductivity. The thermal conductivity of refractory and carbon brick is generally higher than fireclay type. Table 7.2 provides the minimum brick thickness at various operating temperatures to protect a membrane with a maximum allowable temperature of 250°F/121°C. Table 7.3 provides the K factor (thermal conductivity) of the various corrosion resistant bricks.

Red Shale Brick

Red shale brick is the most frequently used brick in chemically resistant masonry construction. It meets type L in ASTM C-279 specification.

TABLE 7.1 Relative Properties of Corrosion Resistant Brick

Type of brick	Cost	Temperature limit, °F/°C	Resistance to[a]				
			Acids	Caustics	Fluorides	Abrasion	Thermal shock
Red shale	low	100/88	E	P	P	H	P
Fireclay	M	1000/538	E	P	P	M	M
Refractory	M	3000/1649	E	P	P	P	M
Carbon	High	[b]	E	E	E	M	H

[a] E = excellent; P = poor; H = high; M = moderate.
[b] Contact manufacturer

TABLE 7.2 Minimum Brick Thickness for Protection of Membrane with 250°F/121°C Maximum Allowable Temperature

Process temperature, °F/°C	Minimum brick thickness, in.
250/121	1.5
300/149	3.75
400/201	7.5
500/260	11.25

TABLE 7.3 Thermal Conductivity of Corrosion Resistant Brick

Type of brick	Thermal conductivity, Btu/ft²-in.-h	Coefficient of thermal elongation per °F $\times$ 10^{-6}	
Red shale	8	3–3.5	
Fireclay	9	2.4–3.3	
Carbon			
Domestic	36–46	1.7–1.9	
Foreign	11.5–20	2.8–3.6	
Refractory Mullite			
500°F/260°C	12	0°F/−17.78°C	1.5
1000°F/538°C	11.5	750°F/399°C	2.3
1500°F/815°C	11.2	1470°F/799°C	2.6
2000°F/1093°C	11.5	2200°F/1200°C	2.7
2500°F/1371°C	12.0		
High fired alumina			
250°F/121°C	35.0	0°F/−17.78°C	3.9
500°F/260°C	29.5	750°F/399°C	4.8
1000°F/538°C	22.0	1470°F/799°C	6.1
1500°F/815°C	18.0	2200°F/1204°C	6.5
2000°F/1093°C	16.0	3000°F/1649°C	6.6
2300°F/1260°C	17.0		
2500°F/1371°C	17.8		
Silicon carbide	20	3.6	
Silica	6–6.5	400°F/204°C	0.07%
Porcelain 68°F/20°C	−91	4.3	
Granite block	6–7.24	5.9	
Foamed glass 12 lb/ft³			
100–400°F/38–204°C	0.58–0.75	1.6	

These brick have a lower absorption rate than fireclay brick and are usually somewhat more brittle. In addition, they are usually less dimensionally true than fireclay brick.

Applications for these brick include areas where lowest absorption rates are desired but where thermal shock is not a major factor. Such areas include process vessels.

Typical physical properties are:

Weight loss in 5 h boil test	2.8%
Modulus of rupture (min.) average of 5 pieces	3300 psi
Individual min.	2500 psi
Absorption (max.) average of 5	0.7%
Individual	1.2%
Compressive strength	10000 psi
Weight, lb/ft^3	145

Fireclay Brick

Fireclay brick is the second most frequently used brick. It is usually described as meeting type H in ASTM C-279. They contain a higher proportion of alumina and lower percentages of silica and iron than do shale brick. Fireclay bricks have a higher absorption rate than shale bricks, although some manufacturers will provide a denser brick that will meet type L for absorption. These brick are usually selected for outdoor exposures where rapid thermal changes occur since they are less brittle than the shale brick. Since they have a low iron content, they are used in process equipment where this characteristic is important in maintaining product purity.

Fireclay bricks are more dimensionally true than red shale bricks and are available in many shapes. Because of this they are preferred in work where uniformity of joints in the masonry is important.

The following are typical physical properties of fireclay brick:

Weight loss, 5 h boil	<2%
Absorption, max.	4%
Modulus of rupture (average of 5)	3500 psi
Compressive strength	7000 + psi
Weight, lb/ft^3	135–145

Carbon Brick

Carbon brick are used in areas exposed to strong alkali (pH 12.5+) and hydrofluoric acid or fluoride salts in acid medium. These brick have a much higher absorption rate than either shale or fireclay brick, and lower coefficients of expansion. Carbon bricks are more shock resistant than either shale or fireclay brick. This permits them to be used in areas where pressure changes in equipment are rapid, a condition that can cause shale or fireclay brick to spall because of the inability of the denser brick to lower its internal buildup pressures to keep pace with the pressure drop in the interior of the brick lined vessel, when such pressure changes are sudden and sharp.

Greater thicknesses of carbon brick are required than of red shale or fireclay brick to reach thermal drops equivalent to those of shale or fireclay bricks. This is the result of the much greater heat conductivity of carbon.

Following are the typical physical properties of carbon brick:

	Domestic	Foreign
Modulus of rupture, psi	2600	1500
Absorption, %	15	17
Compressive strength, psi	8030–8800	10,000
Tensile strength, psi	800	1000
Weight, lb/ft^3	96.7	93

Refractory Brick

For very high temperature operations refractory brick must be used. These brick are made from alumina clays, of which there are two general classifications: mullite or high alumina. High alumina bricks may vary in alumina content from 90 to 99% Al_2O_3.

Alumina brick, although perfectly satisfactory for the high temperature operations, generally does not have as good a chemical resistance as other types. These brick should be used with caution when exposed to free chlorine, phosphoric acid, and alkali hydroxides. The manufacturer should be consulted before specifying this brick. Typical physical properties of 92.9% high fired alumina brick are as follows:

Modulus of rupture, psi	4000–5000
Absorption, %	2.0–7.2
Compressive strength, psi	10,000–15,000
Weight, lb/ft^3	198–200

Mullite brick contains very little free Al_2O_3 therefore its chemical resistance is superior to that of the high alumina brick. Typical physical properties are as follows:

Modulus of rupture, psi	2500
Absorption, %	6–7
Linear thermal expansion, %/°F	3.5

Compressive strength, psi	10,000–14,000
Apparent porosity (% by ASTM C20)	14.0–18.0
Weight, lb/ft^3	155–162

Silicon Carbide Brick

Silicon carbide brick are suitable for high temperature applications. They are also a high abrasion resistant brick. These brick are made to fireclay sizes.

Silica Brick

This brick is the choice when handling high concentrations of phosphoric and other acids. They are made to fireclay brick sizes.

Porcelain

Porcelain is one of the hardest and densist of chemical resistant bricks. It is resistant to a wide range of chemicals up to 1500°F/815°C and is composed of 85% alumina. Outstanding features are its ease of cleaning and nil absorption.

Foamed Glass

These brick, or blocks, are produced from borosilicate glass. They offer a wide range of corrosion resistance, except for acid fluorides and strong alkalies. Because of its higher insulating qualities, thinner linings can be utilized, resulting in reduced lining weight. Operating temperature of up to 960°F/515°C may be used. Typical physical properties of a 12 lb/ft^3 block are:

Compressive strength, psi	200
Tensile strength, psi	72
Absorption	Nil

CELLULAR BOROSILICATE GLASS

Cellular borosilicate glass is similar to foamed glass, except for having a borosilicate composition. Probably the best known plain borosilicate glass is Pyrex. Cellular borosilicate glass offers advantages over the heavy weight of brick linings required for high temperatures. These cellular glass blocks provide superior chemical resistance; greater durability, strength, and dimensional stability; and a lower coefficient of thermal expansion. Disadvantages include a higher cost and the fact that it will be fractured or crushed by a heavy load, although not as readily as foamed glass.

Because of its structure and composition the cellular glass is an excellent insulating material for the inside hot surfaces of process equipment. It has a weight of only 12 lb/ft^3 (0.19 g/cm^3). For example, a process vessel handling 65% sulfuric acid at 350°F/177°C would have an outer shell of mild steel, an inner membrane of sheet rubber or vinyl, and an inner insulation of four 4 in. courses (layers) of acid brick laid up with a silicate mortar. As a result of the extra thickness of the lining, the outer steel walls and vessel supports would have to be heavy.

If cellular borosilicate glass blocks were to be used, a single 2 in. thick lining could be substituted for the acid brick. The advantages are an improved insulation at a labor cost below that for two courses of brick lining and only slightly heavier than a single course of brick. The protection against abrasion and impact are the same as that provided by the inner course of acid brick.

The chemical stability of borosilicate glass is one of the most comprehensive of any known construction material. It is highly resistant to water, acids, salt solutions, organic substance, and even halogens like chlorine and bromine. Only hydrofluoric acid, phosphoric acid, with fluorides, or strong alkalies at temperatures above 102°F/49°C can visibly affect the glass surface. Refer to Table 7.4 for the compatibility of borosilicate glass with selected corrodents.

MORTARS

A brick/membrane liner system, if properly designed, generally does not need mortar to hold the bricks together. For example, in a round vessel the bricks are keyed in (the brick is wider on the membrane side than on the process side) so that the wall cannot fall inward. However, mortar does provide the advantages of easing the installation, improving thermal and chemical protection of the membrane, and provides additional structural integrity in case some bricks crack during installation, operation, or maintenance.

The mortar must have a heavy enough consistency to support the weight of the brick without being squeezed from the joints while the joint is curing. Application is made by buttering each joint. Joints are usually ⅛ in. (3 mm) wide.

Both inorganic and organic based mortars are used. Inorganic mortars have operating temperatures over 1000°F/538°C, resist weak caustics and most strong acids (except hydrofluoric), and are not affected by organics.

Organic mortars have only moderate operating temperatures but resist corrosives over a wider pH range than do inorganic mortars. They also have greater tension and compression strengths. Refer to Table 7.5.

TABLE 7.4 Compatibility of Borosilicate Glass with Selected Corrodents[a]

Chemical	°F	°C	Chemical	°F	°C
Acetaldehyde	450	232	Barium sulfate	250	121
Acetamide	270	132	Barium sulfide	250	121
Acetic acid 10%	400	204	Benzaldehyde	200	93
Acetic acid 50%	400	204	Benzene	200	93
Acetic acid 80%	400	204	Benzene sulfonic acid 10%	200	93
Acetic acid, glacial	400	204	Benzoic acid	200	93
Acetic anhydride	250	121	Benzyl alcohol	200	93
Acetone	250	121	Benzyl chloride	200	93
Adipic acid	210	99	Borax	250	121
Allyl alcohol	120	49	Boric acid	300	149
Allyl chloride	250	121	Bromine gas, moist	250	121
Alum	250	121	Bromine liquid	90	32
Aluminum chloride, aqueous	250	121	Butadiene	90	32
Aluminum chloride, dry	180	82	Butyl acetate	250	121
Aluminum fluoride	x	x	Butyl alcohol	200	93
Aluminum hydroxide	250	121	Butyric acid	200	93
Aluminum nitrate	100	38	Calcium bisulfite	250	121
Aluminum oxychloride	190	88	Calcium carbonate	250	121
Aluminum sulfate	250	121	Calcium chlorate	200	93
Ammonium bifluoride	x	x	Calcium chloride	200	93
Ammonium carbonate	250	121	Calcium hydroxide 10%	250	121
Ammonium chloride 10%	250	121	Calcium hydroxide, sat.	x	x
Ammonium chloride 50%	250	121	Calcium hypochlorite	200	93
Ammonium chloride, sat.	250	121	Calcium nitrate	100	38
Ammonium fluoride 10%	x	x	Carbon bisulfide	250	121
Ammonium fluoride 25%	x	x	Carbon dioxide, dry	160	71
Ammonium hydroxide 25%	250	121	Carbon dioxide, wet	160	71
Ammonium hydroxide, sat.	250	121	Carbon disulfide	250	121
Ammonium nitrate	200	93	Carbon monoxide	450	232
Ammonium persulfate	200	93	Carbon tetrachloride	200	93
Ammonium phosphate	90	32	Carbonic acid	200	93
Ammonium sulfate 10–40%	200	93	Cellosolve	160	71
Amyl acetate	200	93	Chloracetic acid, 50% water	250	121
Amyl alcohol	250	121	Chloracetic acid	250	121
Amyl chloride	250	121	Chlorine gas, dry	450	232
Aniline	200	93	Chlorine gas, wet	400	204
Antimony trichloride	250	121	Chlorine, liquid	140	60
Aqua regia 3:1	200	93	Chlorobenzene	200	93
Barium carbonate	250	121	Chloroform	200	93
Barium chloride	250	121	Chlorosulfonic acid	200	93
Barium hydroxide	250	121	Chromic acid 10%	200	93

TABLE 7.4 Continued

Chemical	°F	°C	Chemical	°F	°C
Chromic acid 50%	200	93	Methyl ethyl ketone	200	93
Citric acid 15%	200	93	Methyl isobutyl ketone	200	93
Citric acid, concentrated	200	93	Nitric acid 5%	400	204
Copper chloride	250	121	Nitric acid 20%	400	204
Copper sulfate	200	93	Nitric acid 70%	400	204
Cresol	200	93	Nitric acid, anhydrous	250	121
Cupric chloride 5%	160	71	Oleum	400	204
Cupric chloride 50%	160	71	Perchloric acid 10%	200	93
Cyclohexane	200	93	Perchloric acid 70%	200	93
Cyclohexanol			Phenol	200	93
Dichloroacetic acid	310	154	Phosphoric acid 50–80%	300	149
Dichloroethane (ethylene dichloride)	250	121	Picric acid	200	93
			Potassium bromide 30%	250	121
Ethylene glycol	210	99	Silver bromide 10%		
Ferric chloride	290	143	Sodium carbonate	250	121
Ferric chloride 50% in water	280	138	Sodium chloride	250	121
Ferric nitrate 10–50%	180	82	Sodium hydroxide 10%	x	x
Ferrous chloride	200	93	Sodium hydroxide 50%	x	x
Fluorine gas, dry	300	149	Sodium hydroxide, concentrated	x	x
Fluorine gas, moist	x	x	Sodium hypochlorite 20%	150	66
Hydrobromic acid, dilute	200	93	Sodium hypochlorite, concentrated	150	66
Hydrobromic acid 20%	200	93			
Hydrobromic acid 50%	200	93	Sodium sulfide to 50%	x	x
Hydrochloric acid 20%	200	93	Stannic chloride	210	99
Hydrochloric acid 38%	200	93	Stannous chloride	210	99
Hydrocyanic acid 10%	200	93	Sulfuric acid 10%	400	204
Hydrofluoric acid 30%	x	x	Sulfuric acid 50%	400	204
Hydrofluoric acid 70%	x	x	Sulfuric acid 70%	400	204
Hydrofluoric acid 100%	x	x	Sulfuric acid 90%	400	204
Hypochlorous acid	190	88	Sulfuric acid 98%	400	204
Iodine solution 10%	200	93	Sulfuric acid 100%	400	204
Ketones, general	200	93	Sulfurous acid	210	99
Lactic acid 25%	200	93	Thionyl chloride	210	99
Lactic acid, concentrated	200	93	Toluene	250	121
Magnesium chloride	250	121	Trichloroacetic acid	210	99
Malic acid	160	72	White liquor	210	99
Methyl chloride	200	93	Zinc chloride	210	99

[a] The chemicals listed are in the pure state or in a saturated solution unless otherwise indicated. Compatibility is shown to the maximum allowable temperature for which data are available. Incompatibility is shown by an x. A blank space indicates that data are unavailable.
Source: Ref. 1.

TABLE 7.5 General Properties of Mortars

Mortar type	Compression/tensile strength, psi	Maximum temperature, °F/°C	Resistance to[a]		
			Acids	Caustics	Organics
Phenolic	12000/1200	305/151	NO	W	M
Furan	12000/1400	380/193	NO	E	M
Sulfurcarbon	6000/600	300/149	E	X	M
Polyester	13000/1500	250/121	M	W	M
Potassium silicate	3000/400	1500/815	E-HF	W	E
Aluminum	3000/400	2700/1482	E-HF	W	E

[a] NO = not oxidizing; M = most; W = weak only; E = resistant to all; X = not resistant; E-HF = resistant to all except hydrofluoric acid and fluorides.

Chemically resistant mortars are formulated using an inorganic binder or liquid resin system, fillers such as silica, carbon, or combinations thereof, and a hardener or catalyst system.

Carbon is the most inert of the fillers, having a wide resistance to most chemicals; therefore, it is the filler most often used. It is resistant to strong alkalies, hydrochloric acid, and fluorine chemicals. The general resistance of these fillers is shown in Table 7.6.

Inorganic Mortars

The inorganic mortars are the original mortars. They are commonly referred to as acid proof mortars, primarily because they are limited in application to a maximum pH of 7. They cannot be used in alkaline or alternate alkaline-acid service. There are two general types of inorganic mortars: the hot pour sulfur and the ambiently mixed and applied silicate mortars.

Sulfur Mortars

Sulfur mortars are hot melt compounds, available in flake, powder, and ingot forms. They must be heated to a temperature of 250°F/120°C and poured into the joints while hot. Sulfur mortars are particularly useful against oxidizing acids. When they are carbon filled, they are suitable for use against combinations of oxidizing acids and hydrofluoric acid. Chemical resistance to strong alkaline solutions and certain organic solvents is poor.

The sulfur mortars possess certain advantages over some of the resin mortars, primarily their resistance to oxidizing, nonoxidizing, and mixed acids, ease of use, resistance to thermal shock, high early strength, unlimited shelf life, and economy.

TABLE 7.6 General Resistance of Mortar Fillers

Corrodent at 20% concentration	Filler[a]		
	Carbon	Silica	Carbon-silica
Hydrochloric acid	R	R	R
Hydrofluoric acid	R	N	N
Sulfuric acid	R	R	R
Potassium hydroxide	R	N	N
Sodium hydroxide	R	N	N
Neutral salts	R	R	R
Solvents, conc.	R	R	N

[a]R = recommended; N = not recommended.

Sodium Silicate Mortars

Sodium silicate mortars are available as either two component systems, which consist of the liquid sodium solution and the filler powder containing settling agents and selected aggregates, or they may be a one part system in powder form to be mixed with water when used. There are some differences in the chemical resistance between the two types. The differences in the chemical resistance of the two types of mortars are shown in Table 7.7. Sodium silicate mortars are useful in the pH range of 0–6, except where sulfuric acid exposures exist in the vapor phase, wet-dry exposures, or in concentrations above 93%.

Potassium Silicate Mortars

Potassium silicate mortars are preferable to sodium silicate mortars. They have better workability because of their smoothness and lack of tackiness. They do

TABLE 7.7 Corrosion Resistance of Sodium Silicate Mortars

Corrodent at room temp.	Type of mortar[a]	
	Normal	Water resistant
Acetic acid, glacial	P	P
Chlorine dioxide, water sol.	N	N
Hydrogen peroxide	N	R
Nitric acid 5%	C	R
Nitric acid 20%	C	R
Nitric acid, over 20%	R	R
Sodium bicarbonate	N	N
Sodium sulfite	R	R
Sulfates, aluminum	R	R
Sulfates, copper	P	P
Sulfates, iron	P	P
Sulfates, magnesium	P	P
Sulfates, nickel	P	P
Sulfates, zinc	P	P
Sulfuric acid, to 93%	P	P
Sulfuric acid, over 93%	P	P

[a] R = recommended; N = not recommended; P = potential failure; C = conditional.

not run or flow from the brickwork, and they do not stick to the trowel. The potassium silicate mortars have a greater resistance to strong acid solutions, as well as to sulfation. These mortars are available with halogen-free hardening systems which eliminate the remote possibility of catalyst poisoning in certain chemical operations. The general corrosion resistance of the two types of potassium silicate mortars is shown in Table 7.8.

Silica Mortars

The silica type mortars consist of a colloidal silica binder with quartz fillers. The main difference compared with other mortars is total freedom from metal ions that could contribute to sulfation hydration within the mortar joints in high concentrations of sulfuric acid. This is a unique system. It can be used up to 2000°F/ 1093°C. The silica type mortars used in the pH range of 0–7 are resistant to all materials except hydrofluoric acid and acid fluorides.

TABLE 7.8 General Corrosion Resistance of Potassium Silicate Mortars

	Type of mortar[a]	
Corrodent at room temp.	Normal	Halogen free
Acetic acid, glacial	R	R
Chlorine dioxide, water sol.	R	R
Hydrogen peroxide	N	N
Nitric acid, 5%	R	R
Nitric acid, 20%	R	R
Nitric acid, over 20%	R	R
Sodium bicarbonate	N	N
Sodium sulfite	N	N
Sulfates, aluminum	R	R
Sulfates, copper	R	R
Sulfates, iron	R	R
Sulfates, magnesium	R	R
Sulfates, nickel	R	R
Sulfates, zinc	R	R
Sulfuric acid, to 93%	R	R
Sulfuric acid, over 93%	R	R

[a] R = recommended; N = not recommended.

ORGANIC MORTARS

Epoxy

The three main epoxy resins used in the formulation of corrosion resistant mortars are based on Bisphenol A, Bisphenol F, and epoxy phenol. The corrosion resistance as well as the physical and mechanical properties will be determined by the type of hardener used.

The three main types of hardeners used with the Bisphenol A resin are aliphatic amines, modified aliphatic amines, and aromatic amines. Silica is the filler most often used with epoxy mortars. This prohibits the use of epoxy mortars with hydrofluoric acid, other fluorine chemicals, and strong, hot alkalies. Carbon fillers can be substituted but with some sacrifice to the working properties.

TABLE 7.9   General Corrosion Resistance of Epoxy Mortars

	Hardeners[a]			
			Aromatic amines	
			Bisphenol	
	Aliphatic	Modified aliphatic		
Corrodent at room temp.	amines	amines	A	F
Acetic acid 5–10%	C	U	R	
Acetone	U	U	U	U
Benzene	U	U	R	R
Butyl acetate	U	U	U	R
Butyl alcohol	R	R	R	R
Chromic acid 5%	U	U	R	R
Chromic acid 10%	U	U	U	R
Formaldehyde 35%	R	R	R	R
Gasoline	R	R	R	R
Hydrochloric acid to 36%	U	U	R	R
Nitric acid 30%	U	U	U	U
Phosphoric acid 50%	U	U	R	R
Sulfuric acid 25%	R	U	R	R
Sulfuric acid 50%	U	U	R	R
Sulfuric acid 75%	U	U	U	U
Trichloroethylene	U	U	U	R

[a] R = recommended; U = unsatisfactory.

The Bisphenol F series of mortars are similar to the Bisphenol A series. They both use alkaline hardeners and the same fillers. The main advantage of the Bisphenol F mortars are their improved resistance to aliphatic and aromatic solvents, and higher concentrations of oxidizing and nonoxidizing acids.

The general corrosion resistance is shown in Table 7.9.

Phenolic Mortars

The phenolic mortars provide resistance to high concentrations of acids and to sulfuric acid at elevated temperatures. Fillers for the phenolic mortars are 100% carbon, 100% silica, or part carbon and part silica. For high concentration of sulfuric acid, silica is the filler of choice. Carbon fillers are used where resistance to high concentrations of hydrofluoric acid is required. Table 7.10 lists some typical compatibilities of phenolic mortars.

Furan Mortars

The furan mortars are resistant to most nonoxidizing organic and inorganic acids, alkalies, salts, oils, greases, and solvents to temperatures of 360°F/182°C. Fillers

TABLE 7.10 Typical Compatibilities of Phenolic Materials

Corrodent at room temp.	Filler[a]	
	Carbon	Silica
Amyl alcohol	R	R
Chromic acid 10%	U	U
Gasoline	R	R
Hydrofluoric acid to 50%	R	U
Hydrofluoric acid 93%	R	U
Methyl ethyl ketone	R	R
Nitric acid 10%	U	U
Sodium hydroxide to 5%	U	U
Sodium hydroxide 30%	U	U
Sodium hypochlorite 5%	U	U
Sulfuric acid to 50%	R	R
Sulfuric acid 93%	R	R
Xylene	R	R

[a] R = recommended; U = unsatisfactory.

are either 100% carbon, 100% silica, or part carbon and part silica. The 100% carbon filled mortar provides the widest range of corrosion resistance. Typical corrosion compatibilities are shown in Table 7.11.

Polyester Mortars

The polyester mortars were originally developed to resist chlorine dioxide. There are a number of types of polyester resins available. The ones most commonly used are the isophthalic, chlorendic, and Bisphenol A fumurate. Depending on the application, the polyester mortars can be formulated to incorporate carbon and silica fillers. One hundred percent carbon fillers are used to resist hydrofluoric acid, fluorine chemicals, and strong alkalies such as sodium and potassium hydroxide. The chlorendic and Bisphenol A fumurate resins have improved chemi-

TABLE 7.11 Typical Compatibilities of Furan Mortars

Corrodent at room temp.	100% carbon filler[a]	Part carbon filler part silica[a]
Acetic acid, glacial	R	R
Benzene	R	R
Cadmium salts	R	R
Chlorine dioxide	U	U
Chromic acid	U	U
Copper salts	R	R
Ethyl acetate	R	R
Ethyl alcohol	R	R
Formaldehyde	R	R
Fatty acids	R	R
Gasoline	R	R
Hydrochloric acid	R	R
Hydrofluoric acid	R	U
Iron salts	R	R
Lactic acid	R	R
Methyl ethyl ketone	R	R
Nitric acid	U	U
Phosphoric acid	R	R
Sodium chloride	R	R
Sodium hydroxide to 20%	R	U
Sodium hydroxide 40%	R	U
Sulfuric acid 50%	R	R
Sulfuric acid 80%	U	U
Trichloroethylene	R	R

[a] R = recommended; N = unsatisfactory.

TABLE 7.12 Comparison of Corrosion Resistance of Polyester Mortars

Corrodent at room temp.	Polyster[a]	
	Chlorendic	Bisphenol A fumarate
Acetic acid, glacial	U	U
Benzene	U	U
Chlorine dioxide	R	R
Ethyl alcohol	R	R
Hydrochloric acid 36%	R	R
Hydrogen peroxide	R	U
Methanol	R	R
Methyl ethyl ketone	U	U
Motor oil and gasoline	R	R
Nitric acid 40%	R	U
Phenol 5%	R	R
Sodium hydroxide 50%	U	R
Sulfuric acid 75%	R	U
Toluene	U	U
Triethanolamine	U	R
Vinyl toluene	U	U

[a] R = recommended; U = unsatisfactory.

TABLE 7.13 Comparative Resistance of Vinyl Ester Mortars

Corrodent	Vinyl ester[a]	
	Vinyl ester	Novolac
Acetic acid, glacial	U	R
Benzene	R	R
Chlorine dioxide	R	R
Ethyl alcohol	R	R
Hydrochloric acid 36%	R	R
Hydrogen peroxide	R	R
Methanol	U	R
Methyl ethyl ketone	U	U
Motor oil and gasoline	R	R
Nitric acid 40%	U	R
Phenol 5%	R	R
Sodium hydroxide 50%	R	R
Sulfuric acid, 75%	R	R
Toluene	U	R
Triethanolamine	R	R
Vinyl toluene	U	R
Max. temp. °F/°C	220/104	230/110

[a] R = recommended; U = unsatisfactory.

TABLE 7.14 Compatibility of Various Mortars with Selected Corrodents[a]

Mortars	Acetic acid 10%
Silicate	U
Sodium silicate	R———————————————————————————————
Potassium silicate	R———————————————————————————————
Silica	R———————————————————————————————
Sulfur	R———————————
Furan resin	R—————————————————————————
Polyester	R—————————————
Epoxy	R————————

°F	60	80	100	120	140	160	180	200	220	240	260	280	300	320	340	360	380	400	420	440	460
°C	15	26	38	49	60	71	82	93	104	116	127	138	149	160	171	182	193	204	216	227	238

Mortars	Acetic acid 50%
Silicate	U
Sodium silicate	R———————————————————————————————
Potassium silicate	R———————————————————————————————
Silica	R———————————————————————————————
Sulfur	U
Furan resin	R—————————————————————————————
Polyester	U
Epoxy	U

°F	60	80	100	120	140	160	180	200	220	240	260	280	300	320	340	360	380	400	420	440	460
°C	15	26	38	49	60	71	82	93	104	116	127	138	149	160	171	182	193	204	216	227	238

Mortars	Acetic acid 80%
Silicate	U
Sodium silicate	R———————————————————————————————
Potassium silicate	R———————————————————————————————
Silica	R———————————————————————————————
Sulfur	U
Furan resin	R———————————————————————————
Polyester	U
Epoxy	U

°F	60	80	100	120	140	160	180	200	220	240	260	280	300	320	340	360	380	400	420	440	460
°C	15	26	38	49	60	71	82	93	104	116	127	138	149	160	171	182	193	204	216	227	238

TABLE 7.14 Continued

Mortars	Acetic acid, glacial
Silicate	U
Sodium silicate	R———————————————————————————————
Potassium silicate	R———————————————————————————————
Silica	R———————————————————————————————
Sulfur	U
Furan resin	R————————————————————————
Polyester	U
Epoxy	U

	°F	60	80	100	120	140	160	180	200	220	240	260	280	300	320	340	360	380	400	420	440	460
	°C	15	26	38	49	60	71	82	93	104	116	127	138	149	160	171	182	193	204	216	227	238

Mortars	Acetic anhydride
Silicate	
Sodium silicate	R———————————————————————————————
Potassium silicate	R———————————————————————————————
Silica	R———————————————————————————————
Sulfur	U
Furan resin	U
Polyester	U
Epoxy	U

	°F	60	80	100	120	140	160	180	200	220	240	260	280	300	320	340	360	380	400	420	440	460
	°C	15	26	38	49	60	71	82	93	104	116	127	138	149	160	171	182	193	204	216	227	238

Mortars	Aluminum chloride, aqueous
Silicate	R———————————————————————————————
Sodium silicate	R———————————————————————————————
Potassium silicate	R———————————————————————————————
Silica	R———————————————————————————————
Sulfur	R—————————————
Furan resin	R————————————————————————
Polyester	R—————————————————
Epoxy	R————

	°F	60	80	100	120	140	160	180	200	220	240	260	280	300	320	340	360	380	400	420	440	460
	°C	15	26	38	49	60	71	82	93	104	116	127	138	149	160	171	182	193	204	216	227	238

TABLE 7.14 Continued

Mortars	Aluminum fluoride
Silicate	U
Sodium silicate	U
Potassium silicate	U
Silica	U
Sulfur	R—————————
Furan resin	R———————————————————
Polyester	R—————————————
Epoxy	R—————————————

°F: 60 80 100 120 140 160 180 200 220 240 260 280 300 320 340 360 380 400 420 440 460
°C: 15 26 38 49 60 71 82 93 104 116 127 138 149 160 171 182 193 204 216 227 238

Mortars	Ammonium chloride 10%
Silicate	R———————————————————————————————
Sodium silicate	R——
Potassium silicate	R———————————————————————————
Silica	R—————————————————————
Sulfur	R—————————
Furan resin	R———————————————————————————
Polyester	R—————————————
Epoxy	R—————————————

°F: 60 80 100 120 140 160 180 200 220 240 260 280 300 320 340 360 380 400 420 440 460
°C: 15 26 38 49 60 71 82 93 104 116 127 138 149 160 171 182 193 204 216 227 238

Mortars	Ammonium chloride 50%
Silicate	R———————————————————————————————
Sodium silicate	R——
Potassium silicate	R———————————————————————————————
Silica	R———————————————————————————————
Sulfur	R—————————
Furan resin	R———————————————————————————
Polyester	R—————————————
Epoxy	R—————————————

°F: 60 80 100 120 140 160 180 200 220 240 260 280 300 320 340 360 380 400 420 440 460
°C: 15 26 38 49 60 71 82 93 104 116 127 138 149 160 171 182 193 204 216 227 238

TABLE **7.14** Continued

Mortars	Ammonium chloride, saturated
Silicate	R—————————————————————————————
Sodium silicate	R——
Potassium silicate	R————————————————————————
Silica	R————————————————————————
Sulfur	R——————————
Furan resin	R————————————————————
Polyester	R——————————
Epoxy	R——————————

°F	60	80	100	120	140	160	180	200	220	240	260	280	300	320	340	360	380	400	420	440	460
°C	15	26	38	49	60	71	82	93	104	116	127	138	149	160	171	182	193	204	216	227	238

Mortars	Ammonium fluoride 10%
Silicate	U
Sodium silicate	U
Potassium silicate	U
Silica	U
Sulfur	U
Furan resin	R—————————————————————
Polyester	R——————
Epoxy	R——————

°F	60	80	100	120	140	160	180	200	220	240	260	280	300	320	340	360	380	400	420	440	460
°C	15	26	38	49	60	71	82	93	104	116	127	138	149	160	171	182	193	204	216	227	238

Mortars	Ammonium fluoride 25%
Silicate	U
Sodium silicate	U
Potassium silicate	U
Silica	U
Sulfur	U
Furan resin	R—————————————————————
Polyester	R————
Epoxy	R——

°F	60	80	100	120	140	160	180	200	220	240	260	280	300	320	340	360	380	400	420	440	460
°C	15	26	38	49	60	71	82	93	104	116	127	138	149	160	171	182	193	204	216	227	238

TABLE 7.14 Continued

Mortars	Ammonium hydroxide 25%
Silicate	U
Sodium silicate	U
Potassium silicate	U
Silica	U
Sulfur	U
Furan resin	R—————————————————————
Polyester	R—————————————
Epoxy	R——————————

°F: 60 80 100 120 140 160 180 200 220 240 260 280 300 320 340 360 380 400 420 440 460
°C: 15 26 38 49 60 71 82 93 104 116 127 138 149 160 171 182 193 204 216 227 238

Mortars	Ammonium hydroxide, saturated
Silicate	U
Sodium silicate	
Potassium silicate	U
Silica	U
Sulfur	U
Furan resin	R————————————————
Polyester	R———
Epoxy	R———

°F: 60 80 100 120 140 160 180 200 220 240 260 280 300 320 340 360 380 400 420 440 460
°C: 15 26 38 49 60 71 82 93 104 116 127 138 149 160 171 182 193 204 216 227 238

Mortars	Aqua regia 3:1
Silicate	R—————————————————————
Sodium silicate	R—————————————————————
Potassium silicate	R—————————————————————
Silica	R—————————————————————
Sulfur	U
Furan resin	U
Polyester	U
Epoxy	U

°F: 60 80 100 120 140 160 180 200 220 240 260 280 300 320 340 360 380 400 420 440 460
°C: 15 26 38 49 60 71 82 93 104 116 127 138 149 160 171 182 193 204 216 227 238

TABLE 7.14 Continued

Mortars	Bromine gas, dry
Silicate	
Sodium silicate	R———————————————————————————
Potassium silicate	
Silica	
Sulfur	U
Furan resin	U
Polyester	U
Epoxy	U

°F	60	80	100	120	140	160	180	200	220	240	260	280	300	320	340	360	380	400	420	440	460
°C	15	26	38	49	60	71	82	93	104	116	127	138	149	160	171	182	193	204	216	227	238

Mortars	Bromine gas, moist
Silicate	
Sodium silicate	R———————————————————————————
Potassium silicate	
Silica	
Sulfur	U
Furan resin	U
Polyester	U
Epoxy	U

°F	60	80	100	120	140	160	180	200	220	240	260	280	300	320	340	360	380	400	420	440	460
°C	15	26	38	49	60	71	82	93	104	116	127	138	149	160	171	182	193	204	216	227	238

Mortars	Bromine liquid
Silicate	R———————————————————————————
Sodium silicate	
Potassium silicate	R———————————————————————————
Silica	R———————————————————————————
Sulfur	R————————————
Furan resin	U
Polyester	U
Epoxy	U

°F	60	80	100	120	140	160	180	200	220	240	260	280	300	320	340	360	380	400	420	440	460
°C	15	26	38	49	60	71	82	93	104	116	127	138	149	160	171	182	193	204	216	227	238

TABLE 7.14 Continued

Mortars	Calcium hypochlorite
Silicate	
Sodium silicate	U
Potassium silicate	R————————————————————————
Silica	R————————————————————————
Sulfur	U
Furan resin	U
Polyester	
Epoxy	U

°F	60	80	100	120	140	160	180	200	220	240	260	280	300	320	340	360	380	400	420	440	460
°C	15	26	38	49	60	71	82	93	104	116	127	138	149	160	171	182	193	204	216	227	238

Mortars	Carbon tetrachloride
Silicate	R————————————————————————
Sodium silicate	R————————————————————————
Potassium silicate	R————————————————————————
Silica	R————————————————————————
Sulfur	U
Furan resin	R————————————————————
Polyester	R———————
Epoxy	R———————

°F	60	80	100	120	140	160	180	200	220	240	260	280	300	320	340	360	380	400	420	440	460
°C	15	26	38	49	60	71	82	93	104	116	127	138	149	160	171	182	193	204	216	227	238

Mortars	Chlorine gas, dry
Silicate	R————————————————————————
Sodium silicate	R————————————————————————
Potassium silicate	R————————————————————————
Silica	R————————————————————————
Sulfur	U
Furan resin	U
Polyester	U
Epoxy	U

°F	60	80	100	120	140	160	180	200	220	240	260	280	300	320	340	360	380	400	420	440	460
°C	15	26	38	49	60	71	82	93	104	116	127	138	149	160	171	182	193	204	216	227	238

TABLE **7.14** Continued

Mortars	Chlorine gas, wet																				
Silicate	R————————————————————————————																				
Sodium silicate	R————————————————————————————																				
Potassium silicate	R————————————————————————————																				
Silica	R————————————————————————————																				
Sulfur	U																				
Furan resin	U																				
Polyester	U																				
Epoxy	U																				
°F	60	80	100	120	140	160	180	200	220	240	260	280	300	320	340	360	380	400	420	440	460
°C	15	26	38	49	60	71	82	93	104	116	127	138	149	160	171	182	193	204	216	227	238

Mortars	Chlorine liquid																				
Silicate	R————————————————————————————																				
Sodium silicate																					
Potassium silicate	R————————————————————————————																				
Silica	R————————————————————————————																				
Sulfur	U																				
Furan resin	U																				
Polyester	U																				
Epoxy	U																				
°F	60	80	100	120	140	160	180	200	220	240	260	280	300	320	340	360	380	400	420	440	460
°C	15	26	38	49	60	71	82	93	104	116	127	138	149	160	171	182	193	204	216	227	238

Mortars	Chromic acid 10%																				
Silicate	U																				
Sodium silicate	R————————————————————————————																				
Potassium silicate	R————————————————————————————																				
Silica	R————————————————————————————																				
Sulfur	U																				
Furan resin	U																				
Polyester	R———																				
Epoxy	U																				
°F	60	80	100	120	140	160	180	200	220	240	260	280	300	320	340	360	380	400	420	440	460
°C	15	26	38	49	60	71	82	93	104	116	127	138	149	160	171	182	193	204	216	227	238

TABLE 7.14 Continued

Mortars	Chromic acid 50%
Silicate	R———————————————————————
Sodium silicate	
Potassium silicate	
Silica	R———————————————————
Sulfur	U
Furan resin	U
Polyester 30%	R———
Epoxy	U

	°F	60	80	100	120	140	160	180	200	220	240	260	280	300	320	340	360	380	400	420	440	460
	°C	15	26	38	49	60	71	82	93	104	116	127	138	149	160	171	182	193	204	216	227	238

Mortars	Ferric chloride
Silicate	R———————————————————————
Sodium silicate	U
Potassium silicate	R———————————————————
Silica	R———————————————
Sulfur	R————————
Furan resin	R—————————————————
Polyester	R——————————
Epoxy	R—————————

	°F	60	80	100	120	140	160	180	200	220	240	260	280	300	320	340	360	380	400	420	440	460
	°C	15	26	38	49	60	71	82	93	104	116	127	138	149	160	171	182	193	204	216	227	238

Mortars	Ferric chloride 50% in water
Silicate	R———————————————————————
Sodium silicate	U
Potassium silicate	R———————————————————
Silica	R———————————————————
Sulfur	R————————
Furan resin	R—————————————
Polyester	R—————————
Epoxy	R—————————

	°F	60	80	100	120	140	160	180	200	220	240	260	280	300	320	340	360	380	400	420	440	460
	°C	15	26	38	49	60	71	82	93	104	116	127	138	149	160	171	182	193	204	216	227	238

TABLE 7.14 Continued

Mortars	Hydrobromic acid 20%																				
Silicate	R—————————————————————————————————																				
Sodium silicate	R—————————————————————————————————																				
Potassium silicate	R—————————————————————————————————																				
Silica	R—————————————————————————————————																				
Sulfur	R———————————————																				
Furan resin	R——																				
Polyester	R——																				
Epoxy	U																				
°F	60	80	100	120	140	160	180	200	220	240	260	280	300	320	340	360	380	400	420	440	460
°C	15	26	38	49	60	71	82	93	104	116	127	138	149	160	171	182	193	204	216	227	238

Mortars	Hydrobromic acid 50%																				
Silicate	R—————————————————————————————————																				
Sodium silicate	R—————————————————————————————————																				
Potassium silicate	R—————————————————————————————————																				
Silica	R—————————————————————————————————																				
Sulfur	R———————————————																				
Furan resin	R——																				
Polyester	R———																				
Epoxy	U																				
°F	60	80	100	120	140	160	180	200	220	240	260	280	300	320	340	360	380	400	420	440	460
°C	15	26	38	49	60	71	82	93	104	116	127	138	149	160	171	182	193	204	216	227	238

Mortars	Hydrochloric acid 20%																				
Silicate	R—————————————————————————————————																				
Sodium silicate	R—————————————————————————————————																				
Potassium silicate	R—————————————————————————————————																				
Silica	R—————————————————————————————————																				
Sulfur	R———————————																				
Furan resin	R—————————————————————————																				
Polyester	R————																				
Epoxy	U																				
°F	60	80	100	120	140	160	180	200	220	240	260	280	300	320	340	360	380	400	420	440	460
°C	15	26	38	49	60	71	82	93	104	116	127	138	149	160	171	182	193	204	216	227	238

TABLE 7.14 Continued

Mortars	Hydrochloric acid 38%
Silicate	R————————————————————————
Sodium silicate	R————————————————————————
Potassium silicate	R————————————————————————
Silica	R————————————————————————
Sulfur	R—————————
Furan resin	R————————————————————
Polyester	R——
Epoxy	U

°F	60	80	100	120	140	160	180	200	220	240	260	280	300	320	340	360	380	400	420	440	460
°C	15	26	38	49	60	71	82	93	104	116	127	138	149	160	171	182	193	204	216	227	238

Mortars	Hydrofluoric acid 30%
Silicate	U
Sodium silicate	U
Potassium silicate	U
Silica	U
Sulfur	R—————
Furan resin	R————————————————————
Polyester	R——
Epoxy	U

°F	60	80	100	120	140	160	180	200	220	240	260	280	300	320	340	360	380	400	420	440	460
°C	15	26	38	49	60	71	82	93	104	116	127	138	149	160	171	182	193	204	216	227	238

Mortars	Hydrofluoric acid 70%
Silicate	U
Sodium silicate	U
Potassium silicate	U
Silica	U
Sulfur	U
Furan resin	U
Polyester	
Epoxy	U

°F	60	80	100	120	140	160	180	200	220	240	260	280	300	320	340	360	380	400	420	440	460
°C	15	26	38	49	60	71	82	93	104	116	127	138	149	160	171	182	193	204	216	227	238

TABLE 7.14 Continued

Mortars	Hydrofluoric acid 100%																				
Silicate	U																				
Sodium silicate	U																				
Potassium silicate	U																				
Silica	U																				
Sulfur	U																				
Furan resin	U																				
Polyester																					
Epoxy	U																				
°F	60	80	100	120	140	160	180	200	220	240	260	280	300	320	340	360	380	400	420	440	460
°C	15	26	38	49	60	71	82	93	104	116	127	138	149	160	171	182	193	204	216	227	238

Mortars	Magnesium chloride																				
Silicate	R———————————————————————————————————																				
Sodium silicate																					
Potassium silicate	R———————————————————————————————————																				
Silica	R———————————————————————————————————																				
Sulfur	R———————————————																				
Furan resin	R———————————————————————																				
Polyester	R—————————																				
Epoxy	R———————————————																				
°F	60	80	100	120	140	160	180	200	220	240	260	280	300	320	340	360	380	400	420	440	460
°C	15	26	38	49	60	71	82	93	104	116	127	138	149	160	171	182	193	204	216	227	238

Mortars	Nitric acid 5%																				
Silicate	R———————————————————————————————————																				
Sodium silicate	R———————————————————————————————————																				
Potassium silicate	R———————————————————————————————————																				
Silica	R———————————————————————————————————																				
Sulfur	R—————————																				
Furan resin	U																				
Polyester	R—————————																				
Epoxy	U																				
°F	60	80	100	120	140	160	180	200	220	240	260	280	300	320	340	360	380	400	420	440	460
°C	15	26	38	49	60	71	82	93	104	116	127	138	149	160	171	182	193	204	216	227	238

TABLE 7.14 Continued

Mortars	Nitric acid 20%
Silicate	R————————————————————————————————————
Sodium silicate	R————————————————————————————————————
Potassium silicate	
Silica	
Sulfur	R———
Furan resin	U
Polyester	R———
Epoxy	U

°F	60	80	100	120	140	160	180	200	220	240	260	280	300	320	340	360	380	400	420	440	460
°C	15	26	38	49	60	71	82	93	104	116	127	138	149	160	171	182	193	204	216	227	238

Mortars	Nitric acid 70%
Silicate	R————————————————————————————————————
Sodium silicate	R————————————————————————————————————
Potassium silicate	
Silica	
Sulfur	U
Furan resin	U
Polyester	U
Epoxy	U

°F	60	80	100	120	140	160	180	200	220	240	260	280	300	320	340	360	380	400	420	440	460
°C	15	26	38	49	60	71	82	93	104	116	127	138	149	160	171	182	193	204	216	227	238

Mortars	Nitric acid, anhydrous
Silicate	R————————————————————————————————————
Sodium silicate	R————————————————————————————————————
Potassium silicate	
Silica	
Sulfur	U
Furan resin	U
Polyester	U
Epoxy	U

°F	60	80	100	120	140	160	180	200	220	240	260	280	300	320	340	360	380	400	420	440	460
°C	15	26	38	49	60	71	82	93	104	116	127	138	149	160	171	182	193	204	216	227	238

TABLE **7.14** Continued

Mortars	Oleum
Silicate	
Sodium silicate	R———
Potassium silicate	
Silica	
Sulfur	U
Furan resin	U
Polyester	U
Epoxy	U

°F	60	80	100	120	140	160	180	200	220	240	260	280	300	320	340	360	380	400	420	440	460
°C	15	26	38	49	60	71	82	93	104	116	127	138	149	160	171	182	193	204	216	227	238

Mortars	Phosphoric acid 50–80%
Silicate	R————————————————
Sodium silicate	U
Potassium silicate	R————————————————
Silica	R————————————————
Sulfur	R————
Furan resin	R——————————
Polyester	R————
Epoxy	U

°F	60	80	100	120	140	160	180	200	220	240	260	280	300	320	340	360	380	400	420	440	460
°C	15	26	38	49	60	71	82	93	104	116	127	138	149	160	171	182	193	204	216	227	238

Mortars	Sodium chloride
Silicate	U
Sodium silicate	R————
Potassium silicate	R————
Silica	R————
Sulfur	R—————
Furan resin	R——————————
Polyester	R————
Epoxy	R————

°F	60	80	100	120	140	160	180	200	220	240	260	280	300	320	340	360	380	400	420	440	460
°C	15	26	38	49	60	71	82	93	104	116	127	138	149	160	171	182	193	204	216	227	238

TABLE 7.14 Continued

Mortars	Sodium hydroxide 10%
Silicate	U
Sodium silicate	U
Potassium silicate	U
Silica	U
Sulfur	U
Furan resin	R————————————————————————————
Polyester	R———
Epoxy	R———————————

	°F	60	80	100	120	140	160	180	200	220	240	260	280	300	320	340	360	380	400	420	440	460
	°C	15	26	38	49	60	71	82	93	104	116	127	138	149	160	171	182	193	204	216	227	238

Mortars	Sodium hydroxide 50%
Silicate	U
Sodium silicate	U
Potassium silicate	U
Silica	U
Sulfur	U
Furan resin	R————————————————————————————
Polyester	R———
Epoxy	R———————————

	°F	60	80	100	120	140	160	180	200	220	240	260	280	300	320	340	360	380	400	420	440	460
	°C	15	26	38	49	60	71	82	93	104	116	127	138	149	160	171	182	193	204	216	227	238

Mortars	Sodium hydroxide, concentrated
Silicate	U
Sodium silicate	U
Potassium silicate	U
Silica	U
Sulfur	U
Furan resin	
Polyester	
Epoxy	R———

	°F	60	80	100	120	140	160	180	200	220	240	260	280	300	320	340	360	380	400	420	440	460
	°C	15	26	38	49	60	71	82	93	104	116	127	138	149	160	171	182	193	204	216	227	238

TABLE 7.14 Continued

Mortars	Sodium hypochlorite 20%
Silicate	U
Sodium silicate	U
Potassium silicate	U
Silica	U
Sulfur	U
Furan resin	U
Polyester	U
Epoxy	U

°F	60	80	100	120	140	160	180	200	220	240	260	280	300	320	340	360	380	400	420	440	460
°C	15	26	38	49	60	71	82	93	104	116	127	138	149	160	171	182	193	204	216	227	238

Mortars	Sodium hypochlorite, concentrated
Silicate	U
Sodium silicate	U
Potassium silicate	U
Silica	U
Sulfur	U
Furan resin	U
Polyester	U
Epoxy	U

°F	60	80	100	120	140	160	180	200	220	240	260	280	300	320	340	360	380	400	420	440	460
°C	15	26	38	49	60	71	82	93	104	116	127	138	149	160	171	182	193	204	216	227	238

Mortars	Sulfuric acid 10%
Silicate	U
Sodium silicate	R————————————————————————————————
Potassium silicate	
Silica	
Sulfur	R—————————
Furan resin	R————————————————
Polyester	R—————————————
Epoxy	R———

°F	60	80	100	120	140	160	180	200	220	240	260	280	300	320	340	360	380	400	420	440	460
°C	15	26	38	49	60	71	82	93	104	116	127	138	149	160	171	182	193	204	216	227	238

TABLE 7.14 Continued

Mortars	Sulfuric acid 50%
Silicate	U
Sodium silicate	R————————————————————————————————
Potassium silicate	
Silica	
Sulfur	R————
Furan resin	R————————————————————————
Polyester	R————————————
Epoxy	U

°F: 60 80 100 120 140 160 180 200 220 240 260 280 300 320 340 360 380 400 420 440 460
°C: 15 26 38 49 60 71 82 93 104 116 127 138 149 160 171 182 193 204 216 227 238

Mortars	Sulfuric acid 70%
Silicate	U
Sodium silicate	R————
Potassium silicate	R————————
Silica	R————————
Sulfur	R———
Furan resin	R———
Polyester	R———
Epoxy	U

°F: 60 80 100 120 140 160 180 200 220 240 260 280 300 320 340 360 380 400 420 440 460
°C: 15 26 38 49 60 71 82 93 104 116 127 138 149 160 171 182 193 204 216 227 238

Mortars	Sulfuric acid 90%
Silicate	U
Sodium silicate	R——
Potassium silicate	
Silica	
Sulfur	U
Furan resin	U
Polyester	U
Epoxy	U

°F: 60 80 100 120 140 160 180 200 220 240 260 280 300 320 340 360 380 400 420 440 460
°C: 15 26 38 49 60 71 82 93 104 116 127 138 149 160 171 182 193 204 216 227 238

Masonry Linings

215

TABLE 7.14 Continued

Mortars	Sulfuric acid 98%
Silicate	U
Sodium silicate	R———
Potassium silicate	
Silica	
Sulfur	U
Furan resin	U
Polyester	U
Epoxy	U

°F	60	80	100	120	140	160	180	200	220	240	260	280	300	320	340	360	380	400	420	440	460
°C	15	26	38	49	60	71	82	93	104	116	127	138	149	160	171	182	193	204	216	227	238

[a]The table is arranged alphabetically according to corrodent. Unless otherwise noted, the corrodent is considered pure, in the case of liquids, and a saturated aqueous solution in the case of solids. All percentages shown are weight percents.

Corrosion is a function of temperature. When using the tables note that the vertical lines refer to temperatures midway between the temperatures cited. An entry of R indicates that the material is resistant to the maximum temperature shown. An entry of U indicates that the material is unsatisfactory. A blank indicates that no data are available.

Source: Ref. 1.

cal resistance and higher thermal capabilities than the isophthalic resins. The Bisphenol A fumurate resins exhibit greatly improved resistance to strong alkalies. A comparison of the corrosion resistance between the chlorendic and Bisphenol A fumurate resins is shown in Table 7.12.

Vinyl Ester and Vinyl Ester Novolac Mortars

These resins have many of the same properties as the epoxy, acrylic, and Bisphenol A fumurate resins. The vinyl ester resins have replaced polyester resins in mortars for bleach towers in the pulp and paper industry. The major advantage of these resin systems is their resistance to most oxidizing mediums and high concentrations of sulfuric acid, sodium hydroxide, and many solvents. Comparative resistances of the two types of vinyl ester mortars are shown in Table 7.13.

Table 7.14 shows the compatibility of various mortars with selected corrodents. Reference 1 provides a more comprehensive listing.

REFERENCES

1. Schweitzer, Philip A. *Corrosion Resistance Tables*, 4th ed., vols. 1–3, Marcel Dekker, New York, 1995.

2. Boova, AA. Chemical-resistant mortars, grouts, and monolithic surfacings. In PA Schweitzer (ed.), *Corrosion Engineering Handbook*, Marcel Dekker, New York, 1996, pp. 459–487.
3. Sheppard, WL, Jr. *Chemical Resistant Masonry*, 2nd ed., Marcel Dekker, New York, 1982.
4. Schweitzer, Philip A. *Encyclopedia of Corrosion Technology*, Marcel Dekker, New York, 1998.

8

Glass Linings

The bonding of glass to metal is not a recent development. Relics of glass bonded to gold jewelry (enameling) have been found which date back to 400 B.C. Until the early 1800s, enameling continued as an art form. The first application of a glass coating for any purpose other than art form took place when cast iron sanitary ware was first coated.

The brewing industry was responsible for the development of the first large scale glassed steel equipment. It was the result of the need to impose consistency of the quality of the beer. This took place in the early 1800s. Between that time and the start of the Second World War there were no notable developments in glassed steel composites.

With the advent of war the need developed for critical chemicals that were often corrosive and sticky. This sparked development programs to find materials to meet these needs. It was during this time that the importance of characteristics other than corrosion resistance of the composite were recognized, specifically thermal loading and mechanical stressing. Since that time, continued research efforts have culminated in a glassed steel product that is one of the major materials of construction used by the chemical processing industry.

There are three main areas of application where glass-lined equipment is used:

1. Those involving highly corrosive acids and bases
2. Those involving high-purity processes where cleanliness is important
    for the ability to clean as well as to minimize the risk of metal contami-
    nation
3. For polymerization reactions to prevent polymers from sticking on ves-
    sels

The majority of the shell substrates for glassed steel vessels are designed
and manufactured in accordance with the latest edition of the ASME Boiler and
Pressure Vessel Code, Section VIII Division I. This code determines the type of
metal, thicknesses, joining techniques, designs, and other details for the specific
temperature and pressure requirements of the application. However, the code has
no jurisdiction over the glass lining itself; this is governed by the quality standards
of the manufacturer combined with the specific requirements of the application
and the quality specifications of the user.

Carbon steel is the most commonly used shell material with stainless steel
being used for extremely low temperature operation.

GLASS SELECTION

Many advances have been made in the composition of glass used to produce
linings. Various formulations are available from the different manufacturers to
meet specific requirements. Although so-called "standard" formulations are
available, which represents a good balance between chemical and physical property serviceability, these can be modified for specific applications. These systems
are generally recommended for an operating temperature range of -20 to
$+450°F/-29$ to $+232°C$. For example, there are special glass formulations available for:

1. Low temperature service operating down to $-200°F/-129°C$.
2. High temperature service extending up to $650°F/343°C$, the upper tem-
    perature limit of the carbon steel usually specified for the vessel shell.
3. Low product adherence. These glass systems employ a special surface
    layer that is largely free of many microscopic surface imperfections.
    This surface remains effective only until the onset of corrosive etch.

CORROSION OF GLASS

A glass system used to produce a lining can be considered as a three-dimensional
network type structure, consisting of one or more oxide groups. The network
formers are acidic type oxides that form the backbone of the glass structure.
Silicon dioxide (SiO_2) is the primary network former and is obtained from relatively inexpensive beach sand. It is usually present in glasses in amounts ex-

ceeding 50 wt%. The network modifiers are base type oxides but are not part of the network forming structure. They cannot form glasses by themselves, as the name implies, these oxides modify the properties of the network formers. The intermediates are amphoteric in nature and can act as either network formers or modifiers, depending upon the concentration, nature, and amounts of the other constituents. Aluminum oxide and titanium dioxide improve general corrosion resistance while zirconium dioxide improves alkali resistance.

The cover coat systems for glass linings are complex mixtures of up to 15 oxides taken from the above three groups and built around the framework of the silica network.

The corrosion of glass linings takes place by means of two basic mechanisms. One relates to the removal of the modifiers and/or the intermediates, while the other is the removal of the network formers.

Acids (fluorine and phosphorus compounds excepted), small ions, and the first stage of water attack invoke removal of the intermediates and modifiers, by a diffusion controlled ion exchange mechanism in which the small ion, e.g., hydrogen from the acid, exchanges for the larger modifier/intermediate ion. This results in a stress relief cracking in the network former that eventually leads to the dulling of the glass.

Corrosive attack by alkalies, fluorine, and phosphorus containing compounds and the second stage of water attack removes compounds from the network former through a regenerated dissolution type reaction.

Dulling of the glass surface is usually the first sign of corrosion. This may be caused by either the ion exchange or the dissolution type of corrosion reaction and is usually uniform in appearance. The corrosive upset of the glass surface leads to localized stress differences, which in turn lead to varying degrees of stress relief and eventually glass chipping.

One of the most serious types of glass corrosion damage is pitting. This type of attack is both corrosive and glass composition sensitive, and can be caused by fluorine based chemistries and alkalies. Older glass systems and glass/crystal composites are more susceptible to this type of attack.

CORROSION RESISTANCE OF GLASS LININGS

When considering the application of glassed steel equipment to handle a specific corrosive, thought must be given to the effects of the corrodent on any tantalum repair plugs that may be present in the vessel. For example, while sulfuric acid can be handled in a glassed steel vessel, the presence of small amounts of sulfur trioxide would attack any tantalum present. The same would apply to chlorosulfonic acid and oleum. Reference 1 provides a comprehensive listing of the compatibility of tantalum with selected corrodents. Note that glass linings may be specified to be plug free, but they are more expensive.

TABLE 8.1 Compatibility of Glass Lining with Selected Corrodents[1]

Chemical	°F	°C	Chemical	°F	°C
Acrylic acid	302	150	Cyano acetic acid	212	100
Aluminum acetate	392	200	Dichlorobenzene	428	220
Aluminum chlorate aq	230	110	Dichloro-acetic acid	302	150
Aluminum chloride 10%	bp		Diethylamine	212	100
Aluminum potassium	248	120	Diethyl ether	212	100
sulfate 50% aq			Dimethyl sulfate	302	150
Amino-ethanol	338	170	Ethyl acetate	392	200
Aminophenol sulfonic acid	266	130	Ethyl alcohol	392	200
Ammonium carbonate aq	bp		Ethylenediamine 98%	176	80
Ammonium chloride 10%	302	150	Fatty acids	302	150
Ammonium nitrate aq	bp		Ferric chloride 10%	bp	
Ammonium phosphate	bp		Formaldehyde	302	150
Ammonium sulfate	bp		Formic acid 98%	356	180
Ammonium sulfate	x		Fumaric acid	302	150
Ammonium sulfide	176	80	Gallic acid	212	100
Aniline	363	184	Glutamic acid	104	40
Antimony (III) chloride	428	220	Glycerine	212	100
Antimony (IV) chloride	302	150	Glycol	302	150
Aqua regia	302	150	Glycolic acid 57%	302	150
Barium sulfate	302	150	Hydrochloric acid 30%	266	130
Benzaldehyde	302	150	Hydrogen peroxide 30%	158	70
Benzene	482	250	Hydrogen sulfide aq	302	150
Benzoic acid	302	150	Hydroiodic acid 20%	x	
Benzyl chloride	266	130	Iodine	392	200
Boric acid aq	302	150	Isopropyl alcohol	302	150
Bromine	212	100	Lactic acid	bp	
Butanol	284	140	Lead acetate	572	300
Carbon dioxide aq	302	150	Lithium chloride	x	
Carbon dioxide	482	250	Lithium hydroxide conc	x	
Carbon disulfide	392	200	Magnesium carbonate aq	212	100
Carbon tetrachloride	392	200	Magnesium chloride 30%	230	110
Chloride bleaching agent	356	180	Magnesium sulfate aq	302	150
Chlorinated paraffin	356	180	Maleic acid	356	180
Chlorine	392	200	Methanol	392	200
Chlorine water	356	180	Monochloroacetic acid	bp	
Chlorosulfonic acid	302	150	Naphthalene	419	215
Chloropropionic acid	347	175	Nitric oxides	392	200
Chromic acid 30%	212	100	Nitrobenzene	302	150
Chromic acid aq	302	150	Nitric acid 50%	302	150
Citric acid 10%	bp		Oleum 10% SO_3	338	170
Cupric chloride 5%	302	150	Oxalic acid 50%	302	150
Cupric nitrate 50%	212	100	Palmitic acid	230	110
Cupric sulfate aq	302	150	Perchloric acid 70%	bp	

TABLE 8.1   Continued

Chemical	°F	°C	Chemical	°F	°C
Phenol	392	200	Sulfuric acid 20%	284	140
Phthalic anhydride	482	260	Sulfuric acid 60%	320	160
Picric acid	302	150	Sulfuric acid 98%	428	220
Potassium bromide aq	bp		Tannic acid	302	150
Potassium chloride aq	bp		Tartaric acid	284	140
Pyridine	bp		Trichloroethylene	302	150
Sodium bisulfate	572	300	Tin chloride	482	250
Sodium chlorate aq	176	80	Toluene	302	150
Sodium chloride aq	bp		Trichloro-acetic acid	302	150
Sodium nitrate	608	320	Triethanolamine	482	250
Sodium sulfide 4%	x		Triethylamine 30%	176	80
Stearic acid	320	160	Trisodium phosphate 50%	176	80
Succinnic acid 35%	284	140	Urea	302	150
Succinnic acid sat. sol.	x		Zinc bromide aq	bp	
Sulfur	302	150	Zinc chloride melt	626	330
Sulfur dioxide	392	200	Zinc chloride aq	284	140

[a] The chemicals listed are in the pure state or in a saturated solution unless otherwise indicated. Compatibility is shown to the maximum allowable temperature for which data are available. Incompatibility is shown by an x.
bp = boiling point
aq = aqueous

In general, the following corrodents may be handled safely in glassed steel equipment.

Hydrochloric acid up to 300°F/149°C
Chlorides in general
Bromides
Solids
Sulfuric acid up to 450°F/232°C
Chlorosulfonic acid
Acetic acid
Organic compounds

Corrosives that will attack glassed steel are

Alkaline compounds (conditionally).
Salts with small cations, e.g., lithium, magnesium, aluminum in aqueous media, should be used with caution above 150°F/66°C.
Phosphorus compounds frequently contain fluorides, and some phosphorus compounds possess a mutual solubility for glass.

Refer to Table 8.1 for the compatibility of special general purpose glass lining material for use in the chemical process industry.

CAUSES OF GLASSED STEEL DAMAGE

The most common causes of damage to a glass lining are the result of either mechanical or thermal damage during operation or cleaning. Mechanical damage can result from impact, stress, abrasion, or spalling while thermal damage can result from excessive temperature differential (thermal shock), and improper welding procedures.

Impact

Impact damage can be caused by solids, liquids, or vapors. Impacting energy of 9 in.-lb is sufficient to cause damage (e.g., a 1 lb steel ball dropped from a height of 9 in. will produce 9 in.-lb of impacting energy). Failures may result from the interaction of solids, liquids, or vapors with the glass surface, or in the case of solids from the metal backside.

Damage from solids usually results from a moving agitator hitting a stationary particle (e.g., bolt) and forcing it to the sidewall or bottom of the vessel. External blows on the vessel can also result in damage to the lining.

Liquid damage is most often related to the use of high pressure fluids to clean the glass surface from product buildup, e.g., liquid blast or jet cleaning. It is much better to use a suitable solvent to remove the product buildup, one that will not etch the glass, is nontoxic and nonexplosive, etc.

Vapors, or more accurately, the collapse of vapor bubbles (cavitation) can cause liner failure. Calculations have shown that, depending on the hydrostatic head, a steam vapor bubble upon collapse can generate over 100 in.-lb of energy, while glass breaks at 9 in.-lb. Bubble collapse can be the result of three mechanisms:

1. *Condensation.* For example, steam sparging into cold water, vapor bubbles formed by an exothermic addition of a chemical and its collapse at a cooled vessel wall. Since the energy of the bubble collapse is directly related to the bubble size, the problem can be eliminated by the use of small holed spargers directed away from the sidewall.
2. *Pressure buildup.* When low boiling, high vapor pressure components are present, vapor bubbles can form at the backside of the agitator blade and collapse at the front. Slower speed agitation and higher pressure operation are possible corrective measures.
3. *Chemical reaction.* Neutralization operations, e.g., ammonia in an acidic solution. Rate of addition and agitation shear control can help reduce the problem.

Stress

Pressure, vibration, and nozzle loadings can produce tension type stresses that are capable of offsetting the residual compressive strain of the glassed steel composite.

Code rated pressure vessels are not deliberately operated above the rated operating pressure. However, clogged pressure relief nozzles combined with a pressure generating reaction can result in overpressuring.

Vibration

Vibration problems are usually the result of steam/water hammer, improper baffle positioning, misalignment of the agitator, and malfunctioning drives/pumps. Good design practices should correct these problems. Vibration by itself will not usually cause glass failure; however, combined with other loading stresses, failure could result.

Nozzle Loading

Nozzles are the most trouble-prone part of glassed steel equipment. Most damage results from bending loads resulting from nozzle appendages which are not properly supported. A good rule of thumb is that all appendages should be aligned properly and independently supported.

Another important factor is gasket selection. It is essential that proper gaskets, designed for use with glassed steel equipment be used. These gaskets are designated as being a CRT type. It is also important that all bolts be tightened to the proper torque and in the proper sequence. Details are available from the manufacturers.

Abrasion

A number of factors affect abrasion, many of which are interrelated. Among these are particle type (as it relates to hardness), particle size and distribution, shape, density, concentration, velocity, and solvent.

The use of the Moh hardness scale for minerals may be useful. This is a listing in which a material lower in the table with a higher number will scratch the material above it. Refer to Table 8.2.

Since glass has a 5.5 rating, any material lower in the table will scratch it. Moh hardness for other materials can be found in handbooks.

Keeping the particle size small, the density light, the velocity low, and the solvent "lubricating" can help to reduce potential damage.

Spalling

This is a case of delayed fracturing which is the result of a large subsurface bubble that may form during the glassing cycle and does not show up in any of the

TABLE 8.2 Moh Hardness Scale

Material	Moh scale
Talc	1
Gypsum	2
Fingernail	2.5
Calcite	3
Fluorite	4
Apatite	5
(Knife, glass)	5.5
Orthoclose	6
(Steel file)	6.5
Quartz (beach sand)	7
Topaz	8
Corundum (aluminium oxide)	9
Diamond	10

quality assurance checks during the final inspection and shipping. The addition of a stress factor, either mechanical or thermal, will result in spalling. Since this type of damage is extremely sensitive to any stressing influence, it usually shows up quickly after initial operation. Obviously this is the manufacturers' responsibility.

Thermal Shock

The two most common causes of thermal shock damage are:

1. Addition of a cold material to the interior of a hot-walled vessel
2. Addition of a hot material into the jacket of a cold-walled vessel

There is no so-called "standard" maximum allowable temperature differential. Over the years most manufacturers have developed several glass lining systems all with different thermal characteristics. Therefore, it is important that the manufacturer be contacted and data on the specific system be gotten. Table 8.3 provides a very general guideline for allowable temperature differentials, containing several averaging assumptions, that provides some insight into the problem.

Welding

Welding on glassed steel shells can be done providing certain restrictions are adhered to. First, all code rated vessels require that a code certified welder must be used in conjunction with an inspector. Never weld more than a ½ in. long bead in any one area. Allow the area to cool to the point where the hand can be

TABLE 8.3 Allowable Temperature Differentials

Glass wall temperature °F/°C	Maximum allowable differential °F/°C
−20 to 250/−29 to 121	260/127
300/149	225/107
350/177	195/90
400/204	170/77
450/227	150/66

comfortably placed on the glass next to the weld. Consult with the manufacturer for specific data.

The use of water in the jacket to cool down (or speed up) the process can be risky and should be avoided.

REFERENCES

1. Schweitzer, Philip A. *Corrosion Resistance Tables*, 4th ed. vols. 1–3, Marcel Dekker, New York, 1995.
2. De Clerck, Donald H. Glass Linings, in *Corrosion Engineering Handbook* (Philip A. Schweitzer, ed.), Marcel Dekker, New York, 1996, pp. 489–544.

9

Comparative Corrosion Resistance of Lining Materials

Each of the following charts provides the compatibility of the various lining materials with the specific corrodent listed. The chemicals listed are in the pure state or in a saturated solution unless otherwise indicated. Compatibility is shown to the maximum allowable temperature for which data are available. Incompatibility is shown by an ''x.'' A blank space indicates that data are unavailable. Data are taken from Philip A. Schweitzer, *Corrosion Resistance Tables*, 4th ed., vols. 1–3, Marcel Dekker, New York, 1995.

The following abbreviations are used in the charts:

Epoxy GP	Epoxy general purpose
Epoxy CR	Epoxy chemical resistant
Epoxy P	Epoxy polyamide
CPVC	Chlorinated polyvinylchloride
ECTFE	Ethylene chlorotrifluoroethylene
ETFE	Ethylene tetrafluoroethylene
FEP	Fluorinated ethylene propylene
PFA	Perfluoroalkoxy
HMW PE	High molecular weight polyethylene
UHMW PE	Ultra high molecular weight polyethylene
PP	Polypropylene
PVC	Polyvinylchloride

PVDC	Polyvinylidene chloride
PVDF	Polyvinylidene fluoride
PTFE	Polytetrafluoroethylene (Teflon)
CIIR	Chlorobutyl rubber
EPDM	Ethylene-propylene-diene
EPT	Ethylene-propylene terpolymer
FKM	Fluoroelastomers
FPM	Perfluoroelastomers
PE	Polyethylene
EU	Polyether urethane elastomer
T	Polysulfide elastomer
AU	Polyurethane elastomer

Acetaldehyde

Lining material	°F	°C	Lining material	°F	°C
Glass	300	149	PVDC	150	66
Asphalt			PVDF	150	66
Coal tar	x	x	PTFE	450	232
Coal tar epoxy	x	x	Vinyl ester	x	x
Epoxy GP	x	x	Butyl rubber	80	27
Epoxy CR	x	x	CIIR		
Epoxy P	x	x	Hypalon	80	27
Zinc rich	80	27	EPDM	200	93
CPVC	x	x	EPT	200	93
ECTFE			FKM	x	x
ETFE	200	93	Hard rubber		
FEP	200	93	Soft rubber		
Phenolic	x	x	FPM	x	x
PFA			Neoprene	200	93
HMW PE	x	x	Nitrile rubber	x	x
UHMW PE	90	32	PE		
PP	120	49	EU	x	x
PVC Type 1	x	x	T		
PVC Type 2	x	x	AU	x	x
PVC Plastisol	x	x			

Acetic Acid 10%

Lining material	°F	°C	Lining material	°F	°C
Glass	400	204	PVDC	150	66
Asphalt			PVDF	300	149
Coal tar			PTFE	450	232
Coal tar epoxy	100	38	Vinyl ester	200	93
Epoxy GP	x	x	Butyl rubber	150	66
Epoxy CR	90	32	CIIR	160	71
Epoxy P	x	x	Hypalon	200	93
Zinc rich			EPDM	140	60
CPVC	90	32	EPT	x	x
ECTFE	250	121	FKM	190	88
ETFE	250	121	Hard rubber	200	93
FEP	400	204	Soft rubber		
Phenolic	210	99	FPM	200	93
PFA	200	93	Neoprene	160	71
HMW PE	140	60	Nitrile rubber	200	93
UHMW PE	140	60	PE	80	27
PP	220	104	EU	x	x
PVC Type 1	140	60	T	80	27
PVC Type 2	100	38	AU	x	x
PVC Plastisol	100	38			

Acetic Acid 20%

Lining material	°F	°C	Lining material	°F	°C
Glass	400	204	PVDC	120	49
Asphalt			PVDF	300	149
Coal tar			PTFE	450	232
Coal tar epoxy	100	38	Vinyl ester	200	93
Epoxy GP			Butyl rubber	150	66
Epoxy CR	x	x	CIIR	150	66
Epoxy P	x	x	Hypalon	200	93
Zinc rich			EPDM	140	60
CPVC	180	82	EPT	x	x
ECTFE	250	121	FKM	200	93
ETFE	250	121	Hard rubber	210	99
FEP	400	204	Soft rubber		
Phenolic	210	99	FPM	200	93
PFA	200	93	Neoprene	160	71
HMW PE	140	60	Nitrile rubber	200	93
UHMW PE	140	60	PE	80	27
PP	220	104	EU	x	x
PVC Type 1	140	60	T	80	27
PVC Type 2	140	60	AU	80	27
PVC Plastisol	140	60			

Acetic Acid 80%

Lining material	Temp. °F	°C	Lining material	Temp. °F	°C
Glass	400	204	PVDC	130	54
Asphalt			PVDF	180	82
Coal tar			PTFE	450	232
Coal tar epoxy	x	x	Vinyl ester	150	66
Epoxy GP	x	x	Butyl rubber	110	43
Epoxy CR			CIIR	160	71
Epoxy P	x	x	Hypalon	200	93
Zinc rich			EPDM	140	60
CPVC	x	x	EPT	x	x
ECTFE	150	66	FKM	160	71
ETFE	230	110	Hard rubber	150	66
FEP	400	204	Soft rubber		
Phenolic			FPM	90	32
PFA	200	93	Neoprene	160	71
HMW PE	80	27	Nitrile rubber	210	99
UHMW PE	80	27	PE	80	27
PP	200	93	EU	x	x
PVC Type 1	140	60	T	80	27
PVC Type 2	x	x	AU	x	x
PVC Plastisol	x	x			

Acetic Acid, Glacial

Lining material	Temp. °F	°C	Lining material	Temp. °F	°C
Glass	400	204	PVDC	140	60
Asphalt			PVDF	190	88
Coal tar			PTFE	450	232
Coal tar epoxy	x	x	Vinyl ester	150	66
Epoxy GP	x	x	Butyl rubber	90	32
Epoxy CR	x	x	CIIR	160	71
Epoxy P	x	x	Hypalon	x	x
Zinc rich			EPDM	300	149
CPVC	x	x	EPT	x	x
ECTFE	200	93	FKM	x	x
ETFE	230	110	Hard rubber	200	93
FEP	400	204	Soft rubber	x	x
Phenolic	120	49	FPM	210	99
PFA	250	121	Neoprene	x	x
HMW PE	80	27	Nitrile rubber	90	32
UHMW PE			PE	100	38
PP	180	82	EU	x	x
PVC Type 1	130	54	T	80	27
PVC Type 2	x	x	AU	x	x
PVC Plastisol	x	x			

Acetyl Chloride

Lining material	Temp.		Lining material	Temp.	
	°F	°C		°F	°C
Glass			PVDC	130	54
Asphalt			PVDF	130	54
Coal tar			PTFE	450	232
Coal tar epoxy	x	x	Vinyl ester	x	x
Epoxy GP			Butyl rubber		
Epoxy CR			CIIR		
Epoxy P	x	x	Hypalon	x	x
Zinc rich			EPDM	x	x
CPVC	x	x	EPT	x	x
ECTFE	150	66	FKM	400	204
ETFE	150	66	Hard rubber		
FEP	400	204	Soft rubber		
Phenolic	x	x	FPM	210	99
PFA	200	93	Neoprene	x	x
HMW PE	x	x	Nitrile rubber	x	x
UHMW PE			PE		
PP	x	x	EU	x	x
PVC Type 1	x	x	T		
PVC Type 2	x	x	AU	x	x
PVC Plastisol	x	x			

Acrylic Acid

Lining material	Temp.		Lining material	Temp.	
	°F	°C		°F	°C
Glass			PVDC		
Asphalt			PVDF	150	66
Coal tar			PTFE	450	232
Coal tar epoxy	x	x	Vinyl ester	100	38
Epoxy GP	x	x	Butyl rubber		
Epoxy CR	90	32	CIIR		
Epoxy P	100	38	Hypalon		
Zinc rich	x	x	EPDM		
CPVC	x	x	EPT 75%	150	66
ECTFE			FKM	x	x
ETFE			Hard rubber		
FEP	200	93	Soft rubber		
Phenolic			FPM	210	99
PFA			Neoprene	x	x
HMW PE			Nitrile rubber	x	x
UHMW PE			PE		
PP	x	x	EU		
PVC Type 1	x	x	T 75%	80	27
PVC Type 2	x	x	AU		
PVC Plastisol	x	x			

Alum

Lining material	Temp. °F	Temp. °C	Lining material	Temp. °F	Temp. °C
Glass	170	77	PVDC	180	82
Asphalt			PVDF	210	99
Coal tar			PTFE	450	232
Coal tar epoxy	100	38	Vinyl ester	240	116
Epoxy GP			Butyl rubber	190	88
Epoxy CR	200	93	CIIR		
Epoxy P	100	38	Hypalon	200	93
Zinc rich			EPDM	200	93
CPVC	200	93	EPT	140	60
ECTFE	300	149	FKM	190	88
ETFE	300	149	Hard rubber	190	88
FEP	400	204	Soft rubber		
Phenolic			FPM	210	99
PFA	200	93	Neoprene	200	93
HMW PE	140	60	Nitrile rubber	200	93
UHMW PE	140	60	PE		
PP	220	104	EU		
PVC Type 1	140	60	T		
PVC Type 2	140	60	AU		
PVC Plastisol	140	60			

Aluminum Bromide

Lining material	Temp. °F	Temp. °C	Lining material	Temp. °F	Temp. °C
Glass			PVDC		
Asphalt			PVDF	250	121
Coal tar			PTFE	250	121
Coal tar epoxy	160	71	Vinyl ester	160	71
Epoxy GP			Butyl rubber		
Epoxy CR	160	71	CIIR		
Epoxy P	160	71	Hypalon		
Zinc rich			EPDM		
CPVC	170	77	EPT	140	60
ECTFE			FKM	170	77
ETFE			Hard rubber		
FEP			Soft rubber		
Phenolic			FPM	210	99
PFA			Neoprene		
HMW PE			Nitrile rubber	140	60
UHMW PE			PE		
PP	170	77	EU		
PVC Type 1	140	60	T		
PVC Type 2	140	60	AU		
PVC Plastisol	140	60			

Aluminum Chloride, Aqueous

Lining material	Temp. °F	Temp. °C	Lining material	Temp. °F	Temp. °C
Glass	250	121	PVDC	150	66
Asphalt			PVDF	300	149
Coal tar			PTFE	450	232
Coal tar epoxy	100	38	Vinyl ester	260	127
Epoxy GP	90	32	Butyl rubber	150	66
Epoxy CR	220	104	CIIR	200	93
Epoxy P	110	43	Hypalon	250	121
Zinc rich	x	x	EPDM	300	149
CPVC	200	93	EPT	180	82
ECTFE	300	149	FKM	400	204
ETFE	300	149	Hard rubber	200	93
FEP	260	127	Soft rubber	140	60
Phenolic	300	149	FPM	210	99
PFA	200	93	Neoprene	200	93
HMW PE	140	60	Nitrile rubber	200	93
UHMW PE	140	60	PE		
PP	200	93	EU	130	54
PVC Type 1	140	60	T		
PVC Type 2	140	60	AU		
PVC Plastisol	140	60			

Aluminum Sulfate, Saturated

Lining material	Temp. °F	Temp. °C	Lining material	Temp. °F	Temp. °C
Glass	250	121	PVDC	180	82
Asphalt			PVDF	300	149
Coal tar			PTFE	400	232
Coal tar epoxy	100	38	Vinyl ester	250	121
Epoxy GP			Butyl rubber	190	88
Epoxy CR	200	93	CIIR	180	82
Epoxy P	100	38	Hypalon	160	71
Zinc rich			EPDM	300	149
CPVC	200	93	EPT	140	60
ECTFE	300	149	FKM	300	149
ETFE	300	149	Hard rubber	190	88
FEP	400	204	Soft rubber	140	60
Phenolic	290	143	FPM	210	99
PFA	220	104	Neoprene	200	93
HMW PE	140	60	Nitrile rubber	200	93
UHMW PE	140	60	PE		
PP	220	104	EU	120	49
PVC Type 1	140	60	T		
PVC Type 2	140	60	AU		
PVC Plastisol	140	60			

Ammonium Chloride 28%

Lining material	°F	°C	Lining material	°F	°C
Glass	250	121	PVDC		
Asphalt			PVDF	280	138
Coal tar			PTFE	450	232
Coal tar epoxy	100	38	Vinyl ester	200	93
Epoxy GP			Butyl rubber	190	88
Epoxy CR	200	93	CIIR	200	93
Epoxy P	100	38	Hypalon	200	93
Zinc rich			EPDM	210	99
CPVC	180	82	EPT	200	93
ECTFE	300	149	FKM	300	149
ETFE	300	149	Hard rubber	200	93
FEP	400	204	Soft rubber	140	60
Phenolic			FPM	210	99
PFA	200	93	Neoprene	200	93
HMW PE	140	60	Nitrile rubber	200	93
UHMW PE	140	60	PE	80	27
PP	180	82	EU	130	54
PVC Type 1	140	60	T	150	66
PVC Type 2			AU	90	32
PVC Plastisol					

Ammonium Chloride, Saturated

Lining material	°F	°C	Lining material	°F	°C
Glass	250	121	PVDC	160	71
Asphalt			PVDF	280	138
Coal tar			PTFE	450	232
Coal tar epoxy			Vinyl ester	200	93
Epoxy GP			Butyl rubber	190	88
Epoxy CR	200	93	CIIR	200	93
Epoxy P	100	38	Hypalon	200	93
Zinc rich			EPDM	300	149
CPVC	270	132	EPT	180	82
ECTFE	300	149	FKM	300	149
ETFE	300	149	Hard rubber	200	93
FEP	400	204	Soft rubber	140	60
Phenolic	80	27	FPM	210	99
PFA	200	93	Neoprene	200	93
HMW PE	140	60	Nitrile rubber	200	93
UHMW PE	140	60	PE	80	27
PP	200	93	EU	120	49
PVC Type 1	140	60	T	150	66
PVC Type 2	140	60	AU	90	32
PVC Plastisol	140	60			

Ammonium Fluoride 25%

Lining material	°F	°C	Lining material	°F	°C
Glass	x	x	PVDC	90	32
Asphalt			PVDF	280	138
Coal tar			PTFE	450	232
Coal tar epoxy	x	x	Vinyl ester	220	104
Epoxy GP			Butyl rubber	150	66
Epoxy CR			CIIR		
Epoxy P	x	x	Hypalon		
Zinc rich			EPDM	300	149
CPVC	200	93	EPT	140	60
ECTFE	300	149	FKM	140	60
ETFE	300	149	Hard rubber	200	93
FEP	400	204	Soft rubber	x	x
Phenolic			FPM	140	60
PFA	200	93	Neoprene	200	93
HMW PE	140	60	Nitrile rubber	120	49
UHMW PE	140	60	PE		
PP	200	93	EU		
PVC Type 1	140	60	T		
PVC Type 2	90	32	AU		
PVC Plastisol	90	32			

Aniline Hydrochloride

Lining material	°F	°C	Lining material	°F	°C
Glass	250	121	PVDC	80	27
Asphalt			PVDF	120	49
Coal tar			PTFE	450	232
Coal tar epoxy	x	x	Vinyl ester	140	60
Epoxy GP			Butyl rubber	x	x
Epoxy CR			CIIR		
Epoxy P	x	x	Hypalon	200	93
Zinc rich			EPDM	210	99
CPVC	x	x	EPT	140	60
ECTFE	80	27	FKM	180	82
ETFE 10%	150	66	Hard rubber	x	x
FEP	400	204	Soft rubber	x	x
Phenolic	160	71	FPM	210	99
PFA			Neoprene	x	x
HMW PE	x	x	Nitrile rubber	x	x
UHMW PE			PE	x	x
PP	170	77	EU	x	x
PVC Type 1	x	x	T		
PVC Type 2	x	x	AU		
PVC Plastisol	x	x			

Antimony Trichloride

Lining material	Temp. °F	Temp. °C	Lining material	Temp. °F	Temp. °C
Glass	250	121	PVDC	160	71
Asphalt			PVDF	150	66
Coal tar			PTFE	450	232
Coal tar epoxy	140	60	Vinyl ester	220	104
Epoxy GP			Butyl rubber	150	66
Epoxy CR			CIIR	140	60
Epoxy P	140	60	Hypalon	140	60
Zinc rich			EPDM	300	149
CPVC	200	93	EPT	x	x
ECTFE	100	38	FKM	190	88
ETFE	210	99	Hard rubber		
FEP	400	204	Soft rubber		
Phenolic			FPM	210	99
PFA	200	93	Neoprene	140	60
HMW PE	140	60	Nitrile rubber		
UHMW PE	140	60	PE	120	49
PP	180	82	EU		
PVC Type 1	140	60	T		
PVC Type 2	140	60	AU		
PVC Plastisol	140	60			

Aqua Regia 3; 1

Lining material	Temp. °F	Temp. °C	Lining material	Temp. °F	Temp. °C
Glass	200	93	PVDC	120	49
Asphalt			PVDF	170	77
Coal tar			PTFE	450	232
Coal tar epoxy	x	x	Vinyl ester	x	x
Epoxy GP	x	x	Butyl rubber		
Epoxy CR	x	x	CIIR		
Epoxy P	x	x	Hypalon		
Zinc rich			EPDM	x	x
CPVC	80	27	EPT	x	x
ECTFE	250	121	FKM	190	88
ETFE	210	99	Hard rubber	x	x
FEP	400	204	Soft rubber		
Phenolic	x	x	FPM	210	99
PFA	240	116	Neoprene	x	x
HMW PE	130	54	Nitrile rubber	x	x
UHMW PE	130	54	PE	x	x
PP	x	x	EU		
PVC Type 1	x	x	T		
PVC Type 2	x	x	AU	x	x
PVC Plastisol	x	x			

Barium Chloride, Saturated

Lining material	Temp. °F	°C	Lining material	Temp. °F	°C
Glass	250	121	PVDC	180	82
Asphalt			PVDF	280	138
Coal tar			PTFE	450	232
Coal tar epoxy	140	60	Vinyl ester	190	88
Epoxy GP			Butyl rubber	190	88
Epoxy CR	200	93	CIIR	160	71
Epoxy P	140	60	Hypalon	250	121
Zinc rich			EPDM	300	149
CPVC	180	82	EPT	180	82
ECTFE	300	149	FKM	400	204
ETFE	300	149	Hard rubber	190	88
FEP	400	204	Soft rubber	140	60
Phenolic			FPM to 20%	210	99
PFA	200	93	Neoprene	200	93
HMW PE	140	60	Nitrile rubber	200	93
UHMW PE	140	60	PE	100	38
PP	220	104	EU		
PVC Type 1	140	60	T		
PVC Type 2	140	60	AU	90	32
PVC Plastisol	140	60			

Benzaldehyde

Lining material	Temp. °F	°C	Lining material	Temp. °F	°C
Glass	250	121	PVDC	x	x
Asphalt			PVDF	120	49
Coal tar	x	x	PTFE	450	232
Coal tar epoxy	x	x	Vinyl ester	x	x
Epoxy GP	x	x	Butyl rubber	90	32
Epoxy CR	x	x	CIIR		
Epoxy P	x	x	Hypalon	x	x
Zinc rich	x	x	EPDM	300	149
CPVC	x	x	EPT	x	x
ECTFE	150	66	FKM	x	x
ETFE	210	99	Hard rubber	x	x
FEP	400	204	Soft rubber		
Phenolic	80	27	FPM	210	99
PFA	200	93	Neoprene	x	x
HMW PE	x	x	Nitrile rubber	x	x
UHMW PE	x	x	PE	x	x
PP	80	27	EU	x	x
PVC Type 1	x	x	T		
PVC Type 2	x	x	AU	x	x
PVC Plastisol	x	x			

Boric Acid

Lining material	°F	°C	Lining material	°F	°C
Glass	300	149	PVDC	170	77
Asphalt			PVDF	280	138
Coal tar			PTFE	450	232
Coal tar epoxy	100	38	Vinyl ester	200	93
Epoxy GP	90	27	Butyl rubber	190	38
Epoxy CR	200	93	CIIR 10%	250	121
Epoxy P	140	60	Hypalon	290	143
Zinc rich			EPDM	300	149
CPVC	210	99	EPT	140	60
ECTFE	300	149	FKM	400	204
ETFE	300	149	Hard rubber	200	93
FEP	400	204	Soft rubber	140	60
Phenolic	300	149	FPM	210	99
PFA	200	93	Neoprene	200	93
HMW PE	140	60	Nitrile rubber	180	82
UHMW PE	140	60	PE	100	38
PP	220	104	EU	130	54
PVC Type 1	140	60	T		
PVC Type 2	140	60	AU	90	32
PVC Plastisol	140	60			

Brine, Acid

Lining material	°F	°C	Lining material	°F	°C
Glass	200	93	PVDC	140	60
Asphalt			PVDF	280	138
Coal tar			PTFE	450	232
Coal tar epoxy	130	54	Vinyl ester	220	104
Epoxy GP			Butyl rubber		
Epoxy CR			CIIR		
Epoxy P	130	54	Hypalon	180	82
Zinc rich			EPDM	300	149
CPVC	180	82	EPT	210	99
ECTFE	300	149	FKM	400	204
ETFE	300	149	Hard rubber		
FEP	400	204	Soft rubber		
Phenolic	160	71	FPM	210	99
PFA			Neoprene	160	71
HMW PE			Nitrile rubber	180	82
UHMW PE	140	60	PE		
PP	230	104	EU		
PVC Type 1	140	60	T		
PVC Type 2	140	60	AU		
PVC Plastisol	140	60			

Comparative Corrosion Resistance of Lining Materials

239

Bromine Liquid

Lining material	°F	°C	Lining material	°F	°C
Glass	90	32	PVDC	x	x
Asphalt			PVDF	140	60
Coal tar			PTFE	450	232
Coal tar epoxy			Vinyl ester	x	x
Epoxy GP			Butyl rubber		
Epoxy CR			CIIR		
Epoxy P			Hypalon	80	27
Zinc rich			EPDM	x	x
CPVC	x	x	EPT	x	x
ECTFE	150	66	FKM	400	204
ETFE			Hard rubber		
FEP	400	204	Soft rubber		
Phenolic			FPM	140	60
PFA			Neoprene	x	x
HMW PE	x	x	Nitrile rubber	x	x
UHMW PE	x	x	PE	x	x
PP	x	x	EU		
PVC Type 1	x	x	T		
PVC Type 2	x	x	AU	x	x
PVC Plastisol	x	x			

Bromine Water, Dilute

Lining material	°F	°C	Lining material	°F	°C
Glass			PVDC	80	27
Asphalt			PVDF	210	99
Coal tar			PTFE	450	232
Coal tar epoxy	x	x	Vinyl ester	180	82
Epoxy GP	x	x	Butyl rubber		
Epoxy CR	x	x	CIIR		
Epoxy P	x	x	Hypalon	80	27
Zinc rich			EPDM	x	x
CPVC	x	x	EPT	x	x
ECTFE	250	121	FKM	200	93
ETFE 10%	230	110	Hard rubber		
FEP	400	204	Soft rubber		
Phenolic			FPM		
PFA 25%	200	93	Neoprene	x	x
HMW PE	80	27	Nitrile rubber	x	x
UHMW PE	x	x	PE 5%	100	38
PP	x	x	EU		
PVC Type 1	140	60	T	80	27
PVC Type 2	x	x	AU	x	x
PVC Plastisol	x	x			

Bromine Water, Saturated

Lining material	°F	°C	Lining material	°F	°C
Glass	200	93	PVDC	x	x
Asphalt			PVDF	210	99
Coal tar			PTFE	450	232
Coal tar epoxy	x	x	Vinyl ester		
Epoxy GP	x	x	Butyl rubber		
Epoxy CR	x	x	CIIR		
Epoxy P	x	x	Hypalon	80	27
Zinc rich			EPDM	x	x
CPVC	x	x	EPT	x	x
ECTFE	250	121	FKM	210	99
ETFE			Hard rubber		
FEP	400	204	Soft rubber		
Phenolic			FPM		
PFA			Neoprene	x	x
HMW PE			Nitrile rubber	x	x
UHMW PE	x	x	PE		
PP	x	x	EU		
PVC Type 1	140	60	T	80	27
PVC Type 2	x	x	AU	x	x
PVC Plastisol	x	x			

Butyric Acid

Lining material	°F	°C	Lining material	°F	°C
Glass	250	121	PVDC	80	27
Asphalt			PVDF	210	99
Coal tar			PTFE	450	232
Coal tar epoxy	x	x	Vinyl ester	130	54
Epoxy GP	x	x	Butyl rubber	x	x
Epoxy CR	90	32	CIIR		
Epoxy P	x	x	Hypalon	x	x
Zinc rich			EPDM	140	60
CPVC	140	60	EPT	x	x
ECTFE	250	121	FKM	120	49
ETFE	250	121	Hard rubber	x	x
FEP	400	204	Soft rubber	x	x
Phenolic	160	71	FPM	210	99
PFA	210	99	Neoprene	x	x
HMW PE	x	x	Nitrile rubber	x	x
UHMW PE	130	54	PE	100	38
PP	180	82	EU		
PVC Type 1	80	27	T		
PVC Type 2	x	x	AU		
PVC Plastisol	x	x			

Calcium Chloride, Saturated

Lining material	Temp. °F	Temp. °C	Lining material	Temp. °F	Temp. °C
Glass	250	121	PVDC	180	82
Asphalt			PVDF	280	138
Coal tar			PTFE	450	232
Coal tar epoxy	110	43	Vinyl ester	180	82
Epoxy GP	80	27	Butyl rubber	190	88
Epoxy CR	200	90	CIIR	160	71
Epoxy P	110	43	Hypalon	200	93
Zinc rich	x	x	EPDM	210	99
CPVC	180	82	EPT	180	82
ECTFE	300	149	FKM	300	149
ETFE	300	149	Hard rubber	190	88
FEP	400	204	Soft rubber	140	60
Phenolic	200	93	FPM	210	99
PFA	200	93	Neoprene	200	93
HMW PE	140	60	Nitrile rubber	180	82
UHMW PE	140	60	PE	100	38
PP	220	104	EU	130	54
PVC Type 1	140	60	T	150	66
PVC Type 2	140	60	AU	80	27
PVC Plastisol	140	60			

Calcium Hypochlorite, Saturated

Lining material	Temp. °F	Temp. °C	Lining material	Temp. °F	Temp. °C
Glass	250	121	PVDC	120	49
Asphalt			PVDF	280	138
Coal tar			PTFE	450	232
Coal tar epoxy	x	x	Vinyl ester	180	82
Epoxy GP	80	27	Butyl rubber	150	66
Epoxy CR	100	32	CIIR		
Epoxy P	x	x	Hypalon	250	121
Zinc rich			EPDM	300	149
CPVC	200	93	EPT	x	x
ECTFE	300	149	FKM	400	204
ETFE	300	149	Hard rubber	200	93
FEP	400	204	Soft rubber	x	x
Phenolic	x	x	FPM	210	99
PFA	200	93	Neoprene	80	27
HMW PE	140	60	Nitrile rubber	x	x
UHMW PE	140	60	PE	x	x
PP	210	99	EU		
PVC Type 1	140	60	T		
PVC Type 2	140	60	AU	x	x
PVC Plastisol	140	60			

Chlorine Dioxide, 15%

Lining material	Temp. °F	Temp. °C	Lining material	Temp. °F	Temp. °C
Glass	80	27	PVDC	120	49
Asphalt			PVDF	200	93
Coal tar			PTFE	450	232
Coal tar epoxy	x	x	Vinyl ester	180	82
Epoxy GP	x	x	Butyl rubber	x	x
Epoxy CR	x	x	CIIR		
Epoxy P	x	x	Hypalon		
Zinc rich	140	60	EPDM		
CPVC	250	121	EPT	x	x
ECTFE	250	121	FKM		
ETFE	400	204	Hard rubber		
FEP			Soft rubber		
Phenolic			FPM	210	99
PFA			Neoprene	x	x
HMW PE			Nitrile rubber	x	x
UHMW PE			PE	100	38
PP	x	x	EU		
PVC Type 1	140	60	T		
PVC Type 2			AU		
PVC Plastisol					

Chlorine Water, Saturated

Lining material	Temp. °F	Temp. °C	Lining material	Temp. °F	Temp. °C
Glass	200	93	PVDC	180	82
Asphalt			PVDF	220	104
Coal tar	x	x	PTFE	450	232
Coal tar epoxy	x	x	Vinyl ester	180	82
Epoxy GP	x	x	Butyl rubber	x	x
Epoxy CR	x	x	CIIR		
Epoxy P	x	x	Hypalon	90	32
Zinc rich			EPDM	80	27
CPVC	210	99	EPT	80	27
ECTFE	250	121	FKM	190	88
ETFE	100	38	Hard rubber	150	66
FEP	400	204	Soft rubber		
Phenolic			FPM	210	99
PFA			Neoprene	x	x
HMW PE	80	27	Nitrile rubber	x	x
UHMW PE	140	60	PE	100	38
PP	140	60	EU		
PVC Type 1	140	60	T	x	x
PVC Type 2	140	60	AU	x	x
PVC Plastisol	140	60			

Chloroacetic Acid

Lining material	Temp. °F	Temp. °C	Lining material	Temp. °F	Temp. °C
Glass	250	121	PVDC	120	49
Asphalt			PVDF	200	93
Coal tar	x	x	PTFE	450	232
Coal tar epoxy	x	x	Vinyl ester	200	93
Epoxy GP	x	x	Butyl rubber	160	71
Epoxy CR	x	x	CIIR to 25%	160	71
Epoxy P	x	x	Hypalon	x	x
Zinc rich	x	x	EPDM	300	149
CPVC	250	121	EPT	x	x
ECTFE	250	121	FKM	x	x
ETFE	250	121	Hard rubber	x	x
FEP	400	204	Soft rubber	x	x
Phenolic	x	x	FPM	210	99
PFA	200	93	Neoprene	x	x
HMW PE	x	x	Nitrile rubber	x	x
UHMW PE	x	x	PE	x	x
PP	180	82	EU	x	x
PVC Type 1	140	60	T		
PVC Type 2	x	x	AU	x	x
PVC Plastisol	x	x			

Chromic Acid 50%

Lining material	Temp. °F	Temp. °C	Lining material	Temp. °F	Temp. °C
Glass	250	121	PVDC	180	82
Asphalt			PVDF	250	121
Coal tar			PTFE	450	232
Coal tar epoxy			Vinyl ester	x	x
Epoxy GP	x	x	Butyl rubber	x	x
Epoxy CR	x	x	CIIR	x	x
Epoxy P			Hypalon	160	71
Zinc rich			EPDM	x	x
CPVC	210	99	EPT	x	x
ECTFE	250	121	FKM	350	177
ETFE	150	66	Hard rubber	x	x
FEP	400	204	Soft rubber	x	x
Phenolic	x	x	FPM	210	99
PFA	200	93	Neoprene	100	38
HMW PE	90	32	Nitrile rubber	190	88
UHMW PE	90	32	PE		
PP	150	66	EU	x	x
PVC Type 1	x	x	T	x	x
PVC Type 2	x	x	AU	x	x
PVC Plastisol	x	x			

Citric Acid

Lining material	°F	°C	Lining material	°F	°C
Glass	200	93	PVDC	180	82
Asphalt			PVDF	250	121
Coal tar			PTFE	450	232
Coal tar epoxy	100	38	Vinyl ester	200	93
Epoxy GP	90	32	Butyl rubber	190	88
Epoxy CR	200	93	CIIR		
Epoxy P	100	38	Hypalon	250	121
Zinc rich			EPDM	300	149
CPVC	180	82	EPT	180	82
ECTFE	300	149	FKM	400	204
ETFE			Hard rubber	150	66
FEP	400	204	Soft rubber	x	x
Phenolic	160	71	FPM	210	99
PFA	200	93	Neoprene	200	93
HMW PE	140	60	Nitrile rubber	180	82
UHMW PE	140	60	PE	100	38
PP	220	104	EU		
PVC Type 1	140	60	T	x	x
PVC Type 2	140	60	AU		
PVC Plastisol	140	60			

Ferrous Chloride, Saturated

Lining material	°F	°C	Lining material	°F	°C
Glass	250	121	PVDC	130	54
Asphalt			PVDF	280	138
Coal tar			PTFE	450	232
Coal tar epoxy			Vinyl ester	200	93
Epoxy GP			Butyl rubber	190	88
Epoxy CR	80	27	CIIR to 50%	160	71
Epoxy P			Hypalon	250	121
Zinc rich			EPDM	200	93
CPVC	210	99	EPT	180	82
ECTFE	300	149	FKM	180	82
ETFE	300	149	Hard rubber	200	93
FEP	400	204	Soft rubber	140	60
Phenolic 40%	300	149	FPM	210	99
PFA	200	93	Neoprene	90	32
HMW PE	140	60	Nitrile rubber	200	93
UHMW PE	140	60	PE		
PP	210	99	EU		
PVC Type 1	140	60	T		
PVC Type 2	140	60	AU		
PVC Plastisol	140	60			

Formic Acid 10–85%

Lining material	Temp. °F	Temp. °C	Lining material	Temp. °F	Temp. °C
Glass	250	121	PVDC	150	66
Asphalt			PVDF	250	121
Coal tar			PTFE	450	232
Coal tar epoxy	x	x	Vinyl ester	100	38
Epoxy GP	x	x	Butyl rubber	150	66
Epoxy CR	x	x	CIIR	150	66
Epoxy P	x	x	Hypalon	200	93
Zinc rich			EPDM	300	149
CPVC	140	60	EPT	200	93
ECTFE	250	121	FKM	190	88
ETFE	270	132	Hard rubber	200	93
FEP	400	204	Soft rubber	x	x
Phenolic	200	93	FPM	210	99
PFA	200	93	Neoprene	160	71
HMW PE	140	60	Nitrile rubber	x	x
UHMW PE	140	60	PE	80	27
PP	230	110	EU	x	x
PVC Type 1	110	43	T		
PVC Type 2	90	32	AU	x	x
PVC Plastisol	90	32			

Hydrobromic Acid 20%

Lining material	Temp. °F	Temp. °C	Lining material	Temp. °F	Temp. °C
Glass	250	121	PVDC	120	49
Asphalt			PVDF	280	138
Coal tar			PTFE	450	232
Coal tar epoxy	x	x	Vinyl ester	180	82
Epoxy GP	x	x	Butyl rubber	160	71
Epoxy CR	80	27	CIIR	120	49
Epoxy P	x	x	Hypalon	100	38
Zinc rich			EPDM	300	149
CPVC	180	82	EPT	140	60
ECTFE	300	149	FKM	400	204
ETFE	300	149	Hard rubber	200	93
FEP	400	204	Soft rubber	140	60
Phenolic	200	93	FPM	210	99
PFA	200	93	Neoprene	x	x
HMW PE	140	60	Nitrile rubber	x	x
UHMW PE	140	60	PE	100	38
PP	200	93	EU	x	x
PVC Type 1	140	60	T		
PVC Type 2	140	60	AU		
PVC Plastisol	140	60			

Hydrobromic Acid 50%

Lining material	°F	°C	Lining material	°F	°C
Glass	250	121	PVDC	130	54
Asphalt			PVDF	280	138
Coal tar			PTFE	450	232
Coal tar epoxy	x	x	Vinyl ester	200	93
Epoxy GP	x	x	Butyl rubber	110	43
Epoxy CR	80	27	CIIR	120	49
Epoxy P	x	x	Hypalon	100	38
Zinc rich			EPDM	300	149
CPVC	190	88	EPT	140	60
ECTFE	300	149	FKM	400	204
ETFE	300	149	Hard rubber	200	93
FEP	400	204	Soft rubber	140	60
Phenolic	200	93	FPM	210	99
PFA	200	93	Neoprene	x	x
HMW PE	140	60	Nitrile rubber	x	x
UHMW PE	140	60	PE	100	38
PP	190	88	EU	x	x
PVC Type 1	140	60	T		
PVC Type 2	140	60	AU		
PVC Plastisol	140	60			

Hydrochloric Acid 20%

Lining material	°F	°C	Lining material	°F	°C
Glass	250	121	PVDC	180	82
Asphalt			PVDF	280	138
Coal tar			PTFE	450	232
Coal tar epoxy	x	x	Vinyl ester	220	104
Epoxy GP	x	x	Butyl rubber	x	x
Epoxy CR	80	27	CIIR	x	x
Epoxy P	x	x	Hypalon	160	71
Zinc rich			EPDM	300	149
CPVC	180	82	EPT	x	x
ECTFE	300	149	FKM	350	177
ETFE	300	149	Hard rubber	200	93
FEP	400	204	Soft rubber	140	60
Phenolic	300	149	FPM	210	99
PFA	250	121	Neoprene	180	82
HMW PE	140	60	Nitrile rubber	130	54
UHMW PE	140	60	PE	100	38
PP	220	104	EU	x	x
PVC Type 1	140	60	T	x	x
PVC Type 2	140	60	AU	x	x
PVC Plastisol	140	60			

Hydrochloric Acid 35%

Lining material	Temp. °F	Temp. °C	Lining material	Temp. °F	Temp. °C
Glass	250	121	PVDC	180	82
Asphalt			PVDF	280	138
Coal tar			PTFE	450	232
Coal tar epoxy	x	x	Vinyl ester	180	82
Epoxy GP	x	x	Butyl rubber	x	x
Epoxy CR	80	27	CIIR	x	x
Epoxy P	x	x	Hypalon	150	66
Zinc rich			EPDM	300	149
CPVC	150	66	EPT	x	x
ECTFE	300	149	FKM	350	177
ETFE	300	149	Hard rubber	200	93
FEP	400	204	Soft rubber	140	60
Phenolic	300	149	FPM	210	99
PFA	250	121	Neoprene	180	82
HMW PE	140	60	Nitrile rubber	x	x
UHMW PE	140	60	PE	100	38
PP	220	104	EU	x	x
PVC Type 1	140	60	T	x	x
PVC Type 2	140	60	AU	x	x
PVC Plastisol	140	60			

Hydrochloric Acid 38%

Lining material	Temp. °F	Temp. °C	Lining material	Temp. °F	Temp. °C
Glass	250	121	PVDC	180	82
Asphalt			PVDF	280	138
Coal tar			PTFE	450	232
Coal tar epoxy	x	x	Vinyl ester	180	82
Epoxy GP	x	x	Butyl rubber	x	x
Epoxy CR	80	27	CIIR	x	x
Epoxy P	x	x	Hypalon	150	66
Zinc rich			EPDM	300	149
CPVC	170	77	EPT	x	x
ECTFE	300	149	FKM	350	177
ETFE	300	149	Hard rubber	200	93
FEP	400	204	Soft rubber	140	60
Phenolic	300	149	FPM	210	99
PFA	200	93	Neoprene	90	32
HMW PE	140	60	Nitrile rubber	x	x
UHMW PE	140	60	PE	100	38
PP	200	93	EU	x	x
PVC Type 1	140	60	T	x	x
PVC Type 2	140	60	AU	x	x
PVC Plastisol	140	60			

Hydrochloric Acid Fumes

Lining material	°F	°C	Lining material	°F	°C
Glass	200	93	PVDC		
Asphalt			PVDF		
Coal tar			PTFE	450	232
Coal tar epoxy			Vinyl ester	350	177
Epoxy GP			Butyl rubber	90	32
Epoxy CR	80	27	CIIR		
Epoxy P			Hypalon	90	32
Zinc rich			EPDM	90	32
CPVC			EPT	80	27
ECTFE			FKM	90	32
ETFE	300	149	Hard rubber	90	32
FEP	400	204	Soft rubber		
Phenolic			FPM	210	99
PFA			Neoprene	90	32
HMW PE			Nitrile rubber	90	32
UHMW PE			PE	100	38
PP			EU		
PVC Type 1			T		
PVC Type 2			AU		
PVC Plastisol					

Hydrofluoric Acid 30%

Lining material	°F	°C	Lining material	°F	°C
Glass	x	x	PVDC	160	71
Asphalt			PVDF	260	127
Coal tar			PTFE	450	232
Coal tar epoxy	x	x	Vinyl ester	x	x
Epoxy GP			Butyl rubber	350	177
Epoxy CR	80	27	CIIR	160	71
Epoxy P	x	x	Hypalon	90	32
Zinc rich			EPDM	140	60
CPVC	x	x	EPT	140	60
ECTFE	250	121	FKM	210	99
ETFE	270	132	Hard rubber	200	93
FEP	400	204	Soft rubber		
Phenolic	x	x	FPM	210	99
PFA	200	93	Neoprene	200	93
HMW PE	140	60	Nitrile rubber	x	x
UHMW PE	140	60	PE	x	x
PP	180	82	EU	x	x
PVC Type 1	130	54	T	x	x
PVC Type 2	130	54	AU		
PVC Plastisol	130	54			

Hydrofluoric Acid 70%

Lining material	Temp. °F	Temp. °C	Lining material	Temp. °F	Temp. °C
Glass	x	x	PVDC		
Asphalt			PVDF	210	99
Coal tar			PTFE	450	232
Coal tar epoxy	x	x	Vinyl ester	x	x
Epoxy GP	x	x	Butyl rubber	150	66
Epoxy CR	80	27	CIIR	x	x
Epoxy P	x	x	Hypalon	160	71
Zinc rich			EPDM	x	x
CPVC	x	x	EPT	x	x
ECTFE	240	116	FKM	350	177
ETFE	250	121	Hard rubber	x	x
FEP	400	204	Soft rubber	x	x
Phenolic	x	x	FPM	210	99
PFA	200	93	Neoprene	200	93
HMW PE	x	x	Nitrile rubber	x	x
UHMW PE	x	x	PE	x	x
PP	200	93	EU	x	x
PVC Type 1	80	27	T	x	x
PVC Type 2			AU		
PVC Plastisol					

Hydrofluoric Acid 100%

Lining material	Temp. °F	Temp. °C	Lining material	Temp. °F	Temp. °C
Glass	x	x	PVDC	x	x
Asphalt			PVDF	200	93
Coal tar			PTFE	450	232
Coal tar epoxy	x	x	Vinyl ester	x	x
Epoxy GP	x	x	Butyl rubber	x	x
Epoxy CR	80	27	CIIR	x	x
Epoxy P	x	x	Hypalon	160	71
Zinc rich			EPDM	x	x
CPVC	x	x	EPT	x	x
ECTFE	240	116	FKM	x	x
ETFE	230	110	Hard rubber	x	x
FEP	400	204	Soft rubber	x	x
Phenolic	x	x	FPM	210	99
PFA	200	93	Neoprene	x	x
HMW PE	x	x	Nitrile rubber	x	x
UHMW PE	x	x	PE	x	x
PP	200	93	EU	x	x
PVC Type 1			T	x	x
PVC Type 2			AU		
PVC Plastisol					

Hydrogen Peroxide 30%

Lining material	Temp. °F	Temp. °C	Lining material	Temp. °F	Temp. °C
Glass	250	121	PVDC	120	49
Asphalt			PVDF	250	121
Coal tar			PTFE	450	232
Coal tar epoxy	x	x	Vinyl ester	170	77
Epoxy GP	x	x	Butyl rubber	x	x
Epoxy CR	x	x	CIIR	x	x
Epoxy P	x	x	Hypalon		
Zinc rich			EPDM	300	149
CPVC	180	82	EPT	x	x
ECTFE	270	132	FKM	350	177
ETFE	250	121	Hard rubber	x	x
FEP	400	204	Soft rubber	x	x
Phenolic			FPM	210	99
PFA	200	93	Neoprene	x	x
HMW PE	140	60	Nitrile rubber	90	32
UHMW PE	140	60	PE	100	38
PP	100	38	EU		
PVC Type 1	140	60	T	x	x
PVC Type 2	x	x	AU		
PVC Plastisol	x	x			

Hydrogen Peroxide 90%

Lining material	Temp. °F	Temp. °C	Lining material	Temp. °F	Temp. °C
Glass	250	121	PVDC	120	49
Asphalt			PVDF	120	49
Coal tar			PTFE	450	232
Coal tar epoxy	x	x	Vinyl ester	150	66
Epoxy GP	x	x	Butyl rubber	x	x
Epoxy CR	x	x	CIIR	x	x
Epoxy P	x	x	Hypalon		
Zinc rich			EPDM	300	149
CPVC	180	82	EPT	x	x
ECTFE	150	66	FKM	250	121
ETFE	150	66	Hard rubber	x	x
FEP	400	204	Soft rubber	x	x
Phenolic	80	27	FPM	210	99
PFA	200	93	Neoprene	x	x
HMW PE	140	60	Nitrile rubber	x	x
UHMW PE	80	27	PE		
PP	110	43	EU		
PVC Type 1	140	60	T	x	x
PVC Type 2	x	x	AU		
PVC Plastisol	x	x			

Hypochlorous Acid

Lining material	Temp. °F	Temp. °C	Lining material	Temp. °F	Temp. °C
Glass	170	77	PVDC	120	49
Asphalt			PVDF	280	138
Coal tar			PTFE	450	232
Coal tar epoxy	x	x	Vinyl ester	150	66
Epoxy GP			Butyl rubber	x	x
Epoxy CR			CIIR		
Epoxy P	x	x	Hypalon	x	x
Zinc rich			EPDM	300	149
CPVC	180	82	EPT	140	60
ECTFE	300	149	FKM	400	204
ETFE	300	149	Hard rubber	150	66
FEP	400	204	Soft rubber		
Phenolic			FPM	210	99
PFA	200	93	Neoprene	x	x
HMW PE	150	66	Nitrile rubber	x	x
UHMW PE			PE	x	x
PP	140	60	EU		
PVC Type 1	140	60	T		
PVC Type 2	140	60	AU		
PVC Plastisol	140	60			

Jet Fuel JP-4

Lining material	Temp. °F	Temp. °C	Lining material	Temp. °F	Temp. °C
Glass	180	82	PVDC	80	27
Asphalt			PVDF	250	121
Coal tar			PTFE	450	232
Coal tar epoxy	x	x	Vinyl ester	180	82
Epoxy GP	90	32	Butyl rubber	x	x
Epoxy CR	110	43	CIIR		
Epoxy P	100	38	Hypalon	x	x
Zinc rich	160	71	EPDM	x	x
CPVC	140	60	EPT	x	x
ECTFE	300	149	FKM	400	204
ETFE	230	110	Hard rubber		
FEP	400	204	Soft rubber		
Phenolic	160	71	FPM	210	99
PFA	200	93	Neoprene	x	x
HMW PE			Nitrile rubber	200	93
UHMW PE			PE	100	38
PP	x	x	EU		
PVC Type 1	140	60	T	80	27
PVC Type 2	140	60	AU	x	x
PVC Plastisol	140	60			

Jet Fuel JP-5

Lining material	Temp. °F	Temp. °C	Lining material	Temp. °F	Temp. °C
Glass	180	82	PVDC	90	32
Asphalt			PVDF	250	121
Coal tar			PTFE	450	232
Coal tar epoxy	x	x	Vinyl ester	120	49
Epoxy GP	100	38	Butyl rubber	x	x
Epoxy CR			CIIR		
Epoxy P	100	38	Hypalon	x	x
Zinc rich			EPDM	x	x
CPVC	140	60	EPT	x	x
ECTFE	300	149	FKM	400	204
ETFE	230	110	Hard rubber		
FEP	400	204	Soft rubber		
Phenolic			FPM	210	99
PFA	200	93	Neoprene	x	x
HMW PE			Nitrile rubber	200	93
UHMW PE			PE	100	38
PP	x	x	EU		
PVC Type 1	140	60	T	80	27
PVC Type 2	140	60	AU	x	x
PVC Plastisol	140	60			

Lactic Acid 25%

Lining material	Temp. °F	Temp. °C	Lining material	Temp. °F	Temp. °C
Glass	200	93	PVDC		
Asphalt			PVDF	120	49
Coal tar			PTFE	450	232
Coal tar epoxy	x	x	Vinyl ester	210	99
Epoxy GP	x	x	Butyl rubber	120	49
Epoxy CR	x	x	CIIR	120	49
Epoxy P	x	x	Hypalon	200	93
Zinc rich			EPDM	210	99
CPVC	180	82	EPT	210	99
ECTFE	150	66	FKM	300	149
ETFE	250	121	Hard rubber	150	66
FEP	400	204	Soft rubber	x	x
Phenolic	160	71	FPM	210	99
PFA	200	93	Neoprene	220	104
HMW PE	150	66	Nitrile rubber	x	x
UHMW PE	140	60	PE	100	38
PP	150	66	EU	x	x
PVC Type 1	140	60	T	x	x
PVC Type 2	140	60	AU		
PVC Plastisol	140	60			

Lactic Acid, Concentrated

Lining Material	°F	°C	Lining material	°F	°C
Glass	200	93	PVDC	80	27
Asphalt			PVDF	110	43
Coal tar			PTFE	450	232
Coal tar epoxy	x	x	Vinyl ester	200	93
Epoxy GP	x	x	Butyl rubber	120	49
Epoxy CR	x	x	CIIR	120	49
Epoxy P	x	x	Hypalon	200	93
Zinc rich			EPDM	210	99
CPVC	100	38	EPT	210	99
ECTFE	150	66	FKM	400	204
ETFE	200	93	Hard rubber	150	66
FEP	400	204	Soft rubber	x	x
Phenolic			FPM	210	99
PFA	200	93	Neoprene	200	93
HMW PE			Nitrile rubber	x	x
UHMW PE	140	60	PE	100	38
PP	150	66	EU	x	x
PVC Type 1	80	27	T	x	x
PVC Type 2	80	27	AU		
PVC Plastisol	80	27			

Magnesium Chloride

Lining material	°F	°C	Lining material	°F	°C
Glass	250	121	PVDC	180	82
Asphalt			PVDF	280	138
Coal tar			PTFE	450	232
Coal tar epoxy	90	32	Vinyl ester	260	127
Epoxy GP	90	32	Butyl rubber	200	93
Epoxy CR	200	93	CIIR	200	93
Epoxy P	110	43	Hypalon	250	121
Zinc rich	x	x	EPDM	310	154
CPVC	230	116	EPT	180	82
ECTFE	300	149	FKM	400	204
ETFE	300	149	Hard rubber	200	93
FEP	400	204	Soft rubber	140	60
Phenolic			FPM	210	99
PFA	200	93	Neoprene	210	99
HMW PE	140	60	Nitrile rubber	220	104
UHMW PE	140	60	PE	80	27
PP	210	99	EU	130	54
PVC Type 1	140	60	T		
PVC Type 2	140	60	AU	80	27
PVC Plastisol	140	60			

Methylene Chloride

Lining material	Temp.		Lining material	Temp.	
	°F	°C		°F	°C
Glass	250	121	PVDC	x	x
Asphalt			PVDF	120	49
Coal tar			PTFE	450	232
Coal tar epoxy	x	x	Vinyl ester	x	x
Epoxy GP	x	x	Butyl rubber	x	x
Epoxy CR	x	x	CIIR		
Epoxy P	x	x	Hypalon	x	x
Zinc rich	90	32	EPDM	x	x
CPVC	x	x	EPT	x	x
ECTFE	x	x	FKM	x	x
ETFE	210	99	Hard rubber		
FEP	400	204	Soft rubber		
Phenolic			FPM	210	99
PFA	200	93	Neoprene	x	x
HMW PE	x	x	Nitrile rubber	x	x
UHMW PE	x	x	PE	x	x
PP	x	x	EU		
PVC Type 1	x	x	T		
PVC Type 2	x	x	AU	x	x
PVC Plastisol	x	x			

Nitric Acid 20%

Lining material	Temp.		Lining material	Temp.	
	°F	°C		°F	°C
Glass	400	204	PVDC	150	66
Asphalt			PVDF	180	82
Coal tar			PTFE	450	232
Coal tar epoxy	x	x	Vinyl ester	150	66
Epoxy GP	x	x	Butyl rubber	160	71
Epoxy CR	x	x	CIIR	150	66
Epoxy P	x	x	Hypalon	100	38
Zinc rich			EPDM		
CPVC	160	71	EPT	x	x
ECTFE	250	121	FKM	400	204
ETFE	150	66	Hard rubber	x	x
FEP	400	204	Soft rubber	x	x
Phenolic			FPM	210	99
PFA	200	93	Neoprene	x	x
HMW PE	140	60	Nitrile rubber	x	x
UHMW PE	140	60	PE	x	x
PP			EU		
PVC Type 1	140	60	T	x	x
PVC Type 2	140	60	AU	x	x
PVC Plastisol	140	60			

Nitric Acid 50%

Lining material	Temp. °F	Temp. °C	Lining material	Temp. °F	Temp. °C
Glass	270	132	PVDC	120	49
Asphalt			PVDF	180	82
Coal tar			PTFE	450	232
Coal tar epoxy	x	x	Vinyl ester	x	x
Epoxy GP	x	x	Butyl rubber	x	x
Epoxy CR	x	x	CIIR	x	x
Epoxy P	x	x	Hypalon	80	27
Zinc rich			EPDM	x	x
CPVC	180	82	EPT	x	x
ECTFE	150	66	FKM	400	204
ETFE	150	66	Hard rubber	x	x
FEP	400	204	Soft rubber	x	x
Phenolic		\	FPM	210	99
PFA	200	93	Neoprene	x	x
HMW PE	80	27	Nitrile rubber	x	x
UHMW PE	x	x	PE	x	x
PP	150	66	EU		
PVC Type 1	140	60	T	x	x
PVC Type 2	140	60	AU	x	x
PVC Plastisol	140	60			

Nitric Acid 100% (Anhydrous)

Lining material	Temp. °F	Temp. °C	Lining material	Temp. °F	Temp. °C
Glass	270	132	PVDC	x	x
Asphalt			PVDF	150	66
Coal tar			PTFE	450	232
Coal tar epoxy	x	x	Vinyl ester	x	x
Epoxy GP	x	x	Butyl rubber	x	x
Epoxy CR	x	x	CIIR	x	x
Epoxy P	x	x	Hypalon	x	x
Zinc rich			EPDM	x	x
CPVC	x	x	EPT	x	x
ECTFE	150	66	FKM	190	88
ETFE	x	x	Hard rubber	x	x
FEP	400	204	Soft rubber	x	x
Phenolic	80	27	FPM	x	x
PFA	80	27	Neoprene	x	x
HMW PE	x	x	Nitrile rubber	x	x
UHMW PE	x	x	PE	x	x
PP	x	x	EU		
PVC Type 1	x	x	T	x	x
PVC Type 2	x	x	AU	x	x
PVC Plastisol	x	x			

Oleum

Lining material	Temp. °F	Temp. °C	Lining material	Temp. °F	Temp. °C
Glass	400	204	PVDC	x	x
Asphalt			PVDF	x	x
Coal tar			PTFE	450	232
Coal tar epoxy	x	x	Vinyl ester	x	x
Epoxy GP	x	x	Butyl rubber	x	x
Epoxy CR	x	x	CIIR		
Epoxy P	x	x	Hypalon	x	x
Zinc rich			EPDM	x	x
CPVC	x	x	EPT	x	x
ECTFE	x	x	FKM	190	88
ETFE	150	66	Hard rubber	x	x
FEP	400	204	Soft rubber		
Phenolic			FPM	210	99
PFA	80	27	Neoprene	x	x
HMW PE			Nitrile rubber	x	x
UHMW PE			PE	x	x
PP	x	x	EU		
PVC Type 1	x	x	T		
PVC Type 2	x	x	AU	x	x
PVC Plastisol	x	x			

Perchloric Acid 10%

Lining material	Temp. °F	Temp. °C	Lining material	Temp. °F	Temp. °C
Glass	250	121	PVDC	130	54
Asphalt			PVDF	250	121
Coal tar			PTFE	450	232
Coal tar epoxy	x	x	Vinyl ester	150	66
Epoxy GP	x	x	Butyl rubber	150	66
Epoxy CR	x	x	CIIR		
Epoxy P	x	x	Hypalon	90	32
Zinc rich			EPDM	140	60
CPVC	180	82	EPT	190	88
ECTFE	150	66	FKM	400	204
ETFE	230	110	Hard rubber		
FEP	400	204	Soft rubber		
Phenolic			FPM	210	99
PFA	200	93	Neoprene		
HMW PE	140	60	Nitrile rubber	x	x
UHMW PE	140	60	PE	100	38
PP	150	66	EU	x	x
PVC Type 1	140	60	T		
PVC Type 2	x	x	AU	x	x
PVC Plastisol	x	x			

Perchloric Acid 70%

Lining material	Temp. °F	Temp. °C	Lining material	Temp. °F	Temp. °C
Glass	250	121	PVDC	120	49
Asphalt			PVDF	120	49
Coal tar			PTFE	450	232
Coal tar epoxy	x	x	Vinyl ester	x	x
Epoxy GP	x	x	Butyl rubber		
Epoxy CR	x	x	CIIR		
Epoxy P	x	x	Hypalon	90	32
Zinc rich			EPDM		
CPVC	180	82	EPT	140	60
ECTFE	150	66	FKM	400	204
ETFE	150	66	Hard rubber		
FEP	400	204	Soft rubber		
Phenolic			FPM	210	99
PFA	200	93	Neoprene	x	x
HMW PE	x	x	Nitrile rubber	x	x
UHMW PE	x	x	PE		
PP	x	x	EU	x	x
PVC Type 1	x	x	T		
PVC Type 2	x	x	AU	x	x
PVC Plastisol	x	x			

Phenol

Lining material	Temp. °F	Temp. °C	Lining material	Temp. °F	Temp. °C
Glass	200	93	PVDC	x	x
Asphalt			PVDF	200	93
Coal tar			PTFE	450	232
Coal tar epoxy	x	x	Vinyl ester	x	x
Epoxy GP	x	x	Butyl rubber	150	66
Epoxy CR	x	x	CIIR	150	66
Epoxy P	x	x	Hypalon	x	x
Zinc rich			EPDM	x	x
CPVC	140	60	EPT	80	27
ECTFE	150	66	FKM	210	99
ETFE	210	99	Hard rubber	x	x
FEP	400	204	Soft rubber	x	x
Phenolic	x	x	FPM	210	99
PFA			Neoprene	x	x
HMW PE	100	38	Nitrile rubber	x	x
UHMW PE	100	38	PE	x	x
PP	180	82	EU	x	x
PVC Type 1	x	x	T	x	x
PVC Type 2	x	x	AU	x	x
PVC Plastisol	x	x			

Phosphoric Acid 25–50%

Lining material	Temp. °F	Temp. °C	Lining material	Temp. °F	Temp. °C
Glass			PVDC	120	49
Asphalt			PVDF	250	121
Coal tar			PTFE	450	232
Coal tar epoxy	x	x	Vinyl ester	200	93
Epoxy GP	x	x	Butyl rubber	190	38
Epoxy CR	80	27	CIIR	200	93
Epoxy P	x	x	Hypalon	250	121
Zinc rich			EPDM	300	149
CPVC	180	82	EPT	180	82
ECTFE	300	149	FKM	190	88
ETFE	300	149	Hard rubber	190	88
FEP	400	204	Soft rubber	140	60
Phenolic	200	93	FPM	210	99
PFA	200	93	Neoprene	180	82
HMW PE	140	60	Nitrile rubber	x	x
UHMW PE	140	60	PE	100	38
PP	210	99	EU		
PVC Type 1	140	60	T	x	x
PVC Type 2	140	60	AU		
PVC Plastisol	140	60			

Phosphoric Acid 50–85%

Lining material	Temp. °F	Temp. °C	Lining material	Temp. °F	Temp. °C
Glass			PVDC	130	54
Asphalt			PVDF	250	121
Coal tar			PTFE	450	232
Coal tar epoxy	x	x	Vinyl ester	210	99
Epoxy GP	x	x	Butyl rubber	150	66
Epoxy CR	80	27	CIIR	150	66
Epoxy P	x	x	Hypalon	200	93
Zinc rich			EPDM	300	149
CPVC	180	82	EPT	180	82
ECTFE	250	121	FKM	300	149
ETFE	270	132	Hard rubber	200	93
FEP	400	204	Soft rubber	150	66
Phenolic	x	x	FPM	210	99
PFA	200	93	Neoprene 50%	140	60
HMW PE	100	38	Nitrile rubber	x	x
UHMW PE	100	38	PE		
PP	210	99	EU		
PVC Type 1	140	60	T	x	x
PVC Type 2	140	60	AU		
PVC Plastisol	140	60			

Potassium Chloride

Lining material	Temp. °F	Temp. °C	Lining material	Temp. °F	Temp. °C
Glass	250	121	PVDC	120	49
Asphalt			PVDF	280	138
Coal tar			PTFE	450	232
Coal tar epoxy	100	38	Vinyl ester	200	93
Epoxy GP	100	38	Butyl rubber	180	82
Epoxy CR	200	93	CIIR		
Epoxy P	100	38	Hypalon	250	121
Zinc rich	x	x	EPDM	300	149
CPVC	200	93	EPT	180	82
ECTFE	300	149	FKM	400	204
ETFE	300	149	Hard rubber	200	93
FEP	400	204	Soft rubber	140	60
Phenolic			FPM	210	99
PFA	200	93	Neoprene	160	71
HMW PE	140	60	Nitrile rubber	230	110
UHMW PE	140	60	PE	100	38
PP	210	99	EU	110	43
PVC Type 1	140	60	T		
PVC Type 2	140	60	AU	90	32
PVC Plastisol	140	60			

Sodium Hydroxide 50%

Lining material	Temp. °F	Temp. °C	Lining material	Temp. °F	Temp. °C
Glass	x	x	PVDC		
Asphalt			PVDF	150	66
Coal tar			PTFE	450	232
Coal tar epoxy	100	38	Vinyl ester	220	104
Epoxy GP	90	32	Butyl rubber	190	88
Epoxy CR	140	60	CIIR	180	82
Epoxy P	100	38	Hypalon	250	121
Zinc rich			EPDM	300	149
CPVC	180	82	EPT	200	93
ECTFE	250	121	FKM	x	x
ETFE	230	110	Hard rubber	200	93
FEP	400	204	Soft rubber	x	x
Phenolic	x	x	FPM	210	99
PFA	250	121	Neoprene	200	93
HMW PE	150	66	Nitrile rubber	150	66
UHMW PE	170	77	PE	100	38
PP	220	104	EU		
PVC Type 1	140	60	T	x	x
PVC Type 2	140	60	AU	90	32
PVC Plastisol	140	60			

Sodium Hypochlorite 20%

Lining material	Temp. °F	Temp. °C	Lining material	Temp. °F	Temp. °C
Glass	250	121	PVDC 10%	130	54
Asphalt			PVDF	280	138
Coal tar			PTFE	450	232
Coal tar epoxy	x	x	Vinyl ester	180	82
Epoxy GP	x	x	Butyl rubber	130	54
Epoxy CR	x	x	CIIR		
Epoxy P	x	x	Hypalon	250	121
Zinc rich			EPDM	160	71
CPVC	190	88	EPT	x	x
ECTFE	300	149	FKM	400	204
ETFE	300	149	Hard rubber	200	93
FEP	400	204	Soft rubber	x	x
Phenolic	x	x	FPM	210	99
PFA	200	93	Neoprene	x	x
HMW PE	140	60	Nitrile rubber	x	x
UHMW PE	140	60	PE	100	38
PP	120	49	EU	x	x
PVC Type 1	140	60	T	x	x
PVC Type 2	140	60	AU		
PVC Plastisol	140	60			

Sodium Hypochlorite, Concentrated

Lining material	Temp. °F	Temp. °C	Lining material	Temp. °F	Temp. °C
Glass	150	66	PVDC	120	49
Asphalt			PVDF	280	138
Coal tar			PTFE	450	232
Coal tar epoxy	x	x	Vinyl ester	100	38
Epoxy GP	80	27	Butyl rubber	90	32
Epoxy CR	80	27	CIIR		
Epoxy P	x	x	Hypalon		
Zinc rich			EPDM	140	60
CPVC	180	82	EPT	x	x
ECTFE	300	149	FKM	400	204
ETFE	300	149	Hard rubber	100	38
FEP	400	204	Soft rubber	x	x
Phenolic	x	x	FPM	210	99
PFA			Neoprene	x	x
HMW PE	140	60	Nitrile rubber	x	x
UHMW PE	140	60	PE		
PP	110	43	EU	x	x
PVC Type 1	140	60	T	x	x
PVC Type 2	140	60	AU	x	x
PVC Plastisol	140	60			

Sulfuric Acid 10%

Lining material	°F	°C	Lining material	°F	°C
Glass	400	204	PVDC	120	49
Asphalt			PVDF	250	121
Coal tar			PTFE	450	232
Coal tar epoxy	x	x	Vinyl ester	200	93
Epoxy GP	80	27	Butyl rubber	180	82
Epoxy CR	80	27	CIIR	200	93
Epoxy P	x	x	Hypalon	250	121
Zinc rich			EPDM	140	60
CPVC	180	82	EPT	200	93
ECTFE	250	121	FKM	300	149
ETFE	300	149	Hard rubber	200	93
FEP	400	204	Soft rubber	140	60
Phenolic	300	149	FPM	240	116
PFA	250	121	Neoprene	200	93
HMW PE	140	60	Nitrile rubber	140	60
UHMW PE	140	60	PE	90	32
PP	200	93	EU		
PVC Type 1	140	60	T	x	x
PVC Type 2	140	60	AU	x	x
PVC Plastisol	140	60			

Sulfuric Acid 30%

Lining material	°F	°C	Lining material	°F	°C
Glass	400	204	PVDC	80	27
Asphalt			PVDF	220	104
Coal tar			PTFE	450	232
Coal tar epoxy	x	x	Vinyl ester	180	82
Epoxy GP	x	x	Butyl rubber	180	82
Epoxy CR	90	32	CIIR		
Epoxy P	x	x	Hypalon	250	121
Zinc rich			EPDM	140	60
CPVC	180	82	EPT	140	60
ECTFE	250	121	FKM	350	177
ETFE	300	149	Hard rubber	200	93
FEP	400	204	Soft rubber	x	x
Phenolic	300	149	FPM	230	110
PFA	250	121	Neoprene	200	93
HMW PE	140	60	Nitrile rubber	140	60
UHMW PE	140	60	PE	80	27
PP	200	93	EU		
PVC Type 1	140	60	T	x	x
PVC Type 2	140	60	AU	x	x
PVC Plastisol	140	60			

Sulfuric Acid 50%

Lining material	°F	°C	Lining material	°F	°C
Glass	400	204	PVDC	x	x
Asphalt			PVDF	220	104
Coal tar			PTFE	450	232
Coal tar epoxy	x	x	Vinyl ester	210	99
Epoxy GP	x	x	Butyl rubber	150	66
Epoxy CR	80	27	CIIR		
Epoxy P	x	x	Hypalon	250	121
Zinc rich			EPDM	140	60
CPVC	180	82	EPT	210	99
ECTFE	250	121	FKM	350	177
ETFE	300	149	Hard rubber	200	93
FEP	400	204	Soft rubber	x	x
Phenolic	300	149	FPM	250	121
PFA	250	121	Neoprene	200	93
HMW PE	140	60	Nitrile rubber	200	93
UHMW PE	140	60	PE	x	x
PP	200	93	EU		
PVC Type 1	140	60	T	x	x
PVC Type 2	140	60	AU	x	x
PVC Plastisol	140	60			

Sulfuric Acid 90%

Lining material	°F	°C	Lining material	°F	°C
Glass	400	204	PVDC	x	x
Asphalt			PVDF	210	99
Coal tar			PTFE	450	232
Coal tar epoxy	x	x	Vinyl ester	x	x
Epoxy GP	x	x	Butyl rubber	x	x
Epoxy CR	x	x	CIIR	x	x
Epoxy P	x	x	Hypalon	x	x
Zinc rich			EPDM	x	x
CPVC	x	x	EPT	80	27
ECTFE	150	66	FKM	350	177
ETFE	300	149	Hard rubber	x	x
FEP	400	204	Soft rubber	x	x
Phenolic	80	27	FPM	150	66
PFA	250	121	Neoprene	x	x
HMW PE	x	x	Nitrile rubber	x	x
UHMW PE	x	x	PE	x	x
PP	180	82	EU		
PVC Type 1	140	60	T	x	x
PVC Type 2	x	x	AU	x	x
PVC Plastisol	x	x			

Sulfuric Acid 98%

Lining material	Temp. °F	Temp. °C	Lining material	Temp. °F	Temp. °C
Glass	400	204	PVDC	x	x
Asphalt			PVDF	140	60
Coal tar			PTFE	450	232
Coal tar epoxy	x	x	Vinyl ester	x	x
Epoxy GP	x	x	Butyl rubber	x	x
Epoxy GR	x	x	CIIR	x	x
Epoxy P	x	x	Hypalon	110	43
Zinc rich			EPDM	x	x
CPVE	x	x	EPT	x	x
ECTFE	150	66	FKM	390	199
ETFE	300	149	Hard rubber	x	x
FEP	400	204	Soft rubber	x	x
Phenolic	x	x	FPM	210	99
PFA	200	93	Neoprene	x	x
HMW PE	x	x	Nitrile rubber	x	x
UHMW PE	x	x	PE	x	x
PP	120	49	EU		
PVC Type 1	x	x	T	x	x
PVC Type 2	x	x	AU	x	x
PVC Plastisol	x	x			

Sulfurous Acid

Lining material	Temp. °F	Temp. °C	Lining material	Temp. °F	Temp. °C
Glass	230	110	PVDC	80	27
Asphalt			PVDF	450	232
Coal tar			PTFE	450	232
Coal tar epoxy	100	38	Vinyl ester 10%	120	49
Epoxy GP			Butyl rubber	150	66
Epoxy GR	90	32	CIIR	200	93
Epoxy P	110	43	Hypalon	160	71
Zinc rich			EPDM	x	x
CPVE	180	82	EPT	180	82
ECTFE	250	121	FKM	400	204
ETFE	230	110	Hard rubber	200	93
FEP	400	204	Soft rubber	x	x
Phenolic	160	71	FPM	210	99
PFA	210	99	Neoprene	x	x
HMW PE	140	60	Nitrile rubber	x	x
UHMW PE	140	60	PE	x	x
PP	180	82	EU	x	x
PVC Type 1	140	60	T		
PVC Type 2	140	60	AU		
PVC Plastisol	140	60			

10

Introduction to Coatings

Construction metals are selected because of their mechanical properties and machineability at a low price, while at the same time they should be corrosion resistant. Very seldom can these properties be met in one and the same material. This is where coatings come into play. By applying an appropriate coating, the base metal with the good mechanical properties can be utilized while the appropriate coating provides corrosion protection.

The majority of coatings are applied on external surfaces to protect the metal from natural atmospheric corrosion, and atmospheric pollution. On occasion it may also be necessary to provide protection from accidental spills and/or splashes. In some instances coatings are applied internally in vessels for corrosion resistance.

There are basically four different classes of coatings:

Organic	Inorganic	Conversion	Metallic*
Coal tars	Silicates	Anodizing	Galvanizing
Phenolics	Ceramics	Phosphating	Vacuum vapor deposition
Vinyls	Glass	Chromate	Electroplating
Acrylics		Molybdate	Diffusion
Epoxy			
Alkyds			
Urethanes			

* These are processes rather than individual coatings since many metals may be applied by each process. The process and item to be coated will determine which metal will be used.

PRINCIPLES OF CORROSION PROTECTION

Most metals used for construction purposes are unsuitable in the atmosphere. These unstable metals are produced by reducing ores artificially; therefore, they will return to their original ores or to similar metallic compounds when exposed to the atmosphere. For example, metallic iron is oxidized to ferric oxyhydride in a thermodynamically stable state (iron in the higher level of free energy is changed to lepidocrocite, γ-FeOOH, in the lower level):

$$4Fe + 3O_2 + 2H_2O \rightarrow 4FeOOH$$

This reaction of a metal in a natural environment is called corrosion. By means of a coating a longer period of time is required for rust to form on the substrate, as shown in Figure 10.1. Therefore, it is important that the proper coating material be selected for application in the specific environment.

For a coating to be effective, it must isolate the base metal from the environment. The service life of a coating is dependent upon the thickness and the chemical properties of the coating layer. The latter determines the durability of a coating material in a specific environment which is the corrosion resistance of a metal coating or the stability of its organic and inorganic compounds. In order to be effective, the coating's durability must be greater than that of the base metal or it must be maintained by some means. In addition, a coating is often required to protect the base metal with its original pore and crack, or with a defect which may have resulted from mechanical damage and/or pitting corrosion.

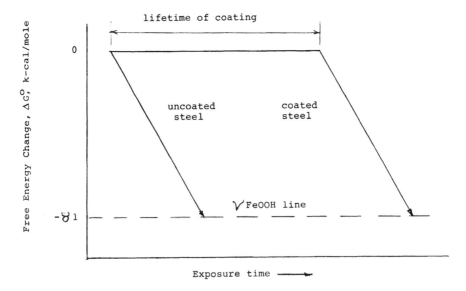

FIGURE 10.1 Role of corrosion resistant coating.

Coatings are classified according to the electrochemical principle upon which they operate to provide protection. These categories are:

1. EMF control
2. Cathodic control protection
3. Anodic control protection
4. Mixed control protection
5. Resistance control protection

The mechanism of the corrosion cell can explain the theories upon which these five categories operate.

CORROSION CELL

A corrosion cell is formed on a metal surface when oxygen and water are present. Refer to Figure 10.2. The electrochemical reactions taking place in the corrosion cell are:

Anodic reaction (M = metal):

$$M \rightarrow M^{n+} + ne$$

(10.1)

Cathodic reaction in acidic solution:

$$2H^+ + 2e \rightarrow H_2$$

(10.2)

Cathodic reaction in neutral and alkaline solutions:

$$O_2 + 2H_2O + 4e \rightarrow 4OH^-$$

(10.3)

The Evans diagram shown in Figure 10.3 represents the mechanism of the corrosion cell. The cathodic current is expressed in the same direction as the anodic current.

In Figure 10.3, E_c shows the single potential for H_2/H^+ or for O_2/OH^- at the cathode and E_a shows the single potential for metal/metal in equilibria at the anode. The single potential is given by the Nernst equation:

$$E = E^\circ + \frac{RT}{nF} \ln a$$

(10.4)

where

E = single potential
E° = standard single potential
R = absolute temperature
n = charge on an ion
F = Faraday constant
a = activity of the ion

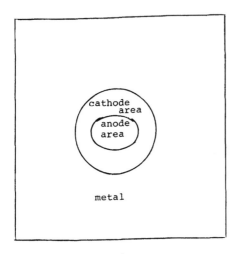

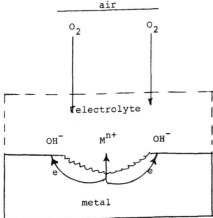

FIGURE 10.2   Structure of a corrosion cell.

When $a = 1$, E is equal to $E°$. The standard single potential $E°$ shows the degree
of activity of the metal and gas.

The electrochemical series consists of the arrangement of metals in order
of electrode potential. The more negative the single potential is, the more active
the metal. Table 10.1 provides the standard single potentials of various metals
and nonmetal reactants.

When the electromotive force $(E_c - E_a)$ is supplied, the corrosion cell is
formed with the current flowing between the anode and the cathode. The cathodic
electrode potential is shifted to the less noble direction. The shifting of potentials

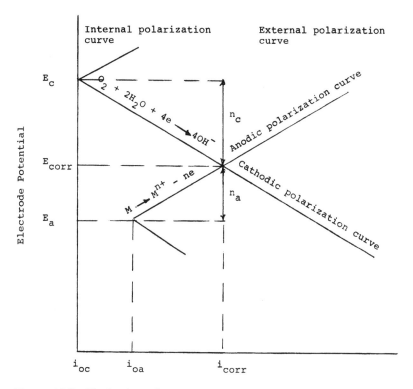

FIGURE 10.3 Mechanism of a corrosion cell.

is called cathodic and anodic polarization. The reaction rate curves $(E - i)$ are known as cathodic or anodic polarization curves. The corrosion potential E_{corr}^c and the corrosion current i_{corr} are given by the intersection of the cathodic and anodic polarization curves—indication that both electrodes react at the same rate in the corrosion process.

The polarization curves in the current density range greater than i_{corr} are called external polarization curves, and those in the current density range less than i_{corr} are called internal polarization curves. By sweeping the electrode potential from the corrosion potential to the cathodic or anodic direction, the external polarization curve can be determined. The internal polarization curve cannot be measured directly by the electrochemical technique since it is impossible to pick up the current separately from the anode and cathode, which exist in the electrode. By analyzing the metallic ions dissolved and the oxidizer reaction, the internal polarization curves can be determined.

Anodic or cathodic overpotential is represented by the difference in potential between E_{corr} and E_a or E_{corr} and E_c and is expressed as n_a or n_c, where

TABLE 10.1 Standard Single Potentials, $E°$ (V, SHE.25°C)

Active		Inert	
Electrode	$E°$	Electrode	$E°$
Li/Li⁺	−3.01	Mo/Mo³⁺	−0.2
Rb/Rb⁺	−0.298	Sn/Sn²⁺	−0.140
Cs/Cs⁺	−0.292	Pb/Pb²⁺	−0.126
K/K⁺	−2.92	H₂/H⁺	±0
Ba/Ba²⁺	−2.92	Bi/BiO	+0.32
Sr/Sr²⁺	−2.89	Cu/Cu²⁺	+0.34
Ca/Ca⁺	−2.84	Rh/Rh²⁺	+0.6
Na/Na⁺	−2.71	Hg/Hg⁺	+0.798
Mg/Mg²⁺	−2.38	Ag/Ag⁺	+0.799
Th/Th⁴⁺	−2.10	Pd/Pd²⁺	+0.83
Ti/Ti²⁺	−1.75	Ir/Ir³⁺	+1.0
Be/Be²⁺	−1.70	Pt/Pt²⁺	+1.2
Al/Al³⁺	−1.66	Au/Au³⁺	+1.42
V/V²⁺	−1.5	Au/Au⁺	+1.7
Mn/Mn²⁺	−1.05		
Zn/Zn²⁺	−0.763	O₂/OH	+0.401
Cr/Cr³⁺	−0.71	I₂/I⁻	+0.536
Fe/Fe²⁺	−0.44	Br₂/Br⁻	+1.066
Cd/Cd	−0.402	Cl₂/Cl⁻	+1.256
In/In³⁺	−0.34	F₂/F⁻	+2.85
Ti/Ti⁺	−0.335	S/S²⁻	−0.51
Co/Co²⁺	−0.27	Se/Se²⁺	−0.78
Ni/Ni²⁺	−0.23	Te/Te²⁺	−0.92

$$n_a = E_{corr} - E_a \qquad n_a > 0 \qquad (10.5)$$

$$n_c = E_{corr} - E_c \qquad n_c < 0 \qquad (10.6)$$

The anodic and cathodic resistance is represented by n_a/i_{corr}.

As soon as the cell circuit is formed, the corrosion reaction starts.

$$E_c - E_a = [n_c] - i_{corr} R \qquad (10.7)$$

where R is the resistance of the electrolyte between the anode and cathode. As the current passes through three processes (the anodic process, the cathodic process, and the transit process in the electrolytes), the electromotive force of a corrosion cell is dissipated.

When the electrode is polarized, the overpotential n is composed of the activation overpotential n^a and the concentration overpotential n^c:

$$n = n^a + n^c \tag{10.8}$$

The activation overpotential n^a results from the potential energy barrier to be overcome for a charge to cross the double layer at the interface ($M = M^{n+} + ne$) and is given as follows:

In the anodic reaction:

$$n_a^a = \beta_a \log \frac{i_a}{i_{oa}} \quad \text{(Tafel equation)} \tag{10.9}$$

$$\beta_a = 2.3 \frac{RT}{\alpha n F} \tag{10.10}$$

In the cathodic reaction:

$$n_c^a = \beta_c \log \frac{i_c}{i_{oc}} \quad \text{(Tafel equation)} \tag{10.11}$$

$$\beta_c = \frac{2.3\,RT}{(1 - \alpha)nF} \tag{10.12}$$

where

$n_a^a =$ activation overpotential in the anodic reaction
$n_c^a =$ activation overpotential in the cathodic reaction
$\beta_a =$ anodic Tafel coefficient
$\beta_c =$ cathodic Tafel coefficient
$\alpha =$ transfer coefficient
$i_a =$ anodic current density
$i_c =$ cathodic current density
$i_{oa} =$ exchange current density of anode
$i_{oc} =$ exchange current density of cathode

The degree of contribution of electrical energy for the activation energy in the electrode reaction ($0 < \alpha < 1$) is indicated by the energy transfer factor α, which in most cases is in the range of 0.3–0.7. The exchange current density i_a or i_c is the flux of charge that passes through the electrical double layer at the single-equilibrium potential E_a or E_c. A linear relationship exists between n^a and $\log i_a$ or $\log i_c$. The Tafel coefficient β_a or β_c is the slope dn^a/d ($\log i_a$ or $\log i_c$) of the polarization curve. Therefore β is one of the important factors controlling the corrosion rate.

The electrode reaction at the low reaction rate is controlled by the activation overpotential. One of the processes controlled by the activation overpotential is the cathodic reaction (in the acid solution):

$$2H^+ + 2e = H_2 \qquad (10.13)$$

Table 10.2 shows the hydrogen overpotentials of various metals. The activation overpotential varies with the kind of metal and the electrolyte condition. Metal dissolution and metal ion deposition are usually controlled by the activation overpotential.

The anodic overpotential is given by

$$n_a = \beta_a \log \frac{i_a}{i_{oa}} \qquad (10.14)$$

At high reaction rates the concentration overpotential n^c becomes the controlling factor in the electrode reaction. In this case the electrode reaction is controlled by mass transfer process, which is the diffusion rate of reactive species. The diffusion current i is given as follows:

$$i = \frac{nFD(C - C_0)}{\delta} \qquad (10.15)$$

TABLE 10.2 Hydrogen Overpotentials of Various Metals

Metal	Temp., °C	Solutions	Hydrogen overpotential $\|n^o\|$, V/mA/ cm^{-2}	Tafel coefficient $\|\beta_c\|$, V	Exchange current density $\|i_{oc}\|$, A/cm^2
Pt smooth	20	1 N HCl	0.00	0.03	10^{-3}
Mo	20	1 N HCl	0.12	0.04	10^{-6}
Au	20	1 N HCl	0.15	0.05	10^{-6}
Ag	20	0.1 N HCl	0.30	0.09	5×10^{-7}
Ni	20	0.1 N HCl	0.31	0.10	8×10^{-7}
Bi	20	1 N HCl	0.40	0.10	10^{-7}
Fe	16	1 N HCl	0.45	0.15	10^{-6}
Cu	20	0.1 N HCl	0.44	0.12	2×10^{-7}
Al	20	2 N H$_2$SO$_4$	0.70	0.10	10^{-10}
Sn	20	1 N HCl	0.75	0.15	10^{-8}
Cd	16	1 N HCl	0.80	0.20	10^{-7}
Zn	20	1 N H$_2$SO$_4$	0.94	0.12	1.6×10^{-11}
Pb	20	0.01–8 N HCl	1.16	0.12	2×10^{13}

where

i = current density
D = diffusion coefficient
C = concentration of reactive species in the bulk solution
C_0 = concentration of reactive species at the interface
δ = thickness of the diffusion layer

When the concentration of the reactive species at the interface is zero, $C_0 = 0$, and the current density reaches a critical value, i_L called the limiting current density:

$$i_L = \frac{nFDC}{\delta} \tag{10.16}$$

From Eqs. (10.15) and (10.16),

$$\frac{C_0}{C} = 1 = \frac{i}{i_L} \tag{10.17}$$

The concentration overpotential is given as follows:

$$n^c = \left[2.3\,\frac{RT}{nF}\right]\log\left[\frac{C_0}{C}\right] \tag{10.18}$$

From Eqs. (10.17) and (10.18),

$$n^c = \left[\frac{2.3\,RT}{nF}\right]\log\left[1 - \frac{i}{i_L}\right] \tag{10.19}$$

As seen in Eq. (10.19), the concentration overpotential increases rapidly as i approaches i_L.

The cathodic reaction is controlled by the activation overpotential n_c^a and the concentration overpotential n_c^c.

The cathodic overpotential is

$$n_c = n_c^a + n_c^c \tag{10.20}$$

The cathodic overpotential n_c can be written in the form:

$$n_c = \beta_c \log\left[\frac{i_c}{i_{oc}}\right] + \frac{2.3RT}{nF}\log\left[1 - \frac{i_c}{i_{cL}}\right] \tag{10.21}$$

See Eqs. (10.12), (10.19), and (10.20). In most cases the corrosion rate can be determined from the anodic and cathodic overpotentials since the rate determining process is determined by the slopes of the two polarization curves.

As mentioned previously, the role of the coating is to isolate the substrate from the environment. The coating accomplishes this based on two characteristics of the coating material: (1) the corrosion resistance of the coating material when the coating is formed by the defect free continuous layer, and (2) the electrochemical action of the coating material when the coating layer has some defect, such as a pore or crack. The mechanism of the corrosion cell can explain the action required of the coating layer. For better understanding Eq. (10.7) is rewritten as follows:

$$i_{corr} = \frac{(E_c - E_a) - |n_c|n_a}{R} \tag{10.22}$$

A corrosion resistant coating is achieved by one of the five different methods to decrease i_{corr} based on Eq. (10.22):

1. *EMF control protection*: Decrease in electromotive force ($E_c - E_a$)
2. *Cathodic control protection*: Increase in cathodic overpotential $|n_c|$
3. *Anodic control protection*: Increase in anodic overpotential $|n_a|$
4. *Mixed control protection*: Increase in both anodic overpotential $|n_a|$ and cathodic overpotential $|n_c|$
5. *Resistance control protection*: Increase in resistance of corrosion cell R

EMF CONTROL PROTECTION

The difference in potential between the anode and the cathode ($E_c - E_a$) is the EMF of the corrosion cell. It is also the degree of thermodynamic instability of the surface metal for the environment. In other words, the less the EMF, the lower the corrosion rate. By covering the surface of the active metal with a continuous layer of a more stable metal, the active metal surface becomes more thermodynamically stable. Dissolved oxygen and hydrogen ions, which are the reactants in the cathodic reaction, are normal oxidizers found in natural environments. In the natural atmosphere the single potential of dissolved oxygen is nearly constant. Because of this, metals with more noble electrode potentials are used as coating materials. These include copper, silver, platinum, gold, and their alloys. A copper coating system provides excellent corrosion resistance under the condition that the defect-free continuous layer covers the surface of the iron substrate. In so doing, the EMF of the iron surface is decreased by the copper coating. The corrosion potential is changed from E_{corr} of uncoated iron to E_{corr} of copper by coating with copper. Under this condition the iron corrodes at the low rate of i_{corr} of copper. However, if the iron substrate is exposed to the environment, as the result of mechanical damage, the substrate is corroded predominantly at the rate of i_{corr} of exposed copper by coupling iron and copper (galvanic corrosion).

Organic coatings and paints are also able to provide EMF control protection. Surface conditions are converted to more stable states by coating with organic compounds. These coatings delay the generation of electromotive force, causing the corrosion of the substrate.

How long an organic coating will be serviceable is dependent upon the durability of the coating itself and its adhesive ability on the base metal. The former is the stability of the coating layer as exposed to various environmental factors, and the latter is determined by the condition of the interface between the organic film and the substrate.

EMF can also be decreased by the use of glass lining, porcelain enameling, and temporary coating with greases and oils.

CATHODIC CONTROL PROTECTION

Cathodic control protection protects the substrate by coating with a less noble metal, for which the slopes of the cathodic polarization curves are steep. The cathodic overpotential of the surface is increased by the coating; therefore, the corrosion potential becomes more negative than that of the substrate. Coating materials used for this purpose are zinc, aluminum, manganese, cadmium, and their alloys. The electrode potential of these metals are more negative than those of iron and steel. When exposed to the environment, these coatings act as sacrificial anodes for the iron and steel substrates.

The protective abilities of coatings are as follows:

1. Original barrier action of coating layer
2. Secondary barrier action of corrosion product layer
3. Galvanic action of coating layer as sacrificial anode

Barrier coatings 1 and 2 predominate as the protective ability even though a sacrificial metal coating is characterized by galvanic action.

Initially the substrate is protected against corrosion by the barrier action of the coating, followed by the barrier action of the corrosion product layer.

Upon exposure to air, aluminum forms a chemically inert Al_2O_3 oxide film that is a rapidly forming self-healing film. Therefore, the passive film on aluminum, as well as the corrosion product layer, is a main barrier and leads to a resistant material in natural environments.

On the other hand, the surface oxide film that forms on zinc is not as an effective barrier as the oxide film on aluminum.

Upon exposure in the natural atmosphere many corrosion cells are formed on the surface of a sacrificial metal coating, thereby accelerating the corrosion rate. During this period corrosion products are gradually formed and converted to a stable layer. This period may last for several months, after which the corrosion rate becomes constant. These corrosion products form the second barrier and are amorphous Al_2O_3 on aluminum, and $Zn(OH)_2$ and basic zinc salts on zinc. ZnO, being electrically conductive, loosens the corrosion product layer and therefore does not contribute to formation of the barrier. When materials such as CO_2, NaCl, and SO_x are present, basic zinc salts are formed, for example, $2ZnCO_3 \cdot 3Zn(OH)_2$ in mild atmospheres, $ZnCl_2 \cdot 6Zn(OH)_2$ in chloride atmospheres, and $ZnSO_4 \cdot 4Zn(OH)_2$ in SO_x atmospheres. How stable each basic zinc salt will be is dependent on the pH and anion concentration of the electrolyte on the zinc. Zinc carbonate forms an effective barrier on steel in mild atmospheres, while basic zinc sulfate and chloride dissolve with decreasing pH of the electro-

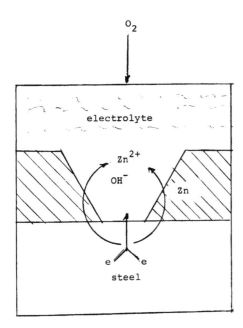

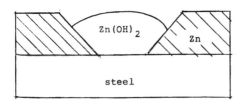

FIGURE 10.4   Schematic illustration of galvanic action sacrificial metal coating.

lyte. The basic zinc sulfate is restricted under atmospheric conditions in its effort to act as a barrier since the pH value of rain in an SO_x atmosphere is usually low, in the area of less than 5. In a chloride environment the value of pH in the electrolyte is not as low as in the SO_x atmosphere; therefore a secondary barrier will form. However, in a severe chloride environment the zinc coating layer will corrode in spite of the existence of basic zinc chloride on the surface.

Galvanic Action of Coating Layer

Sacrificial metal coatings protect the substrate metal by means of galvanic action. When the base metal is exposed to the atmosphere as a result of mechanical damage or the like, the exposed portion of the base metal is polarized cathodically to the corrosion potential of the coating layer. As a result, little corrosion takes place on the exposed metal. A galvanic couple is formed between the exposed part of the base metal and the surrounding coating metal. Sacrificial metals are more negative in electrochemical potential than other metals such as iron or steel. Therefore, the sacrificial metal coating acts as a cathode. This type of reaction of sacrificial metal coatings is known as galvanic or cathodic protection. In addition, the defects are protected by a second barrier of corrosion products of the coating layer. Figure 10.4 schematically illustrates the galvanic action of a sacrificial metal coating.

ANODIC CONTROL PROTECTION

Noble metal coatings provide anodic control protection. They are usually used where corrosion protection and decorative appearance are required. Nickel, chromium, tin, lead, and their alloys are the coating metals that provide anodic protection.

Single Layer Coatings

Single layer metal coatings provide corrosion protection as a result of the original barrier action of the noble metal. With the exception of lead a second barrier of corrosion products is not formed. Noble metals do not provide cathodic protection to steel substances in natural atmospheres because the corrosion potential of the noble metal is more noble than that of the steel. Refer to Table 10.3.

The service life of a single layer coating is affected by any discontinuity in the coating, such as that caused by pores and cracks. For metals to form a protective barrier, the coating thickness must be greater than 30 μm to ensure absence of defects. The surface of a bright nickel coating will remain bright in a clean atmosphere but will change to a dull color when exposed in an SO_x atmosphere.

TABLE 10.3 Corrosion Potentials of Noble Metals

	Corrosion potential (V, SCE)	
pH	2.9	6.5
Chromium	−0.119	−0.186
Nickel	−0.412	−0.430
Tin	−0.486	−0.554
Lead	−0.435	−0.637
Steel	−0.636	−0.505

Chromium coatings are applied as a thin layer to maintain a bright, tarnish-free surface. Cracking of chromium coatings begins at a thickness of 0.5 µm after which a network of fine cracks is formed.

For protection of steel in an SO_x atmosphere, lead and its alloys (5 to 10% tin) coatings are employed. Pitting will occur in the lead coating at the time of

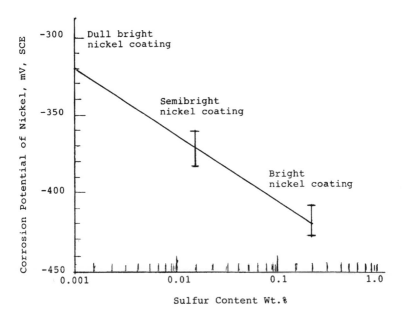

FIGURE 10.5 Effect of sulfur content on the corrosion potential of nickel deposit. (*From Ref. 1.*)

initial exposure, but the pits are self-healing and then the lead surface is protected by the formation of insoluble lead sulfate.

Multilayer Coatings

There are three types of nickel coatings: bright nickel, semibright, and dull. The difference is the quantity of sulfur contained in them as shown below:

Bright nickel deposits $> 0.04\%$ sulfur
Semibright nickel deposits $< 0.005\%$ sulfur
Dull nickel deposits $< 0.001\%$ sulfur

The corrosion potentials of the nickel deposits are dependent on the sulfur content. Figure 10.5 shows the effect of sulfur content on the corrosion potential of a nickel deposit.

As the sulfur content increases, the corrosion potential of a nickel deposit becomes more negative. A bright nickel coating is less protective than a semibright or dull nickel coating. The difference in potential of bright nickel and semibright deposits is more than 50 mV.

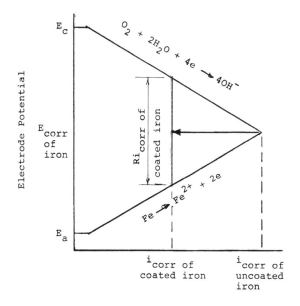

FIGURE 10.6 Resistance control protection. (*From Ref. 1.*)

Use is made in the differences in the potential in the application of multilayer coatings. The more negative bright nickel deposits are used as sacrificial intermediate layers. When bright nickel is used as an intermediate layer, the corrosion behavior is characterized by a sideways diversion. Pitting corrosion is diverted laterally when it reaches the more noble semibright deposit. Thus the behavior of bright nickel prolongs the time for pitting penetration to reach the base metal.

The most negative of all nickel deposits is trinickel. In the triplex layer coating system a coating of trinickel approximately 1 μm thick, containing 0.1–0.25% sulfur is applied between bright nickel and semibright nickel deposits. The high sulfur nickel layer dissolves preferentially, even when pitting corrosion reaches the surface of the semibright deposit. Since the high sulfur layer reacts with the bright nickel layer, pitting corrosion does not penetrate the high sulfur nickel layer in the tunneling form. The application of a high sulfur nickel strike definitely improves the protective ability of a multistage nickel coating.

RESISTANCE CONTROL PROTECTION

Resistance control protection is achieved by using organic compounds, such as some paints, as coating materials. The coating layer delays the transit of ions to the substrate, thereby inhibiting the formation of corrosion cells. Figure 10.6 illustrates the principles of resistance control protection by an organic coating. The corrosion rate of iron is inhibited by the coating from i_{corr} of uncoated iron to i_{corr} of coated iron.

REFERENCE

1. I. Suzuki, *Corrosion Resistant Coatings Technology*, Marcel Dekker, New York, 1989.

11

Principles of Coating

An understanding of the basic principles that describe and predict liquid flow and interfacial interactions is necessary for the effective formulation and the efficient application of coatings. The two primary sciences of liquid flow and solid-liquid interaction are rheology and surface chemistry. Rheology deals with the science of flow and deformation while surface chemistry deals with the science of wetting and dewetting. The key rheological property of coatings is viscosity, which is simply the resistance of a coating to flow, the ratio of shear stress to shear rate.

During the application of a coating, mechanical forces of various types are exerted. The amount of shear force directly affects the viscosity value for non-Newtonian fluids. Most coatings are subject to some degree of "shear thinning" when worked by mixing. As shear rate is increased, the viscosity drops, in some cases dramatically.

RHEOLOGY

As indicated, viscosity, the resistance to flow, is the key property describing the behavior of liquids subjected to forces such as mixing. Viscosity is simply the ratio of shear stress to the shear rate:

$$\eta = \frac{\text{shear stress}}{\text{shear rate}} = \frac{r}{D} \qquad \text{(dynes-sec/cm}^2) \qquad (1)$$

281

TABLE 11.1 Viscosities of Common
Industrial Liquids*

Liquid	Viscosity, cP
Acetone	0.32
Chloroform	0.58
Toluene	0.59
Water (20.20°C)	1.000
Cyclohexane	1.0
Ethyl alcohol	1.2
Turpentine	1.5
Mercury, metal	1.6
Creosote	12.0
Sulfuric acid	25.4
Linseed oil	33.1
Olive oil	84.0
Castor oil	986.0
Glycerine	1490.0

* Values are for approximately 20°C.

The viscosity unit dynes-seconds per square centimeter, or Poise, is a rather small unit for low viscosity fluids such as water (approximately 0.01 P). Therefore, the more common centipoise (0.01 P) is used. Since 100 cP = 1 P, water has a viscosity of approximately 1 cP. Table 11.1 lists the viscosities of some common industrial liquids. A high viscosity liquid requires considerable force (work) to produce a change in shape. For example, high viscosity coatings are more difficult to pump than low viscosity coatings. High viscosity coatings also take longer to flow out when applied.

   Thin or low viscosity liquids flow easily while high viscosity liquids move with considerable resistance. In the ideal or Newtonian case, viscosity is constant over any region of shear. However, very few liquids are truly Newtonian. Most liquids drop in viscosity as shear work is applied. This phenomena is known as shear thinning. A liquid can be affected by the amount of time that force is applied. A shear thinned liquid will tend to return to its initial viscosity over a period of time. Therefore, if viscosity is to be reported accurately, the time under shearing action and the time at rest must also be noted.

VISCOSITY BEHAVIOR

The effect on the viscosity of a fluid varies from fluid to fluid as force is applied. These different effects are described below.

TABLE 11.2 Surface Tensions of Liquids and Polymers

	Surface tension, dynes/cm
Liquid	
Heptane	22.1
Methanol	24.0
Acetone	26.3
Ethylene glycol	48.4
Glycerol	63.1
Water	72.8
Polymer	
Polyethylene	31
Polyvinyl chloride	40
Polyethylene terephthalate (polyester)	43
Nylon	46

Plasticity

Plastic fluids behave more like plastic solids until a specific minimum force is applied to overcome the yield point. Gels and ketchup are extreme examples. Once the yield point has been reached, the liquids begin to approach Newtonian behavior as shear rate is increased. Although plastic behavior is of no benefit to ketchup, it has some benefit in paints. Actually, it is the yield point phenomenon that is of practical value as illustrated by no-drip paints. When the brush stroke force has been removed the paint's viscosity builds quickly until the flow stops. Dripping is prevented because the yield point exceeds the force of gravity.

Pseudoplasticity

The viscosity of pseudoplastic fluids drop as force is applied. However, there is no yield point. The more energy applied, the more the thinning. When shear rate is reduced, the viscosity increases at the same rate by which the force is diminished. There is no hysteresis; the shear stress–shear rate curve is the same in both directions, as shown in Figure 11.1. Figure 11.2 compares pseudoplastic behavior using viscosity–shear rate curves.

Many coatings exhibit this type of behavior, but with time dependency. There is a pronounced delay in viscosity increase after the force has been removed. This form of pseudoplasticity with a hysteresis loop is called thixotropy. Pseudoplasticity is useful in coatings, but thixotropy is more useful.

Thixotropy

Some coatings take advantage of thixotropic behavior to overcome the problem of having a sufficiently low viscosity that levels well and a sufficiently high

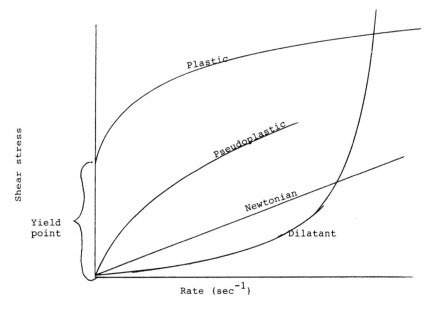

FIGURE 11.1   Shear stress–shear rate curves.

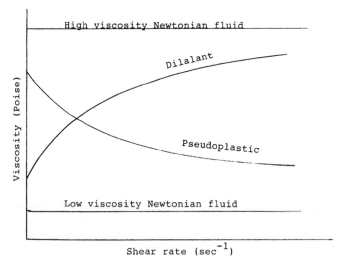

FIGURE 11.2   Viscosity–shear rate curves.

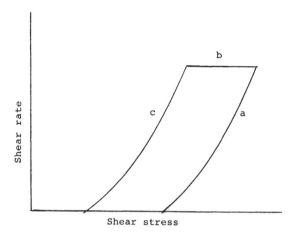

FIGURE 11.3   Thixotropic loop.

viscosity that does not sag. Thixotropy is the dependence of viscosity on time. These coatings retain a low viscosity for a short time after shearing, thus permitting good leveling, but thicken fast enough to prevent sagging.

This thixotropic behavior of a coating is shown in Figure 11.3. The coating is initially sheared at an increasing shear rate, producing curve *a*. Then the coating is sheared at a constant shear rate until constant viscosity (curve *b*) is reached. The shear rate is then gradually reduced, producing curve *c*. The degree of thixotropy is indicated by the enclosed area of the thixotropic loop. Dripless paints owe their driplessness to thixotropy. The paint begins as a moderately viscous material that stays on the brush. It quickly drops in viscosity under the stress of brushing for easy smooth application. A return to higher viscosity when shearing (brushing) stops prevents dripping and sagging.

Dilatancy

Dilatants are liquids whose viscosity increases as shear is applied. Very few liquids possess this property. This property should not be confused with the increase in viscosity resulting from the loss of solvent. True dilatancy takes place without solvent loss.

Changes After Application

The viscosity of a fluid coating starts to increase after it has been applied to a substrate. Several factors are responsible for this increase, as illustrated in Figure

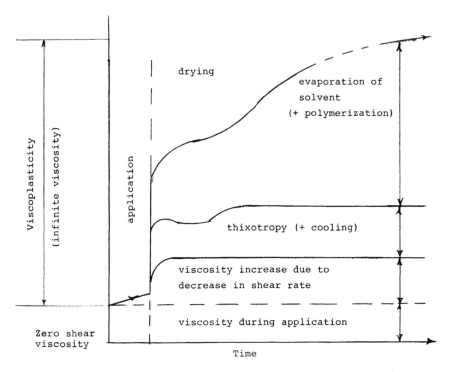

FIGURE 11.4   Change in coating viscosity during application and film formation.

11.4. The curves shown in Figure 11.4 are typical for a coating formulation with a low solid content. Coatings with a high solid content and powder coatings will have curves of different relative magnitudes. The principal increase in viscosity of powder coatings will be due to freezing, as the temperature approaches the melting point.

   As the viscosity increases with time, various coating phenomena are abated. Leveling and sagging can only take place as long as the coating is fluid. As the viscosity increases, these phenomena can no longer take place. The measured time dependency on the viscosity is used to estimate the time taken to solidify. In general, when the viscosity is higher than 100,000 P, leveling and sagging occur to a negligible extent.

Effect of Temperature

Viscosity is extremely sensitive to temperature changes. All comparative measurements should be taken at the same temperature (usually 73.4°F/23°C). A viscosity value without a temperature notation is useless.

Each fluid is affected differently by a temperature change, but the change per degree is usually constant for a specific liquid. In general, a coating's viscosity may be reduced by an increase in temperature and increased by a reduction in temperature.

Effect of Solvents

Higher solution viscosity results from higher resin solids, whereas an increase in solvent volume reduces viscosity. Soluble resins (polymers) produce more pronounced viscosity changes than do insoluble pigments or plastic particles. A plastisol suspension (plastic particles in a liquid plasticizer) may have a medium viscosity at 80% solids, whereas a coating may be highly viscous with a 50% solid concentration. The specific solvent will also have an effect on the viscosity, depending whether they are true solvents, latent solvents, or nonsolvents. Refer to Refs. 1 and 2 for more detail.

Yield Value

The yield value is the shear stress in a viscosity measurement, but one taken at a very low shear. It is the minimum shear stress applied to a liquid that produces flow. When the yield value is greater than shear stress, flow will not take place.

A liquid undergoes deformation without flowing as force is gradually applied. The liquid for all intents and purposes is acting as an elastic solid. Viscosity approaches infinity below the yield value. At a critical force input (the yield value) flow starts.

When the yield value is greater than the shear stress, the liquid behaves as if it were a solid. If a coating is applied at this time, what you deposit is what you get. Leveling will not occur. Coatings that cannot be leveled, even though the apparent viscosity is low, probably have a high yield value. If this is the case, the only solution may be to change the method of application.

SURFACE CHEMISTRY

This is the science that deals with the interface of two materials. The interface may exist between any forms of matter, including a gas phase. However, for the purpose of understanding the interfacial reactions of coating materials, it is only necessary to analyze the liquid-solid interaction. The effect of surface interaction between a liquid coating and the surrounding air is small and may be ignored.

SURFACE TENSION

Surface tension is one of the factors that determines the ability of a coating to wet and adhere to the substrate.

All liquids are composed of molecules, which when close to one another exert attractive forces. It is these mutual attractions that produce the property called surface tension. The units are dynes per centimeter (force per unit length).

When a drop of liquid is suspended in space, it assumes a spherical shape. Since surface molecules are pulled toward those directly beneath them, a minimum surface area (sphere) results. All liquids attempt to form a minimum surface sphere. When a liquid is placed on a solid, a liquid-solid interface develops. Liquid molecules are attracted not only to each other (intramoleculr attraction) but to any solid surface (intermolecular attraction) with which they come into contact. These two interactions are the only ones which must be considered in coating operations.

WETTING

The ability of a liquid to wet a surface is related to its surface tension. By using solvents with lower surface tensions, the ability of a coating to wet a substrate can be improved. When placed on a flat horizontal surface, a liquid will either wet and flow out, or it will dewet to form a semispherical drop. It is also possible for an in-between state to occur in which the liquid neither recedes nor advances but remains stationary.

The angle that the droplet or edge of the liquid makes with the solid substrate is called the contact angle θ. The smaller the contact angle, the better the wetting. Refer to Figure 11.5. A wetting condition takes place when the contact angle is $\theta°$. The liquid's edge continues to advance, even though the rate may be slow for high viscosity materials.

Various interfacial tensions determine the equilibrium contact angle for a liquid drop sitting on an ideally smooth, homogeneous, flat, and nondeformable surface. They are related by Young's equation:

$$\gamma_{LV} \cos \theta = \gamma_{sV} - \gamma_{sL} \tag{2}$$

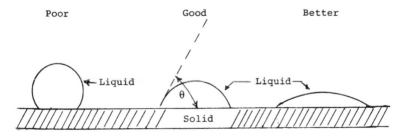

FIGURE 11.5   Schematic illustration of good and poor wetting.

where

> γ_{LV} = surface tension of liquid
> γ_{sv} = surface tension of solid in equilibrium with the saturated vapor of the liquid
> γ_{sL} = interfacial tension between the solid and liquid

From Eq. (2) it can be seen that for spontaneous wetting to occur, the surface tension of the liquid must be greater than the surface tension of the solid. With the application of force it is also possible for a liquid to spread and wet a solid when θ is greater than zero.

COALESCENCE

Coalescence is the fusing of molten particles to form a continuous film, the first step in powder coating. Surface tension, radius of curvature, and viscosity of the molten powder control coalescence. In order to have more time available for leveling, it is desirable to have small particles, low viscosity, and low surface tension.

SAGGING AND SLUMPING

When coatings are applied to inclined or vertical surfaces, it is possible for the coating material to flow downward (under the influence of gravity), which leads to sagging and slumping. Newtonian or shear thinning fluids tend to sag as a result of shear flow. A material with a yield stress will slump.

The velocity (V_0) of the material in flow at the fluid-air interface and the resulting sag or slump length (S) can be calculated for a fluid of index n:

$$V_0 = \left[e \frac{g}{\eta^\circ} \right]^{1/n} \frac{n}{n+1} h^{(n+1)/n} \quad \text{and} \quad S = V_0 t \tag{3}$$

where n° is the zero shear viscosity and h is the film thickness. The special case of Newtonian fluids is obtained by setting $n = 1$ in Eq. (3). The final sag or slump length S is determined by the velocity as well as a time factor t, which is the time interval for which the material remains fluid, or the time it takes for the material to solidify. When everything else is equal, a shear thinning fluid ($n < 1$) will exhibit lower sag/slump velocities. Therefore, a Newtonian or a shear thinning fluid will sag or slump under its own weight until its viscosity increases to the point at which V_0 is negligible. If the material has a yield stress, no sagging will occur if the yield stress σ_y is larger than the force of gravity, pgh. However, if the coating is thick enough (large h), both sagging and slumping can occur if the film thickness is larger than h_s which is given by

$$h_s = \frac{\sigma_y}{Q_g} \tag{4}$$

Between $h = 0$ and $h = h_s$, sagging takes place. The velocity can be determined by substituting $h - h_s$ for h in Eq. (3).

$$V_0 = \left(\frac{Q_g}{\eta_0}\right)^{1/n} \frac{n}{n+1} (h - h_s)^{(n+1)/n} \tag{5}$$

For $h > h_s$ plug flow occurs.

Good sag control and good sprayability of coatings can be maintained with a shear thinning fluid without a yield stress if it has an n value of 0.6.

SURFACTANTS

Surfactants are also known as wetting agents. They are used to lower the surface tension of coatings and paints. Normally, a reduction of 1% or less is sufficient.

Coatings and paints, once modified with a surfactant, are usually permanently changed. This can make them difficult to wet if it is necessary to apply additional coats. The problem can be overcome in several ways. It is desirable to use the smallest amount of the least potent surfactant that will do the job. (Start with the hydrocarbon class.) Also check to see that the substrate is clean before starting.

An alternate possibility is to use a reactive surfactant. These are agents that can react with the coating or binder rendering them less active after curing. Another approach would be to add surfactant to the second material to be applied.

LEVELING

During the coating application imperfections such as waves or furrows appear on the surface. These imperfections must be removed before the wet coating solidifies. Leveling is the critical step to achieve a smooth and uniform coating.

The factors affecting leveling are viscosity, surface tension, yield value, coating thickness, and the degree of wet coating irregularity. These factors are correlated in the leveling equation:

$$a_t = a_0 \frac{\exp(\text{const } \sigma h^3 t)}{3\lambda^4 \eta} \tag{6}$$

where

a_t = height of coating ridge
σ = surface tension of coating
η = coating viscosity
h = coating thickness or height

t = time for leveling
λ = wavelength or distance between ridges

From Eq. (6) we see that leveling is improved by one or more of the following:

1. Longer time t
2. Higher surface tension of coating σ
3. Lower viscosity η
4. Greater coating thickness h
5. Small repeating distance between ridges λ

Since h is raised to the third power, doubling the coating thickness provides an eightfold improvement in leveling. Note that λ (wavelength between ridges) is raised to the fourth power, which indicates that ridges which are far apart create a difficult leveling situation.

ADHESION

Regardless of what excellent properties a coating may possess, it is useless unless it also has good adhesion. The coating's resistance to weather, chemicals, scratches, or impact is only of value while the coating remains on the substrate.

The two most common types of stress under which the bond between a coating and substrate may be placed are tensile and shear stress. Tensile stress is effective perpendicularly to the substrate surface, while shear stress acts along the plane of contact. These stresses may be induced mechanically or thermally. Any difference in the coefficient of expansion between the substrate and coating can result in the paint film becoming detached when the substrate and coating undergo rapid temperature changes.

Failure may also result from chemicals penetrating through the coating, becoming absorbed at the interface causing loss of adhesion. Figure 11.6 illustrates these various interface conditions.

Means have been developed to measure adhesion strength, expressed as force per unit area, and specified as either tensile or shear stress. Consequently, this is the result reported in the various tests for adhesion strength.

Cross-cut Test

In order to determine the adhesion quality of a coating, a lattice pattern is cut into it, penetrating to the substrate. A multiple cutting tool consisting of a set of six "knives" 1 or 2 mm apart is used to yield a uniform pattern. The results are evaluated as shown in Figure 11.7.

It is recommended that this test be considered a "go or no go" result. In this case class "0" would indicate perfect adhesion, whereas even class 1 or 2 would be interpreted as an objectional result.

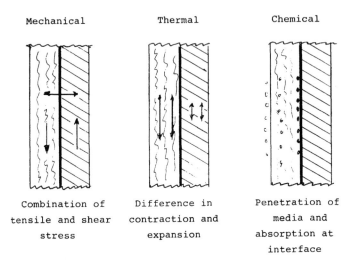

Mechanical	Thermal	Chemical
Combination of tensile and shear stress	Difference in contraction and expansion	Penetration of media and absorption at interface

FIGURE 11.6   Interface condition of polymer layer and substrate.

Tensile Methods

The pull-off method has been standardized internationally. This test utilizes stress patterns caused by loads acting either normal or parallel to the plane of contact. A stud, normally made of steel, is glued with the coating and is subjected to axial tension until separation of the paint film occurs. This determines the maximum tensile stress that is possible at the interface.

When torque is applied about the axis of the stud, the process of separation determines the maximum shear stress that can be attained at the interface, which leads to a measure of adhesion. Refer to Figure 11.8.

Blister Method

Usually the first indication that a coating designed for protection against corrosion is deteriorating is when a blister forms. The following test has been developed to investigate this type of failure.

A hole is bored in the substrate prior to applying the paint coating. The hole is plugged with a nonwetting material such as Teflon. This will permit easy removal of the plug after the substrate has been painted. Refer to Figure 11.9.

Hydrostatic pressure is applied in the hole either with a fluid (oil, mercury, etc.) or with pressurized air. The pressure is the primary measure of the debonding process. The height and diameter of the blister are measured in order to obtain the maximum stress or bonding energy (work of adhesion). From these geometrical data, together with the tensile modulus of the film and its thickness, a critical

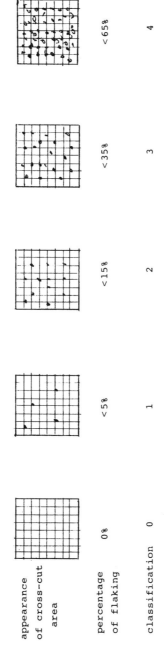

appearance
of cross-cut
area

percentage of flaking	0%	<5%	<15%	<35%	<65%
classification	0	1	2	3	4

FIGURE 11.7   Cross-cut test paint film classification.

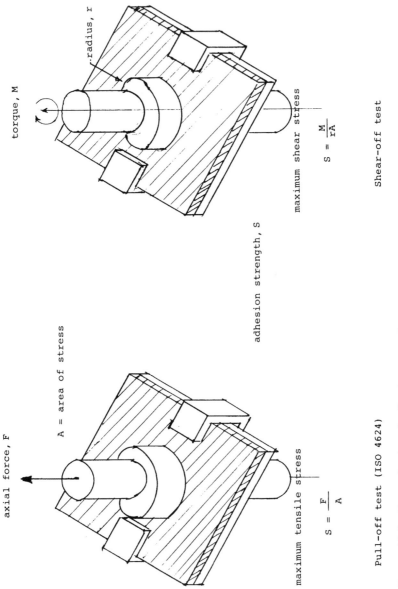

Figure 11.8 Determination of adhesion strength.

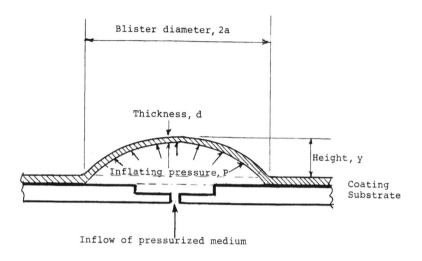

Work of detachment   W = 0.65Pd

$$P = 4.75E \, \frac{dy^3}{a^4}$$

E = Tensile modulus

FIGURE 11.9 Test for measuring adhesion based on blister dimensions.

pressure value can be calculated. Pressure can be used as the basis for determining adhesion strength since it is pressure that causes the blister to grow.

Indentation Debonding

When a needlelike indenter is pressed perpendicularly into the surface of a coating that is bonded to a virtually undeformable substrate, most of the deformation will take place within the film, but there will also be some debonding at the interface.

Referring to Figure 11.10, it is noted that a peeling moment may be calculated which will serve as a measure of the film's ability to withstand delamination in the area of the indentation site.

Optical devices can be used to monitor the gradually increasing area of debonding, especially on thin coatings, on the basis of Newton's rings. Indenters of other typical shapes, besides needlelike, have been used successfully. A 60° angle cone has been proven to be optimum for taking into account boundary conditions at the interface. A particular advantage of the indentation test is that

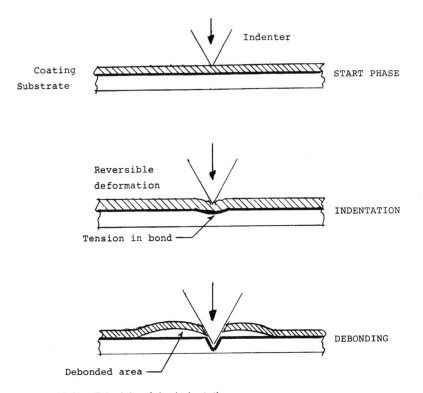

FIGURE 11.10   Principle of the indentation process.

it yields values for bond strengths in absolute terms, as well as information about the durability of the bond between coating and substrate.

Impact Tests

An impact test is used to determine the stonechip resistance of a coating. The value of adhesion at the interface is of primary concern. A steel ball impinging on the test piece can duplicate the situation encountered in actual practice. See Figure 11.11.

As a first approximation the transfer of forces through the film is equal to the case of static loading, and can be calculated in essentially the same way as for an indentation test.

In the debonding area negative tensile (compressive) stress is present in the center of the detachment site and shear stress in the annular region. The maximum diameter of the debonding area can serve as a measure of adhesion at the interface. The diameter and, better yet, the area of the debonding zone can

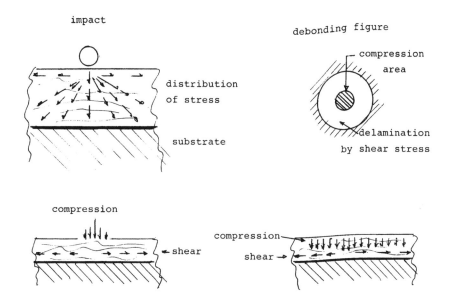

Influence of film thickness

FIGURE 11.11   Impact effect on interface adhesion.

both serve as (reciprocal) measures of adhesion. An extended debonded site is an indication of a low adhesion level.

DETECTION OF ADHESION DEFECTS

Once a coating has been applied and throughout its lifetime, it is necessary to determine any deterioration of the bonding strength between the film and substrate. However, quantitative details of the bonding strength are not required. These tests provide a means of detecting the first signs of adhesion failure.

Ultrasonic Pulse-Echo System

Referring to Figure 11.12, it is seen that an incident of ultrasound will be partially reflected and transmitted at each interface of the test piece, including the backing. It is the pulse partially transmitted at the interface that undergoes more or less total reflection at this free surface.

When the bonding at the interface is intact, the amplitude of the reflected pulse there will be fairly low, in contrast to the amplitude of the transmitted pulse that travels through the substrate and is reflected at its free boundary. If a defect

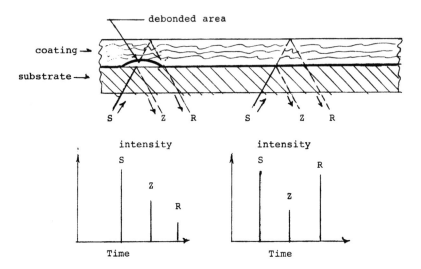

S = start impulse   Z = interface echo   R = rear echo

FIGURE 11.12   Ultrasonic pulse-echo technique.

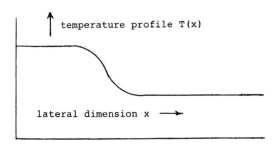

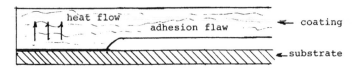

FIGURE 11.13   Thermographic detection of debonding areas.

is present at the interface containing air, or indicating another way that the joint has been disbonded, the amplitude of the related untrasonic pulse will show an increase. This increase is due to the very low accoustic impedance at the site.

Thermographic Detection

Heat flux passing through a coated substrate will provide a uniform temperature on the surface providing there is no flaw or debonded area of the coating. When a flaw or debonded site is present the flow of heat flux is interrupted and a decrease in temperature is detected on the surface. The actual shape of the area of which the temperature decreases is made visible at the surface by an infrared sensor. This can be seen in Figure 11.13.

REFERENCES

1. Temple C. Patton, *Paint Flow and Pigment Dispersion*, 2nd ed., New York, Wiley, 1979.
2. Charles R. Martens, *Technology of Paint, Varnish and Lacquers*, Krieger Publishing Co., New York, 1974.

12

Corrosion Protection
by Organic Coatings

Organic coatings are widely used to protect metal surfaces from corrosion. The effectiveness of such coatings is dependent not only on the properties of the coatings, which are related to the polymeric network and possible flaws in this network, but also on the character of the metal substrate, the surface pretreatment, and the application procedures. Therefore, when considering the application of a coating, it is necessary to take into account the properties of the entire system.

There are three broad classes of polymeric coatings: lacquers, varnishes, and paints. Varnishes are materials that are solutions of either a resin alone in a solvent (spirit varnishes), or an oil and resin together in a solvent (oleo-resinous varnishes). A lacquer is generally considered to be a material whose basic film former is nitrocellulose, cellulose acetate-butyrate, ethyl cellulose, acrylic resin, or another resin that dries by solvent evaporation. The term ''paint'' is applied to more complex formulations of a liquid mixture that dries or hardens to form a protective coating.

Organic coatings provide protection either by the formation of a barrier action from the layer or from active corrosion inhibition provided by pigments in the coating. In actual practice the barrier properties are limited since all organic coatings are permeable to water and oxygen to some extent. The average transmission rate of water through a coating is about 10 to 100 times larger than the water consumption rate of a freely flowing surface, and in normal outdoor conditions an organic coating is saturated with water at least half of its service life.

TABLE 12.1 Diffusion Data for Water Through Organic Films

Polymer	Temp. °C	$p \times 10^9$ [cm³(STP)cm]	$D \times 10^9$ cm²/s
Epoxy	25	10–44	2–8
	40	—	5
Phenolic	25	166	0.2–10
Polyethylene (low density)	25	9	230
Polymethyl methacrylate	50	250	130
Polyisobutylene	30	7–22	—
Polystyrene	25	97	—
Polyvinyl acetate	40	600	150
Polyvinyl chloride	30	13	16
Vinylidene chloride/acrylonitrile copolymer	25	1.7	0.32

Source: Ref. 1.

For the remainder of the time it contains a quantity of water comparable in its behavior to an atmosphere of high humidity. Table 12.1 shows the diffusion data for water through organic films.

It has also been determined that in most cases the diffusion of oxygen through the coating is large enough to allow unlimited corrosion. Taking these factors into account indicates that the physical barrier properties alone do not account for the protective action of coatings. Table 12.2 shows the flux of oxygen through representative free films of paint 100 μm thick.

TABLE 12.2 Flux of Oxygen Through Representative Free Films of Paint, 100 μm Thick

Paint	J, mg/cm² day
Alkyd (15% PVC Fe_2O_3)	0.0069
Alkyd (35% PVC Fe_2O_3)	0.0081
Alkyl-melamine	0.001
Chlorinated rubber (35% PVC Fe_2O_3)	0.017
Cellulose acetate	0.026 (95% RH)
Cellulose nitrate	0.115 (95% RH)
Epoxy melamine	0.008
Epoxy coal-tar	0.0041
Epoxy polyamide (35% PVC Fe_2O_3)	0.0064
Vinyl chloride/vinyl acetate copolymer	0.004 (95% RH)

Source: Ref. 1.

Additional protection may be supplied by resistance inhibition, which is also a part of the barrier mechanism. Retardation of the corrosion action is accomplished by inhibiting the charge transport between cathodic and anodic sites. The reaction rate may be reduced by an increase in the electrical resistance and/or the ionic resistance in the corrosion cycle. Applying an organic coating on a metal surface increases the ionic resistance. The electronic resistance may be increased by the formation of an oxide film on the metal. This is the case for aluminum substrates.

Corrosion of a substrate beneath an organic coating is an electrochemical process which follows the same principles as corrosion of an uncoated substrate. It differs from crevice corrosion since the reactants often reach the substrate through a solid. In addition, during the early stages of corrosion small volumes of liquid are present, resulting in extreme values of pH and ion concentrations. The total corrosion process takes place as follows:

1. Migration through the coating of water, oxygen, and ions
2. Development of an aqueous phase at the coating/substrate interface
3. Activation of the substrate surface for the anodic and cathodic reactions
4. Deterioration of the coating/substrate interfacial bond

COMPOSITION OF ORGANIC COATINGS

The composition of a coating (paint) determines the degree of corrosion protection that will be supplied. A paint formulation is made up of four general classes of ingredients: vehicle (binder, resin component), pigment, filler, and additive. It is the combination of these ingredients that impart the protective properties to the coating.

Resin Component

The resin component is also known as the vehicle or binder, but "resin component" is more descriptive. The resin is the film-forming ingredient of the paint. It forms the matrix of the coating, the continuous polymeric phase in which all other components may be incorporated. Its density and composition largely determine the permeability, corrosion resistance, and UV resistance of the coating.

The most common resins used are vinyls, chlorinated rubber, acrylics, epoxies, urethanes, polyester, autooxidative cross-linking coatings, and water soluble resins.

Vinyl Resins

Vinyl is the general term denoting any compound containing the vinyl linkage ($-CH=CH_2$) group. However, there are many compounds containing this linkage that are not considered as vinyl coatings, such as styrene and propylene.

Vinyl coatings are primarily considered to be copolymers of vinyl chloride and vinyl acetate having the chemical structure shown in Figure 12.1. A relatively large amount of solvent is required to dissolve a vinyl copolymer high in vinyl chloride content. Because of this, the volume of solids in the solution is relatively low. As a result, most vinyl coatings must be applied in thin (1–1.5 mil) coats. Consequently, a vinyl coating system might require five or more coats. The system is considered highly labor intensive, even though in the proper environment excellent protection is provided.

A formulation has been developed that permits 2–2.5 mils to be applied per coat, but this advantage comes at the expense of reduced protection. The thixotropes, fillers, and additives used to permit the greater thicknesses are more susceptible to environmental and moisture penetration.

Chlorinated Rubber

Chlorinated rubber resins include those resins produced by the chlorination of both natural and synthetic rubbers. Chlorine is added to unsaturated double bonds until the final product contains approximately 65% chlorine with the chemical structure shown in Figure 12.2. Since the final product is a hard, brittle material with poor adhesion and elasticity, a plasticizer must be added. The volume of solids of the coating is somewhat higher than that of a vinyl; therefore, a suitably protective chlorinated rubber system often consists of only three coats.

Acrylics

The acrylics can be formulated as thermoplastic resins, thermosetting resins, and as a water emulsion latex. The resins are formed from polymers of acrylate esters, primarily polymethyl methacrylate and polyethyl acrylate. Since the acrylate resins do not contain tertiary hydrogens attached directly to the polymer backbone chain, they are esceptionally stable to oxygen and UV light. The repeating units for the methacrylate and acrylate are as follows:

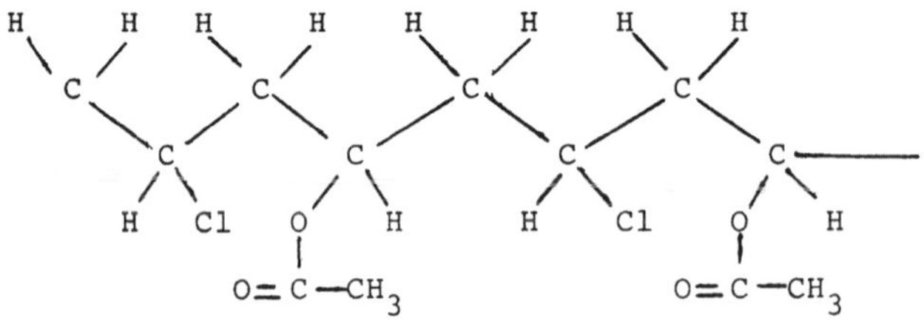

FIGURE 12.1 Chemical structure of vinyl acetate and vinyl chloride copolymer.

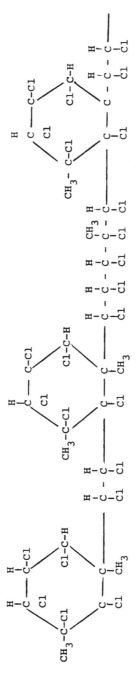

FIGURE 12.2 Chemical structure of chlorinated rubber.

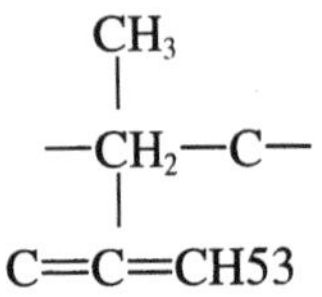

$$CH_3$$
$$|$$
$$-CH_2-C-$$
$$|$$
$$C=C=CH53$$
Polymethacrylate

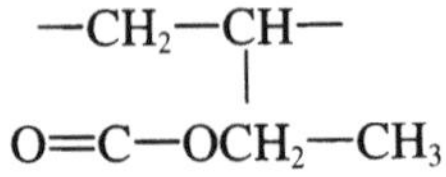

$$-CH_2-CH-$$
$$|$$
$$O=C-OCH_2-CH_3$$
Polyethyl acrylate

Epoxies

Epoxy resins themselves are not suitable for protective coatings because when pigmented and applied they dry to a hard, brittle film with very poor chemical resistance. When copolymerized with other resins, specifically with those of the amine or polyamide family, or with esterfied fatty acids, epoxy resins will form a durable protective coating.

Urethanes

Urethanes are reaction products of isocyanates with materials having hydroxyl groups and simply contain a substantial number of urethane groups ($-N-C-O-$) regardless of what the rest of the molecule may be. The hydroxyl containing side might consist of a variety of compounds including water (moisture cured urethanes), epoxies, polyesters, acrylics, and drying oils. Figure 12.3 shows several typical chemical structures.

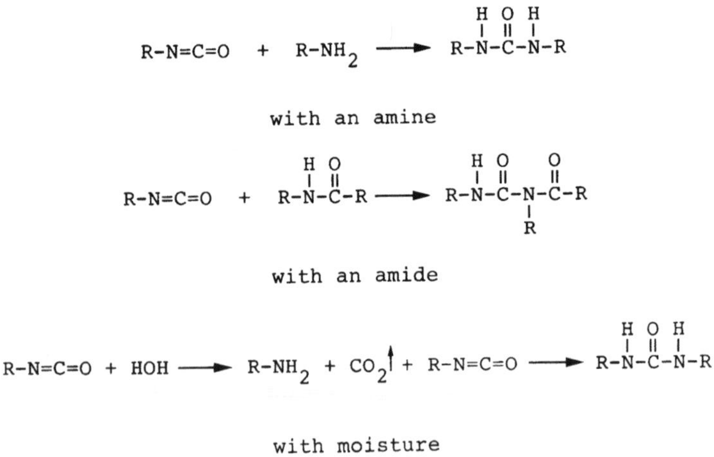

$$R-N=C=O \quad + \quad R-NH_2 \quad \longrightarrow \quad R-N-C-N-R$$

with an amine

$$R-N=C=O \quad + \quad R-N-C-R \longrightarrow R-N-C-N-C-R$$

with an amide

$$R-N=C=O + HOH \longrightarrow R-NH_2 + CO_2\uparrow + R-N=C=O \longrightarrow R-N-C-N-R$$

with moisture

FIGURE 12.3 Typical structures of urethane coatings.

Epoxy and polyester polyols, though more expensive than the acrylic polyols, are more chemically and moisture resistant. However, the acrylic polyol, when suitably reacted to form a urethane coating, is satisfactory for weathering atmospheres.

Coatings formulated with aliphatic isocyanates cost approximately twice as much as similar coatings formulated with epoxy and aromatic isocyanates. However, they have superior color and gloss retention and are resistant to deterioration by sunlight. Aliphatic urethanes will continue to maintain a glossy "wet" look years after application, where as other urethane coatings will chalk or yellow somewhat after exposure to the sun.

Polyesters

In the coating industry, polyesters are characterized by resins based on components that introduce unsaturation ($-C=C-$) directly into the polymer backbone. Typical is the structure of an isophthalic polyester resin, as shown in Figure 12.4. The most common polyester resins are polymerized products of maleic or isophthalic anhydride or their acids. In producing paint, the polyester resin is dissolved in styrene monomer, together with pigment and small amounts of a reaction inhibitor. A free radical initiator, commonly a peroxide, and additional styrene are packaged in another container. When applied, the containers are mixed; sometimes, because of the fast initiating reaction (short pot life), they are mixed in an externally mixing or dual-headed spray gun. After being mixed and applied, a relatively fast reaction takes place, resulting in cross-linking and polymerization of the monomeric styrene with the polyester resin.

Another class of coatings, formulated in a similar manner, are the vinyl esters. The vinyl esters derive from a resin based on a reactive end vinyl group that can open and polymerize. The chemical structure is shown in Figure 12.5.

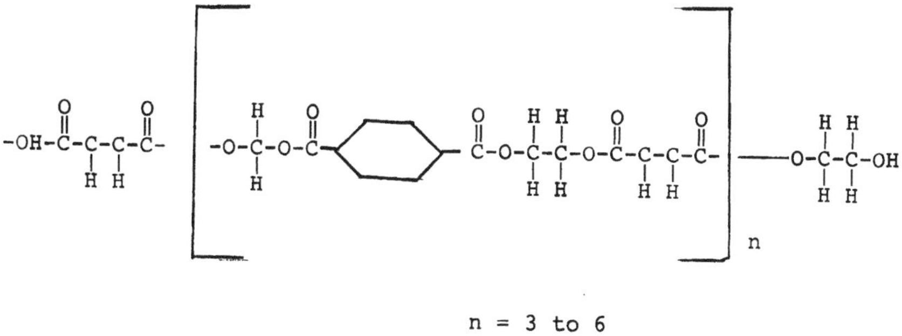

n = 3 to 6

FIGURE 12.4 Chemical structure of polyester coating.

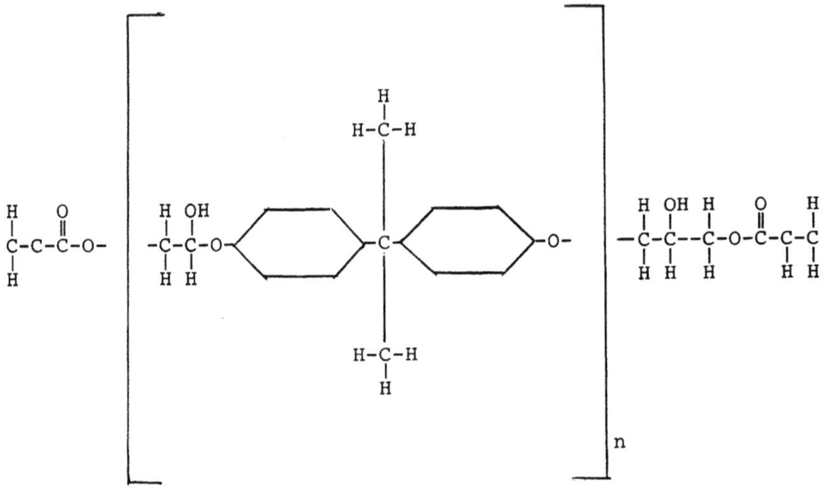

FIGURE 12.5 Chemical structure of vinyl ester coating.

Autooxidative Cross-Linking Coatings

Autooxidative cross-linking coatings rely on a drying oil with oxygen to introduce cross-linking within the resin and attainment of final film properties. The
final coating is formed as a result of the drying oil reacting with the resin, which
is combined with pigments and solvents. The paint is packaged in a single can
which can be opened, mixed, and the paint applied. Since oxygen reacts with
the coating, introducing additional cross-linking, final film properties can take
weeks or even months to attain.

Autooxidative cross-linking type paints include alkyds, epoxies, esters, oil-
modified urethanes, and so on. These are commonly used, when properly formulated, to resist moisture and chemical fume environments. They can be applied
over wood, metal, or masonry substrates.

The advantage of these coatings is their ease of application, excellent adhesion, relatively good environmental resistance (in all but immersion and high
chemical fume environment), and tolerance for poorer surface preparation. Their
major disadvantage, compared with other synthetic resins, is their lessened moisture and chemical resistance.

Water Soluble Resins and Emulsion Coatings

By introducing sufficient carboxyl groups into a polymer, any type of resin can
be made water soluble. These groups are then neutralized with a volatile base

such as ammonia or an amine, rendering the resin a polymeric salt, soluble in water or water/ether-alcohol mixtures. The major disadvantage to such resins is that polymers designed to be dissolved in water will remain permanently sensitive to water. Because of this they are not widely used industrially.

However, water emulsions are widely accepted. The water based latex emulsions are formulated with high molecular weight resins in the form of microscopically fine particles of high molecular weight copolymers of polyvinyl chloride, polyvinyl acetate, acrylic esters, stryren-butadiene, or other resins combined with pigments, plasticizers, UV stabilizers, and other ingredients. Water based epoxy formulations are also available. The epoxy resin and a polyamide copolymer are emulsified and packaged in separate containers. After mixing, coalescence, and ultimate drying, the polyamide reacts with the epoxy to form the final film.

The major advantage of the emulsion coatings is their ease of cleanup. Their disadvantage is the permanent water sensitivity of the coating, making them unsuitable for use in continually wet environments. Water based coatings cannot be applied to blast cleaned steel surfaces with the same confidence as solvent based coatings.

Zinc-Rich Coatings

In all of the above mentioned coatings the final film properties of corrosion protection and environmental resistance have been attributed to organic composition of their constituent resin or binder. The pigment, although playing an important role in the final film properties, is secondary to the resistance of the organic binders.

Conversely, in a zinc-rich coating, the role of the zinc pigment predominates. The high amount of zinc dust metal in the dried film determines the coating's basic property, i.e., galvanic protection. For a coating to be considered zinc-rich, the common rule of thumb is that there must be at least 75% by weight of zinc dust in the dried film. This may change since conductive extenders (notably di-iron phosphide) have been added to improve weldability and burn-through, with supposedly equivalent protection at lower zinc loadings.

The primary advantage of zinc-rich coatings is their ability to protect galvanically. The zinc coating preferentially sacrifices itself in the electrochemical corrosion reaction to protect the underlying steel. This galvanic reaction, together with the filling and sealing effect of zinc reaction products (primarily zinc carbonate, zinc hydroxide, and complex zinc salts), provides more effective corrosion protection to steel substrates than does any other type of coating. Zinc-rich coatings can not be used outside the pH range of 6–10.5.

Pigments

Pigments serve several functions as part of the coating: They reinforce the film structurally, provide color and opacity, and serve special purposes for metal protection.

Typical materials used as color or hiding pigments and to provide aesthetic value, retention of gloss and color, and to help with the film structure and impermeability are iron oxides, titanium dioxide, carbon or lamp black, and others.

Pigments must be somewhat resistant to the environment and must be compatible with the resin. For example, calcium carbonate, which is attacked by acid, should not be used in an acid environment. Water soluble salts are corrosion promoters, so special low salt containing pigments are used in primers for steel.

Primers contain one of three kinds of pigments for special protective properties as follows:

1. *Inert or chemically resistant.* These are used to form barrier coatings in severe environments, such as pH less than 5 or greater than 10. They are also used as nonreactive extender, hiding, or color pigments in neutral environments.
2. *Active.* Leads, chromates, or other inhibitive pigments are used in linseed oil/alkyd primers.
3. *Galvanically sacrificial.* Zinc is employed at high concentrations to obtain electrical contact for galvanic protection in environments between pH 5 and 10.

Table 12.3 shows the type and characteristics of these pigments.

Fillers (Extenders)

In addition to lowering the cost, extenders also provide sag resistance to the liquid paint so that the edges remain covered. When the paint has dried, they reduce the permeability to water and oxygen and provide reinforcing structure within the film. Talc and mica are used extensively as extenders. Mica is limited to approximately 10% of the total pigment. Both talc and mica, but particularly mica, reduce the permeability through the film as platelike particles block permeation, forcing water and oxygen to seek a longer path through the binder around the particle.

Additives

Additives are used in coating formulations for specific purposes and in small, sometimes even trace amounts. Some typical additives and their purposes are as follows:

TABLE 12.3 Characteristics of Pigments for Metal Protective Paints

Pigment	Specific gravity	Color	Opacity	Specific contribution to corrosion resistance
Active pigments				
Red lead	8.8	Orange	Fair	Neutralizes film acids, insolubilizes sulfates and chlorides, renders water noncorrosive
Basic silicon lead chromate	3.9	Orange	Poor	Neutralizes film acids, insolubilizes sulfates and chlorides, renders water noncorrosive
Zinc yellow (chromate)	3.3	Yellow	Fair	Neutralizes film acids, anodic passivator, renders water noncorrosive
Zinc oxide (French process)	5.5	White	—	Neutralizes film acids, renders water noncorrosive
Zinc dust at low concentration in coatings for steel	7.1	Gray	Good	Neutralizes film acids
Galvanically protective pigments				
Zinc dust sacrificial at high concentration	7.1	Gray	Good	Makes electrical contact, galvanically sacrificial
Barrier pigments				
Quartz	2.6	Nil	Translucent	Inert, compatible with vinyl ester additives
Mica	2.8	Nil	Translucent	Impermeability and inertness
Talc	2.8	Nil	Translucent	Impermeability and inertness
Asbestine	2.8	Nil	Translucent	Impermeability and inertness
Barytes	4.1	Nil	Translucent	Impermeability and inertness
Iron oxide	4.1	Red	—	Impermeability and inertness
Iron oxide	4.1	Ochre	—	Impermeability and inertness
Iron oxide black	4.1	Black	—	Impermeability and inertness
Titanium dioxide	4.1	White	Excellent	Impermeability and inertness
Carbon black	1.8	Black	Good	Impermeability and inertness

Note: Titanium dioxide has better "hiding" than any other pigment.
Source: Ref. 2.

Phenylmercury, zinc, cuprous compounds: mildew inhibitors
Cobalt and manganese naphthanates: aid to surface and thorough drying
Zinc oxide: protection of the resin from heat and sun

Latex paints have a number of additives that act as stabilizers, coalescing aids, emulsion stabilizers, freeze-thaw stabilizers, and so on.

Solvents

Organic solvents are usually required only to apply the coating and after application are designed to evaporate from the wet paint film. Water is considered either as a solvent or an emulsifier. The rate at which solvents evaporate influences the application characteristics of the coating. It is imperative that the solvent evaporate completely. If the solvents are partially retained and do not evaporate completely, premature failure by blistering or pinholing is likely to take place.

Most coatings are formulated to be applied at ambient conditions of approximately 75°F/24°C and a 50% relative humidity. If ambient conditions are considerably higher or lower, it is conceivable that solvent release could pose a problem. This potential problem can be eased by changing the "solvent balance." Generally, faster evaporating solvents should be used in colder weather and slower evaporating solvents should be used in hot weather. Classes and characteristics of some common solvents are shown in Table 12.4.

Low viscosity and two component epoxies and powder coatings are examples of paint systems referred to as solvent free. The epoxy coatings may be mixed and applied without the use of a solvent, as the two components typically have a low viscosity.

Powder coatings are cured by sintering. Before application thermoplastic powder coatings consist of a large number of small binder particles. These particles are deposited on a metal surface using special application techniques. Subsequently, the paint is baked in an oven to form a continuous film by sintering.

SURFACE PREPARATION

In order for a paint to provide maximum protection, it is essential that the surface to which the paint is to be applied is properly prepared. The specific preparation system to be specified will be dependent on the coating system to be applied and should usually be in accord with one of the surface preparation specifications defined by the Steel Structures Painting Council which follow:

TABLE 12.4 Characteristics of Solvent Classes

Class	Solvent name	Polarity	Specific gravity	Boiling range °F	Flash point of TCC	Evaporation rate[a]
Aliphatic	VM&P naphtha	Nonpolar	0.74	246–278	52	24.5
	Mineral spirits	Nonpolar	0.76	351–395	128	9.0
Aromatic	Toluene	Intermediate	0.87	230–233	45	4.5
	Xylene	Intermediate	0.87	280–288	80	9.5
	High solvency	Intermediate	0.87	360–400	140	11.6
Ketone	Methyl ethyl ketone (MEK)	High	0.81	172–176	24	2.7
	Methyl isobutyl ketone (MIBK)	High	0.80	252–266	67	9.4
Ester	Cyclohexanone	High	0.95	313–316	112	4.1
	Ethyl acetate	Intermediate	0.90	168–172	26	2.7
Alcohol	Ethanol	Intermediate	0.79	167–178	50	6.8
Unsaturated aromatic	Styrene	Intermediate	0.90			
Glycol	Cellosolve	High	0.93	273–277	110	0.3
Ethers	Butyl cellosolve	High	0.90	336–343	137	0.06

[a] Butyl acetate = 1.

Specification	Description
SP1: Solvent cleaning	Removal of oil, grease, dirt, soil, salts and contaminants by cleaning with solvent, vapor, alkali, emulsion, or steam
SP2: Hand-tool cleaning	Removal of loose rust, mill scale, and paint, to degree specified, by hand chipping and wire brushing
SP3: Power tool cleaning	Removal of loose rust, mill scale, and paint, to degree specified, by power tool chipping, descaling, sanding, wire brushing, and grinding
SP5: White-metal blast cleaning	Removal of all visible rust, mill scale, paint, and foreign matter by blast cleaning, by wheel or nozzle, dry or wet, using sand, grit, or shot for very corrosive atmospheres where high cost of cleaning is warranted
SP6: Commercial blast cleaning	Blast cleaning until at least two-thirds of the surface area is free of all visible residues (for severe exposure)
SP7: Brush-off blast cleaning	Blast cleaning of all except tightly adhering residues of mill scale, rust and coatings, exposing numerous evenly distributed flecks of underlying metal
SP8: Pickling	Complete removal of rust and mill scale by acid pickling, duplex pickling, or electrolyte pickling
SP10: Near white blast cleaning	Blast cleaning to near white metal cleanliness until at least 95% of the surface area is free of all visible residue (for high humidity, chemical atmosphere, marine, and other corrosive environments)

In addition to these mechanical methods, conversion coatings on metals produced by chemical or electrochemical means also increase the adhesive bonding of paint coatings. These types of coatings will be discussed later.

The mechanical processes are used to remove dirt, rust, and mill scale from the surface and are adopted to steel products such as thick plates, construction steels, and steel structures for which chemical processes are not available.

The chemical processes are applied to steel-strip products such as automotive bodies and home electrical appliances, to the internal coating of steel tubes, and to aluminum and zinc and their alloy products.

APPLICATION METHODS

The method of application for a corrosion resistant paint will depend on:

1. Purpose for which coated product is to be used
2. Environment to which the coating will be exposed

3. The type of paint
4. The shape and size of the object to be painted
5. The period of application process
6. Cost.

Brushing

This was once the main coating method, but at the present time spray coating is more widely used. Brush coating has the following advantages:

1. Applicators are simple and inexpensive.
2. Complicated forms and shapes can be coated.
3. Thick films are obtained with one coat.
4. Particularly useful antirusting coating.

The disadvantage of brushing results from the nonuniformity of coating layers, especially coating layers of rapidly drying paints.

Spray

There are two types of spray coatings: air sprays and airless sprays. These coatings are subdivided into hot and cold types.

Air Spray

Air spray is a process for spraying fine paint particles atomized by compressed air. Drying rates are high and a uniform and decorative surface is obtained. This method is used with paints that are not suitable for brushing methods, for mass produced painted products, or for products where surface appearance is important.

Airless Spray

In an airless spray system high pressure is supplied to the paint which is forced through a high pressure resistant hose and airless spray gun. High viscosity paints are warmed before spraying. This technique has the following advantages over the air spray system:

1. The sticking ratio of paint is increased to 25–40%.
2. Thicker films can be applied.
3. The running of paint on the object is reduced.

Steel structures and bridge girders can be coated in the factory by this method since the efficiency of work is several times higher than brushing. However, its paint loss is 30–40% greater than for paints applied by brushing.

Roller Coating

This method is used to coat coils and sheets by passing them through two rollers. The thickness of paint film is controlled by adjusting the rollers. One-side or both-sides coating is possible.

Powder Coating

Because of the absence of organic solvents powder coatings have grown in popularity as antipollution coatings. Coating thicknesses of 25–250 μm can be obtained. Automotive bodies, electric components, housing materials, wires, and cables make use of this method. Polyethylene and epoxy resins are the predominent types of paints used. At the present time the following eleven procedures are used in this coating process:

1. Pouring method (flock coating)
2. Rotational coating of pipes
3. Fluidized bed
4. Dipping in nonfluidizing powders
5. Centrifugal casting
6. Rotational molding
7. Electrostatic powder spraying
8. Electrostatic fluidized bed
9. Pouring or flowing of fluidized powder
10. Electrogasdynamics powder spraying
11. Flame spraying of thermoplastic powders

Electrodeposition

The anodic deposition process for paint coating systems was introduced in the early 1960s, and the cathodic deposition system in 1972. Electrodeposition processes are widely used since they possess the advantages of unmanned coating, automation, energy saving, and lower environmental pollution. This process is used to apply coatings to automotive bodies and parts, domestic electric components, machine parts, and architecturals such as window frames. Schematic illustrations of anodic and cathodic electrodeposition of paints are shown in Figure 12.6.

The primary paints used in the electrodeposition process are anionic type resins with a carboxyl group (RCOOH) polybutadiene resin) and cationic type resin ($R-NH_2$, epoxy resin). Hydrophilic groups and neutralizing agents are added to the water insoluble or undispersed prepolymers to convert them to soluble or dispersed materials.

The dissolution of metal substrate in the cathodic process is much less than that in the anodic process. The primary resins used in the cathodic process are

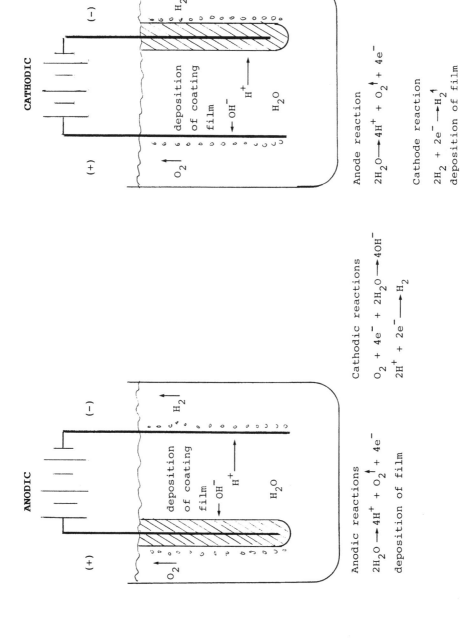

FIGURE 12.6 Electrodeposition of paints.

TABLE 12.5 General Coating Multistage Paint Systems

	System		
Coat	*A*	*B*	*C*
First	Etching primer	Etching primer	Zinc rich primer
Second	Oil corrosion	Oil corrosion	Chlorinated rubber
	Preventative paint	Preventative paint	System primer
Third	Oil corrosion	Oil corrosion	Chlorinated rubber
	Preventative paint	Preventative paint	System primer
Fourth	Long oil alkyd	Phenolic resin	Chlorinated rubber
	Resin paint	System MIO paint	System paint
Fifth	Long oil alkyd	Chlorinated rubber	Chlorinated rubber
	Resin paint	System paint	System paint
Sixth		Chlorinated rubber	
		System paint	

epoxy and since epoxy resins provide good water and alkali resistance as well as adhesion, cationic paint coatings are superior in corrosion protection to anodic paint coatings.

Multilayer Coatings

The protective ability of a coating is also dependent upon its thickness. Thicker coatings provide improved protection. However, single thick paint films tend to crack due to internal stress. When long term protection is required, multilayer paint coating systems are applied.

The multilayer system to be used with a steel structure is determined by the required service life and the environmental conditions under which the steel structure must exist. General coating systems are applied to structures in mild environments while heavy-duty coating systems are applied to these structures in severe environments. Typical general coating systems and heavy duty coating systems are shown in Tables 12.5 and 12.6, respectively.

FACTORS AFFECTING LIFE OF FILM

For a paint film to perform effectively, it must have mechanical resistance to external forces and resistance against environmental factors. As discussed, a paint film is a composite layer that is composed of a vehicle (resin), pigment, and additives, and tends to contain defects in its structure. The common defects existing in a resultant paint film are:

TABLE 12.6 Heavy-Duty Coating Multilayer Paint Systems

Coat	A	B	C	D
			System	
First	Zinc spray or zinc-rich paint	Thick type zinc-rich paint	Thick type zinc-rich paint	Zinc-rich primer
Second	Etching primer	Thick type vinyl or chlorinated rubber system	Thick type epoxy primer	Tar epoxy resin paint
Third	Zinc chromate primer	Thick type vinyl or chlorinated rubber system	Thick type epoxy primer	Tar epoxy resin paint
Fourth	Phenolic resin system MIO paint	Vinyl resin or chlorinated rubber system	Epoxy resin system paint	Tar epoxy resin paint
Fifth	Chlorinated rubber system	Vinyl resin or chlorinated rubber system	Epoxy or polyurethane resin system paint	
Sixth	Chlorinated rubber system paint			

1. Nonuniformity of vehicle molecules
2. Solvent residue
3. Residual stress
4. Differences in expansion coefficients in multilayer coating systems
5. Poor adhesion of polymer to pigment

Strength of Paint Film

Paint films require hardness, flexibility, brittleness resistance, abrasion resistance, mar resistance and sag resistance. Paint coatings are formulated providing a balance of these mechanical properties. The mechanical strength of a paint film is described by the words "hardness and plasticity" which correspond to the modulus of elasticity and to the elongation at break obtained from the stress-strain curve of a paint film. Typical paint films have tensile properties as follows:

Paints	Tensile strength (g/mm)	Elongation at break (%)
Linseed oil	14–492	2–40
Alkyd resin varnish (16% PA)	141–1266	30–50
Amino-alkyd resin varnish ($A/W = 7/3$)	2180–2602	—
NC lacquer	844–2622	2–28
Methyl-n-butyl-meta-acrylic resin	1758–2532	19–49

Mechanical properties of paint coatings vary depending on the type of pigment, baking temperatures, and aging times. As baking temperatures rise, the curing of paint films is promoted and elongation reduced. Tensile strength is improved by curing and the elongation at breaks is reduced with increased drying time.

Structural defects in a paint film causes failures that are determined by environmental conditions such as thermal reaction, oxidation, photooxidation, and photothermal reaction. An important factor in controlling the physical properties of a paint film is the glass transition temperature T_g. In the temperature range higher than T_g the motion of the resin molecules become more active, so that hardness, plasticity, and the permeability of water and oxygen vary greatly. Table 12.7 lists the glass transition temperatures of organic films.

Deterioration of paint films is promoted by photolysis, photooxidation or photo thermal reaction as a result of exposure to natural light. Specifically, UV light (wavelength 40–400 nm) decomposes some polymer structures. Polymer films, such as vinyl chloride resins, are gradually decomposed by absorbing the energy of UV light.

TABLE 12.7 Glass Transition Temperature of Organic Films

Organic film	Glass transition temperature, T_g (°C)
Phthalic acid resin	50
Acrylic lacquer	80–90
Chlorinated rubber	50
Bake type melamine resin	90–100
Anionic resin	80
Catonic resin	120
Epoxy resin	80
Tar epoxy resin	70
Polyurethane resin	40–60
Unsaturated polyester	80–90
Acrylic powder paint	100

The T_g of a polymer is of critical importance in the photolysis process. Radicals formed by photolysis are trapped in the matrix, but they diffuse and react at higher temperatures than T_g. The principal chains of polymers with ketone groups form radicals:

RCOR′   R + COR′   CO + R′

ROCOR′   OCOR′ CO_2 + R′

The resultant radicals accelerate the degradation of the polymer and in some cases HCl (from polyvinyl chloride) or CH_4 is produced.

Adhesion of Paint Film

Paint coated materials are exposed primarily to the natural atmosphere. Consequently, paint adhesion is influenced by atmospheric factors, particularly moisture and water. All organic polymers and organic coatings are permeable to water. They differ only in the degree of permeability. Table 12.1 provides typical values for the permeability and the diffusion coefficient for water in a number of different polymers.

When moisture permeates into the paint substrate interface, blisters form. There are four mechanisms whereby blisters can form:

1. Volume expansion due to swelling
2. Gas inclusion or gas formation

3. Osmotic process due to soluble impurities at the film-substrate interface

4. Electroosmotic blistering

Of the four mechanisms, osmotic blister formation is the most important. Past experience has indicated that 70% of all paint coating failures are the result of poor or inadequate surface preparation prior to application of the coating. Types of osmotically active surfaces at film-substrate interfaces are impurities such as sand particles from incomplete washing after wet sanding, water soluble salt residues from phosphating, or surface nests in a rust layer. It is also possible to entrap hydrophilic solvents to create blisters.

Regardless of the specific driving force, the blister formation process is shown in Figure 12.7 using an example of an intact coating on steel. When the coating is exposed to an aqueous solution, water vapor molecules (and to a lesser extent, oxygen), diffuse into the film toward the coating-substrate interface. Eventually water may accumulate at the interface forming a thin water film of at least several monolayers. The accumulation takes place at sites of poor adhesion or at sites where wet adhesion problems arise. The degree to which permeated water

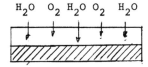

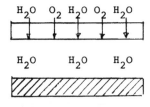

Permeation

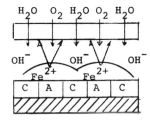

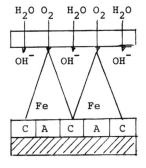

Corrosion initiation

FIGURE 12.7   Initiation of a blister under an intact coating.

may change the adhesion properties of a coated system is referred to as wet adhesion. No initial poorly adhered spots need be present for chemical disbondment to take place. A corrosion reaction can be initiated by the presence of an aqueous electrolyte with an electrochemical double layer, a cathodic species (oxygen) and an anodic species (metal). When the corrosion reaction has started at the interface, the corrosion products can be responsible for the buildup of osmotic pressure, resulting in the formation of macroscopic blisters (see Figure 12.7). Depending on the specific materials and circumstances, the blisters may grow out because of the hydrodynamic pressure in combination with one of the chemical propagation mechanisms, such as cathodic delamination and anodic undercoating.

Oxygen migration through a paint film coating is generally much lower than that of water. In some instances the flow of oxygen increases with the water content of the film, probably because the water acts to swell the polymer.

TYPES OF CORROSION UNDER ORGANIC COATINGS

For corrosion to take place on a metal surface under a coating, it is necessary for an electrochemical double layer to be established. In order for this to take place, it is necessary for the adhesion between the substrate and coating to be broken. This permits a separate thin water layer to be formed at the interface from water which has permeated the coating. As mentioned previously, all organic coatings are permeable to water to some extent. Water permeation occurs under the influence of various driving forces:

1. A concentration gradient, e.g., during immersion or exposure to a humid atmosphere, resulting in true diffusion through the polymer
2. Capillary forces in the coating due to poor curing, improper solvent evaporation, entrapment of air during application, or bad interaction between binder and additives
3. Osmosis due to impurities, or corrosion products at the interface between the metal and coating

Water molecules will eventually reach the coating substrate interface when a coated system has been exposed to an aqueous solution or a humid atmosphere. These molecules can interfere with the bonding between the coating and substrate, eventually resulting in the loss of adhesion and corrosion initiation if a cathodic reaction can take place. A constant supply of a cathodic species, such as water or oxygen, is required for the corrosion reaction to proceed.

Water permeation may also result in the buildup of high osmotic pressures, which are responsible for blistering and delamination.

Cathodic Delamination

When cathodic protection is applied to a coated metal, loss of adhesion between the substrate and paint film, adjacent to defects, often takes place. This loss of adhesion is known as cathodic delamination. Such delamination may also occur in the absence of applied potential. Separation of anodic and cathodic reaction sites under the coating results in the same local driving force as during external polarization. The propagation of a blister due to cathodic delamination under an undamaged coating on a steel substrate is schematically illustrated in Figure 12.8. Under an intact coating, corrosion may be initiated locally at sites of poor adhesion.

A similar situation develops in the case of corrosion under a defective coating. When there is a small defect in the coating, part of the substrate is directly exposed to the corrosive environment. Corrosion products are formed immediately which block the damaged site from oxygen. The defect in the coating is sealed by the corrosion products, after which corrosion propagation takes place, according to the same mechanism as for the initially undamaged coating. Refer to Figure 12.8 for the sequence of events.

Anodic Undermining

Anodic undermining results from the loss of adhesion caused by anodic dissolution of the substrate metal or its oxide. In contrast to cathodic delamination the metal is anodic at the blister edges. Coating defects may cause anodic undermining, but in most cases it is associated with a corrosion sensitive site under the coating, such as a particle from a cleaning or blasting procedure, or a site on the metal surface with potentially increased corrosion activity (scratches). These sites become active once the corrodent has penetrated to the metal surface. The initial corrosion rate is low. However, an osmotic pressure is caused by the soluble corrosion products which stimulates blister growth. Once formed, the blister will grow due to a type of anodic crevice corrosion at the edge of the blister.

Coated aluminum is very sensitive to anodic undermining while steel is more sensitive to cathodic delamination.

Filiform Corrosion

Metals with semipermeable coatings or films may undergo a type of corrosion resulting in numerous meandering threadlike filaments of corrosion beneath the coatings or films. Conditions which promote this type of corrosion are:

1. High relative humidity (60–95% at room temperature).
2. The coating is permeable to water.
3. Contaminants (salts, etc.) are present on or in the coating or at the coating-substrate interface.

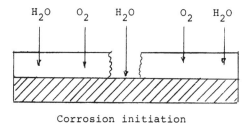

Corrosion initiation

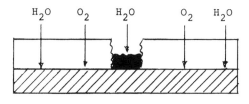

Blocking of pore

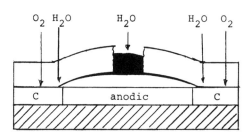

Cathodic delamination

FIGURE 12.8 Blister initiation and propagation under a defective coating (cathodic delamination).

4. The coating has defects (mechanical damage, pores, insufficient coverage of localized areas, air bubbles).

Filiform corrosion under organic coatings is common on steel, aluminum, magnesium, and zinc (galvanized steel). It has also been observed under electroplated silver plate, gold plate, and under phosphate coatings.

This form of corrosion is more prevalent under organic coatings on aluminum than on other metallic surfaces, being a special form of anodic undermining.

A differential aeration cell is the basic driving force. The filaments have considerable length but little width and depth and consist of two parts: a head and a tail. The primary corrosion reactions and subsequently the delamination process of the paint film takes place in the active head, while the tail is filled with the resulting corrosion products. As the head of the filiform moves, the filiform grows in length.

CAUSES OF COATING FAILURE

Coating failures result from two basic causes:

1. Poor or inadequate surface preparation and/or application of the paint to the substrate
2. Atmospheric effects

Coating failures typical from surface preparation and application problems include:

1. Cracking, checking, alligatoring. These type of failures develop with the aging of the paint film. Shrinkage within the film during aging causes cracking and checking. Alligatoring is a film rupture, usually caused by application of a hard brittle film over a more flexible film.
2. Peeling, flaking, delamination. These failures are caused by poor adhesion. When peeling or flaking occurs between coats, it is called delamination.
3. Rusting. Failure of a coated surface may appear as (a) spot rusting at minute areas, (b) pinhole rusting at minute areas, (c) rust nodules breaking through the coating, and (d) underfilm rusting, which eventually causes peeling and flaking of the coating.
4. Lifting and wrinkling. When the solvent of a succeeding coat of paint too rapidly softens the previous coat, lifting results. Rapid surface drying of a coating without uniform drying throughout the rest of the film results in a phenomenon known as wrinkling.
5. Failures around weld areas. Coating adhesion can be hampered by weld flux, which may also accelerate corrosion under the film. Relatively large projections of weld spatter cause possible gaps and cavities that may not be coated sufficiently to provide protection.
6. Edge failures. Edge failures usually take the form of rusting through the film at the edge where the coating is normally the thinnest. This is usually followed by eventual rust creepage under the film.
7. Pinholing, tiny holes which expose the substrate, caused by improper paint spray atomization, or segregation of resin in the coating. If practi-

cal during application, brush out coating. After application and proper cure, apply additional coating.

Polymer coatings are exposed to the environment and therefore are subject to degradation by environmental constituents. The primary factors promoting degradation are thermal, mechanical, radiant, and chemical. Polymers may also be degraded by living organisms such as mildew. Any atmospheric environment is subject to dry and wet cycles. Since water and moisture have a decided effect on the degradation of a coating, the time of wetness of a coating is important. Moisture and water that attack organic films are from rain, fog, dew, snow, and water vapors in the atmosphere. Relative humidity is a particularly important factor. As exposure time is increased in 100% relative humidity, the bond strength of a paint coating is reduced. This is shown in Table 12.8.

Temperature fluctuations and longer time of wetness tends to produce clustered water, which increases the acceleration of degradation of the organic film, particularly in a marine atmosphere. The most severe natural atmosphere for a paint film is that of a seashore environment.

The mode of degradation may involve depolymerization, generally caused by heating, splitting out of constituents in the polymer, chain scission, crosslinking, oxidation, and hydrolysis. Polymers are subject to cracking on the application of a tensile force, particularly when exposed to certain liquid environments. This phenomenon is known as environmental stress cracking, or stress corrosion cracking.

Other factors in the atmosphere that cause corrosion or degradation of the coating are UV light, temperature, oxygen, ozone, pollutants, and wind. Types of failures caused by these factors are:

1. Chalking. UV light, moisture, oxygen, and chemicals degrade the coating resulting in chalk. This may be corrected by providing an additional top coat of material with proper UV inhibitor.
2. Color fading or color change. This may be caused by chalk on the surface or by breakdown of the colored pigments. Pigments can be decomposed or degraded by UV light or reaction with chemicals.
3. Blistering. Blistering may be caused by:
   a. Inadequate release of solvent during both application and drying of the coating system.
   b. Moisture vapor that passes through the film and condenses at a point of low adhesion.
   c. Poor surface preparation.
   d. Poor adhesion of the coating to the substrate or poor intercoat adhesion.
   e. A coat within the paint system that is not resistant to the environment.

TABLE **12.8** Relationship of Bond Strength to Exposure
Time in 100% Relative Humidity

Exposure time (hr)	Bond strength (psi)		
	Epoxy ester	Polyurethane	Thermosetting acrylic
Initial	4790	3410	5700
24	1640	1500	3650
48	1500	1430	3420
120	—	1390	2400
195	1400	1130	1850
500	1390	670	480

    f. Application of a relatively fast drying coating over a relatively porous surface.

    g. Failures due to chemical or solvent attack. When a coating is not resistant to its chemical or solvent environment there is apparent disintegration of the film.

  5. Erosion, coating worn away. Loss of coating due to inadequate erosion protection. Provide material with greater resistance to erosion.

MAINTENANCE OF THE COATING

Any paint job, even if properly done, does not last forever. Within the first 6 months or year after application, spots, inadvertent misses, or weak spots in the coating system can be detected by simple visual inspection. A thorough inspection and repair should be done at this time.

Over a period of time the coating will break down and deteriorate as a result of the effect of the environment. Scheduled inspections should be made to determine the extent and rate of the coating breakdown. Spot touchup repair should be made at localized areas of failure before deterioration of the entire coated surface has taken place.

Extensive costs of total surface preparation (such as complete blast cleaning removal of all old coating) can be avoided if a "planned maintenance approach" of periodic spot touchup and an occasional recoat of the entire surface is followed.

Many coatings that provide good long term protection are more difficult to touch up or repair in the event of physical damage or localized failure. This is particularly true of the thermosetting and zinc-rich coatings. Application of a subsequently applied paint coat to an older aged epoxy, urethane, or other catalyzed coating often results in reduced adhesion, which leads to peeling.

Thermoplastic coatings do not normally present this problem. Solvents of a freshly applied thermoplastic coating soften and allow for intermolecular mixing of the new and old coatings, with good intercoat adhesion.

Heavily pigmented coatings, such as zinc-rich, require agitated pots to keep the pigment in suspension during application. Because of this, touchup and repair of large areas is not recommended using zinc-rich coatings unless it is done by spray using an agitated pot.

For the most part, oil-based coatings (alkyds, epoxy esters, and modifications thereof) have a greater tolerance for poor surface preparation and an ability to wet, penetrate, and adhere to poorly prepared surfaces or old coatings. Consequently, these coatings are often specified for these purposes, even though they do not provide as long a term of corrosion protection.

REFERENCES

1. H. Leidheiser, Jr., Coatings, in F. Mansfield (ed.), *Corrosion Mechanisms*, Marcel Dekker, New York, 1987, pp. 165–209.
2. K. B. Tator, Coating, in P. A. Schweitzer (ed.), *Corrosion and Corrosion Protection Handbook*, 2nd ed., Marcel Dekker, New York, pp. 453–489.

13

Specific Organic Coatings

In order to select a suitable coating system, it is necessary to determine the environmental conditions around the structure to be painted. If the coating is to be applied to an exterior surface, are there nearby chemical plants, pulp and paper mills, or heavy industries that might provide airborne pollutants? Is the objective of the coating to provide an aesthetic appearance or to provide maximum corrosion protection which will limit the colors to the normal grays, whites, and pastels of the most corrosion resistant coatings? Will the surface be predominantly wet (salt or freshwater) or exposed to a chemical contaminant (acid or alkaline)? Is the environment predominantly a weathering environment subject to heat, cold, daily or seasonal temperature changes, precipitation, wind (flexing), exposure to sunlight, or detrimental solar rays? Will the surface be exposed to occasional spells of a corrosive or solvent? After the environmental conditions of the area have been defined, a suitable resistant coating can be selected. Characteristics, corrosion resistance, limitations, and uses of the common generic types of coatings will be discussed.

THERMOPLASTIC RESINS

Protective coatings formulated from these resins do not undergo any chemical change from the time of application until the final properties of a dried protective film have been attained. The film is formed by the evaporation of the solvent, concentrating the binder/pigment solution and causing it to precipitate as a continuous protective film.

Vinyls

These are polyvinyls dissolved in aromatics, ketones, or esters.

1. Resistance: Insoluble in oils, greases, aliphatic hydrocarbons, and alcohols; resistant to water and salt solutions; at room temperature, resistant to inorganic acids and alkalies; fire resistant.
2. Temperature resistance: 180°F/82°C dry; 140°F/60°C wet.
3. Limitations: Dissolved by ketones, aromatics, and ester solvents. Adhesion may be poor until all solvents have vaporized from the coating. Relatively low thickness per coat requiring several coats to provide adequate protection.
4. Features: Tasteless and odorless.
5. Applications: Used on surfaces exposed to potable water, as well as for immersion service, sanitary equipment, and widely used as industrial coatings.

When used for immersion service, it is essential that all solvents have evaporated before placing the vinyl coated object into service. The high polarity of the resin tends to retain the solvents used to dissolve the resin. Solvent evaporation is retarded, with the resultant effect of solvent voids within the coating, pinholes caused by volatilization of retained solvents upon heating or water filled blisters because of hydrogen bonding attraction of water by the retained solvents in the coating.

Chlorinated Rubber

Chlorinated rubber resins include those resins produced by the chlorination of both natural and synthetic rubbers. Chlorine is added to unsaturated double bonds until the final product contains approximately 65% chlorine. A plastisizer must be added since the final product is a hard, brittle material with poor adhesion and elasticity. The volume of solids of the coating dissolved in hydrocarbon solvents is somewhat higher than that of a vinyl; therefore, a suitably protective chlorinated rubber system often consists of only three coats.

1. Resistance: Chemically resistant to acids and alkalies; low permeability to water vapor; abrasion resistant; fire resistant. Nontoxic.
2. Temperature resistance: 200°F/93°C dry; 120°F/49°C wet.
3. Limitations: Degraded by ultraviolet light; attacked by hydrocarbons.
4. Applications: Excellent adherence to concrete and masonry. Used on

structures exposed to water and marine atmospheres (swimming pools, etc.).

When used for water immersion service, the same care must be exercised during evaporation as was described for the vinyls.

Acrylics

The acrylics can be formulated as thermoplastic resins, thermosetting resins, and as a water emulsion latex.

The resins are formed from polymers of acrylate esters, primarily polymethyl methacrylate and polyethyl acrylate. Since the acrylate esters do not contain tertiary hydrogens attached directly to the polymer backbone chain, they are exceptionally stable to oxygen and UV light.

A wide range of monomers is available for use in designing a specific acrylic system. Typically, mixtures of monomers are chosen for the properties they impart to the polymer. The glass transition temperature T_g of the polymer can be varied by selection of the proper monomers. This permits a varied area of application. Table 13.1 illustrates the wide range of T_g resulting from different monomer compositions for emulsion acrylics.

Acrylics may be formulated as lacquers, enamels, and emulsions. Lacquers and baking enamels are used as automotive and appliance finishes. In both these industries acrylics are used as top coats in multicoat finishing systems.

1. Resistance: Somewhat resistant to acids, bases, weak and moderately strong oxidizing agents, and many corrosive industrial gases and fumes. Resistant to weather and UV light.
2. Limitations: Limited penetrating power (water emulsion) because of water surface tension; may flash rust as a primer over bare steel; not suitable for immersion service or strong chemical environments. Soluble in ketones, esters, aliphatic chlorinated hydrocarbons, and aromatic hydrocarbons.

TABLE 13.1 Glass Transition Temperature vs. Application Area

T_g (°C/°F)	Application area
80–100/176–212	High heat resistant coatings
50–65/122–149	Floor care coatings
35–50/95–122	General industrial coatings
10–40/50–100	Decorative paints

3. Application: Compatible with other resins and used to improve application, lightfastness, gloss, or color retention. Used in automotive and appliance finishes.

THERMOSETTING RESINS

After application and solvent evaporation of a thermosetting resin a chemical change takes place. As this chemical reaction is taking place, the coating is said to be "curing." This reaction may take place at room temperature, or in baked coatings at elevated temperatures. The reaction is irreversible. Exposure to high temperature or to solvents does not cause the coating to soften or melt.

Epoxy

Epoxy resins on their own are not suitable for protective coatings because when pigmented and applied, they dry to a hard, brittle film with very poor chemical resistance. By copolymerizing with other resins, the epoxy resins will form a durable protective coating.

Polyamine Epoxy

1. Resistance: Resistant to acids, acid salts, alkalies, and organic solvents.
2. Temperature resistance: 225°F/107°C dry, 190°F/88°C wet.
3. Limitations: Not resistant to UV light, harder and less flexible than other epoxies; less tolerant of moisture during application.
4. Applications: Widest range of chemical and solvent resistance of epoxies; used for piping and vessels.

Polyamide Epoxy

1. Resistance: Resistance inferior to that of polyamine epoxy; only partially resistant to acids, acid salts, alkalies and organic solvents; resistant to moisture.
2. Temperature resistance: 225°F/108°C dry; 150°F/66°C wet.
3. Limitations: Not resistant to UV light; inferior chemical resistance.
4. Applications: Used on wet surfaces under water, as in tidal zone areas of pilings, oil rigs, etc.

Aliphatic Polyamine

1. Resistance: Partially resistant to acids, acid salts, and organic solvents.
2. Temperature resistance: 225°F/108°C dry; 150°F/66°C wet.
3. Limitations: Film formed has greater permeability than other epoxies.
4. Application: Used for protection against mild atmospheric corrosion.

Esters of Epoxies and Fatty Acids

1. Resistance: Resistant to weathering and UV light.
2. Temperature resistance: 225°F/108°C dry; 150°F/66°C wet.
3. Limitations: Chemical resistance generally poor.
4. Applications: Used where the properties of a high quality oil base paint are required.

Coal Tar Plus Epoxy

1. Resistance: Excellent resistance to freshwater, saltwater, and inorganic acids.
2. Temperature resistance: 225°F/108°C dry; 150°F/66°C wet.
3. Limitations: Not resistant to weather and sunlight; attacked by organic solvents.
4. Applications: Used on clean blasted steel for immersion service or below grade service.

Urethanes

Polyurethane-resin based coatings are versatile. They are available as oil modified, moisture curing, blocked, two component, and lacquers. Properties of the various types are shown in Table 13.2.

The urethane coatings possess excellent gloss and color retention and are used as a decorative coating of tank cars and steel in highly corrosive environments. Moisture cured types require humidity during application and may yellow under UV light. They have a temperature resistance of 250°F/121°C dry; 150°F/66°C wet.

Catalyzed (two component) urethanes exhibit very good chemical resistance. They are not recommended for immersion or exposure to strong acids/alkalies. They have a temperature resistance of 225°F/108°C dry; 150°F/66°C wet.

These resins are quite expensive.

Polyesters

In the coating industry, polyesters are characterized by resins based on components that introduce unsaturation ($-C=C-$) directly into the polymer backbone. The most common polyester resins are polymerization products of maleic or isophthalic anhydride or their acids. In producing paint the polyester resin is dissolved in styrene monomer, together with pigment and small amounts of inhibitor. A free radical initiator, commonly a peroxide, and additional styrene are packaged in another container. When applied, the containers are mixed. Sometimes, because of the fast initiating reaction (short pot life) they are mixed in an

TABLE 13.2 Properties of Polyurethane Coatings

Property	One-component Urethane oil	One-component Moisture	One-component Blocked	Two-component	Lacquer
Abrasion resistance	Fair–good	Excellent	Good–excellent	Excellent	Fair
Hardness	Medium	Medium–hard	Medium–hard	Soft–very hard	Soft–medium
Flexibility	Fair–good	Good–excellent	Good	Good–excellent	Excellent
Impact resistance	Good	Excellent	Good–excellent	Excellent	Excellent
Solvent resistance	Fair	Poor–fair	Good	Excellent	Poor
Chemical resistance	Fair	Fair	Good	Excellent	Fair–good
Corrosion resistance	Fair	Fair	Good	Excellent	Good–excellent
Adhesion	Good	Fair–good	Fair	Excellent	Fair–good
Toughness	Good	Excellent	Good	Excellent	Good–excellent
Elongation	Poor	Poor	Poor	Excellent	Excellent
Tensile	Fair	Good	Fair–good	Good–excellent	Excellent
Cure rate	Slow	Slow	Fast	Fast	None
Cure Temperature	Room	Room	300–390°F 149–199°C	212°F/100°C	150–225°F 66–108°C

externally mixing or dual headed spray gun. After being mixed and applied, a relatively fast reaction takes place, resulting in cross-linking and polymerization of the monomeric styrene with the polyester resin.

Polyester coatings exhibit high shrinkage after application. The effect of high shrinkage can be reduced by proper pigmentation which reinforces the coating and reduces the effect of the shrinkage.

1. Resistance: Excellent resistance to acids and aliphatic solvents; good resistance to weathering.
2. Temperature resistance: 180°F/82°C dry and wet.
3. Limitations: Not suitable for use with alkalies and most aromatic solvents since they swell and soften these coatings. They also have a short pot life and must be applied with special equipment.
4. Applications: Coatings for tanks and chemical process equipment.

Vinyl Esters

The vinyl esters are formulated in a manner similar to the polyesters. They are derived from a resin based on a reactive end vinyl group that can open and polymerize. The vinyl esters have better moisture and chemical resistance than the polyesters. They also have good resistance to weathering and acids.

AUTOOXIDATIVE CROSS-LINKING PAINTS

Paints of this type rely on the reaction of a drying oil with oxygen to introduce cross-linking within the resin and attainment of final film properties. The final coating is formed when a drying oil reacts with a resin, which is then combined with pigments and solvents. These paints are packaged in a single can. The can is opened, the paint mixed and applied. As the solvents evaporate, the coating becomes hard. Since oxygen in the atmosphere reacts with the coating to produce additional cross-linking, the final film properties may be formed weeks or months after application.

Among these resin fortified "oil based paints" are included the alkyds, epoxy esters, oil modified urethanes, etc. These paints may be formulated as airdrying or baking types, and with suitable pigments be formulated to be resistant to a variety of moisture and chemical fume environments, as well as application over wood, metal, and masonry substrates. These paints have the following advantages:

Ease of application.
Excellent adhesion.
Relatively good environmental resistance.
Tolerance for poorer surface preparation.

Their major disadvantages are:

> Cannot be used for immersion or in areas of high chemical fume concentrations.
> Reduced moisture and chemical resistance compared to other resin coatings.

Epoxy Esters

Esters of epoxies and fatty acids modified.

1. Resistance: Resistant to weathering; limited resistance to chemicals and solvents.
2. Temperature resistance: 225°F/108°C dry; 150°F/66°C wet.
3. Limitations: Attacked by alkalies. Application time between coats is critical.
4. Applications: Used where the properties of a high quality oil based paint is required.

Alkyd Resin

These paints are the most widely used synthetic resins. They are produced by reactions of phthalic acid with polyhydric alcohols, such as glycerol and vegetable oil (or fatty acid). Classification is by oil content into drying types and baking types. Drying types provide good weathering resistance and good adhesion to a wide variety of substrates, but relatively poor resistance to chemical attack. They have a maximum temperature resistance of 225°F/108°C dry, 150°F/66°C wet.

These paints are used on exterior wood surfaces for primers requiring penetrability and in less severe chemical environments.

Oil Based Paints

These paints are mixtures of pigment and boiled linseed oil or soybean oil, or other similar materials. This paint is used as a ready mixed paint particularly on wood surfaces because of its penetrating power. It is resistant to weather but has relatively poor chemical resistance and will be attacked by alkalies. Temperature limitations are 225°F/108°C dry; 150°F/66°C wet.

Water Emulsion Paints

Water based latex emulsions consist of fine particles of high molecular weight copolymers of polyvinyl chloride, or polyvinyl acetate, acrylic esters, styrene-butadiene, or other resins combined with pigments, plasticizers, UV stabilizers, and other ingredients. A variety of thickeners, coalescing aids, and other additives are present in the water phase.

The paint film is formed by water evaporation. Initial adhesion may be relatively poor as the water continues to evaporate, and coalescing aids, solvents, and surfactants evaporate or are leached from the "curing" film. Final adhesion and environmental resistant properties are reached from as little as a few days to a few months after application.

1. Resistance: Resistant to weather but poor chemical resistance.
2. Temperature resistance: 150°F/66°C dry or wet.
3. Limitations: Not suitable for immersion service.
4. Applications: Used in general decorative applications on wood.

Silicones

There are two types: high temperature; water repellent in water or solvent.

1. Resistance: Excellent resistance to sunlight and weathering; poor resistance to acids and alkalies; resistant to water.
2. Temperature resistance: In aluminum formulation resistant to 1200°F/ 699°C and to weathering at lower temperatures.
3. Limitations: High temperature type requires baking for good cure; not chemically resistant.
4. Applications: Used on surfaces exposed to high temperatures as water repellant; water solvent formulations used on limestone, cement, and nonsilaceous materials; solvent formulations used on bricks and non-calcareous masonry.

BITUMINOUS PAINTS

There are two paints in this category: asphalt and coal tar.

Asphalt Paint

Asphalt paint consists of solids from crude oil refining suspended in aliphatic solvents.

1. Resistance: Good moisture resistance; good resistance to weak acids, alkalies, and salts; better weathering properties than coal tar.
2. Temperature resistance: Softens at 100°F/38°C.
3. Limitations: Heavy dark color hides corrosion under coating.
4. Applications: Used in above ground weathering environments and chemical fume atmospheres.

TABLE 13.3 Properties of Paints

Coating type	Resistance to						Comments
	UV	Weather	Acid	Alkali	Moisture	Salt solutions	
Vinyls dissolved in esters, aromatics, or ketones.	R		R	R	R	R	Adhesion may be poor until all solvents have vaporized from the coating.
Chlorinated rubbers dissolved in hydrocarbon solvents.	N		R	R	R		Excellent adhesion to metals, concrete, and masonry. Used on structures exposed to water and marine atmospheres.
Epoxies, polyamine plus epoxy resin.	N		PR	PR	R	PR	Harder and less flexible than other epoxies. Greatest chemical resistance of the epoxies.
Polyamide plus epoxy resin (polyamide epoxy).	N		PR	PR	R	PR	Chemical resistance inferior to that of the polyamine epoxies.
Aliphatic polyamine.			PR	PR		PR	Flexible film.
Esters of epoxies and fatty acids (epoxy esters).	R	R	PR	N		PR	On surfaces requiring the properties of a high quality oil based paint.
Coal tar plus epoxy resin.	N	N	R		R		Used on clean blast cleaned steel for immersion or below grade service. Attacked by organic solvents.
Oil-based coatings with vehicle (alkyd epoxy, urethane).	R	R		N			Lower cost than most coatings. Used on exterior wood surfaces.

Coating						Remarks
Urethane, moisture cured.	R	R	Weak/R	Weak	R	May yellow under UV light. High gloss and ease of cleaning.
Urethanes catalyzed.	R	R	R	R	R	Expensive. Used as coating on steel in highly corrosive areas.
Silicones, water repellent in water or solvent.	R	R	N	N	R	Used on masonry surfaces.
Silicones, water based aqueous emulsions of polyvinyl acetate, acrylic or styrene-butadiene latex.	R	R	N	N		May flash rust as a primer on steel. Not chemically resistant.
Polyesters, organic acids combined with polybasic alcohols. Styrene is a reaction diluent.	PR	R	R	N		Must be applied with special equipment. Not suitable for use with most aromatic solvents.
Coal tar.	N	N	Weak/R	Weak/R	R	Used on submerged or buried steel.
Asphalt. Solids from crude oil refining in aliphatic solvents.	R	R	Weak/R	Weak/R	Weak/R	Used in above ground weathering environments and chemical fume atmospheres.
Zinc-rich metallic zinc in an organic or inorganic vehicle.						Provides galvanic protection as a primer.
Acrylic-resin water emulsion base.	R	R	Weak/R	Weak	R	Limited penetrating power. May flash rust as a primer over bare steel. Not suitable for immersion service. Soluble in ketones, esters, aliphatic chlorinated hydrocarbons, and aromatic hydrocarbons.

R, resistant; N, not resistant; PR, poor resistance

Coal Tar

Coal tar is a distilled coking by-product in an aromatic solvent.

1. Resistance: Excellent resistance to moisture; good resistance to weak acids and alkalies, petroleum oils, and salts.
2. Temperature resistance: Will soften at 100°F/38°C.
3. Limitations: Will be degraded by exposure to UV light and weathering.
4. Applications: Used on submerged or buried steel and concrete.

ZINC-RICH PAINTS

Zinc-rich paints owe their protection to galvanic action. While all of the preceding coatings owe their final film properties, corrosion resistance, and environmental resistance to the composition of their resin or binder, rather than their pigment, the high amount of zinc dust metal pigment in zinc-rich paints determines the coating's fundamental property: galvanic protection. Many of the previous coatings, chlorinated rubber and epoxies in particular, are formulated as zinc-rich coatings. In so doing, the high pigment content changes the properties of the formulated coating.

Zinc-rich coatings can be classified as organic or inorganic. The organic zinc-riches have organic binders, with polyamide epoxies and chlorinated rubber binders being the most common. Other types such as urethane zinc-rich are also available. These latter coatings are more easily applied than the other zinc-rich coatings.

Inorganic zinc-rich binders are based on silicate solutions, which after drying or curing, crystallize and form an inorganic matrix, holding zinc dust particles together and to the steel substrate.

For a coating to be considered zinc-rich, it must contain at least 75% by weight of zinc dust in the dry film. This may change since conductive extenders (notably di-iron phosphite) have been added to improve weldability and burn-through, with supposedly equivalent protection at lower zinc loadings.

The primary advantage of zinc-rich coatings is their ability to protect galvanically. The zinc pigment in the coating preferentially sacrifices itself in the electrochemical corrosion reaction to protect the underlying steel. This galvanic reaction, together with the filling and sealing effect of zinc reaction products (primarily zinc carbonate, zinc hydroxide, and complex zinc salts), provides more effective corrosion protection to steel substrates than does any other type of coating. Zinc-rich coatings cannot be used outside the pH range of 6–10.5.

Table 13.3 presents a summary of the properties of various paint formulations.

14

Selecting a Paint System

The first step in selecting a paint system for corrosion protection is to determine the environment around the structure or item to be painted. Is the environment predominantly a weathering environment subject to heat, cold, daily or seasonal temperature changes, precipitation, wind (flexing), exposure to sunlight, or detrimental solar rays? If the structure or item is located outdoors, are there chemical plants located nearby, or pulp and paper mills, or other industrial facilities that are apt to discharge airborne pollutants? Is color, gloss, and overall pleasing effect more important than corrosion protection or are the normal grays, whites, and pastels of the more corrosion resistant paints satisfactory? If located in a chemical facility, what specific chemicals are used nearby; is there any chance of a chemical spill on the painted surface?

Since surface preparation is an important factor in the selection of a paint system, the suitability or availability of the surface for specific preparation techniques must be known. In some instances certain types of surface preparation may not be permitted or practical. For example, many companies do not permit open blast cleaning where there is a prevalence of electric motors or hydraulic equipment. Refineries in general do not permit open blast cleaning, or any other method of surface preparation that might result in the possibility of a spark, static electricity buildup, or an explosion hazard.

TABLE 14.1 Multilayer Paint Systems Requiring Commercial Blast (SSPC-SP-6) for Surface Preparation

---

System A. Inorganic Zinc/Epoxy Mastic

Paint layers:
One coat inorganic zinc: 2–3 mils dft (50–75 µm)
One coat epoxy mastic: 4–6 mils dft (100–150 µm)
Properties:
Zinc primer provides outstanding corrosion resistance and undercutting resistance. A barrier protection for the zinc primer is provided by the finish coat of epoxy which also provides a color coat for appearance. Suitable for use on carbon steel only.
Limitations:
A relatively high level of applicator competence required for the primer.

System B. Inorganic Zinc/Epoxy Primer/Polyurethane Finish

Paint layers:
One coat inorganic zinc: 2–3 mils dft (50–75 µm)
One coat epoxy primer: 4–6 mils dft (100–150 µm)
One coat polyurethane finish: 2–4 mils dft (50–100 µm)
Properties:
Zinc primer provides outstanding corrosion resistance and undercutting resistance. The zinc primer is protected by a barrier coating of epoxy primer, while the finish coat of polyurethane provides color and gloss retention. This is a premium industrial finish for steel surfaces. Can only be used on carbon steel.
Limitations:
A relatively high level of applicator competence required for the primer.

System C. Inorganic Zinc/Acrylic Finish

Paint layers:
One coat inorganic zinc: 2–3 mils dft (50–75 µm)
One coat acrylic finish: 2–3 mils dft (50–75 µm)
Properties:
Zinc primer provides outstanding corrosion resistance and undercutting resistance. Water based single package finish has excellent weathering and semigloss appearance.
Limitations:
A relatively high level of applicator competence required for the primer. The finish coat has low temperature curing restrictions.

System D. Aluminum Epoxy Mastic/Epoxy Finish

Paint layers:
One coat aluminum epoxy mastic: 4–6 mils dft (100–150 μm)
One coat epoxy finish: 4–6 mils dft (100–150 μm)

Properties:
Can be used on tight rust and marginally prepared surface. The epoxy finish coat is available in a variety of colors and has good overall chemical resistance. May be used on carbon steel or concrete. Concrete must be clean, rough, and cured at least 28 days. Hand or power tool cleaning, including water blasting, may be used for surface preparation.

System E. Aluminum Epoxy Mastic/Acrylic Finish

Paint layers:
One coat aluminum epoxy mastic: 4–6 mils dft (100–150 μm)
One coat acrylic finish: 2–3 mils dft (50–75 μm)

Properties:
May be used on tight rust and marginally prepared surfaces. The acrylic finish coat is available in a variety of colors and has good overall chemical resistance. This is an excellent maintenance system. Normally used on carbon steel and concrete.

System F. Epoxy Mastic/Epoxy Mastic

Paint layers:
One coat epoxy mastic: 4–6 mils dft (100–150 μm)
One coat epoxy mastic: 4–6 mils dft (100–150 μm)

Properties:
May be used on tight rust and marginally prepared surfaces. The substrate is protected by the formation of a tight barrier stopping moisture from reaching the surface. Normally used on steel or concrete. Concrete must be clean, rough, and cured at least 28 days. If necessary, hand or power tools may be used for cleaning.

System G. Epoxy Primer/Epoxy Finish

Paint layers:
One coat epoxy primer: 4–6 mils dft (100–150 μm)
One coat epoxy finish: 4–6 mils dft (100–150 μm)

Properties:
An easily applied two coat high build barrier protection is provided with ease of application. Used on carbon steel or concrete.

(continued)

TABLE 14.1 Continued

Limitations:
Since these are two component materials, they must be mixed just prior to application. They require additional equipment and more expertise to apply than a single-packed product. Most epoxy finish coats will chalk, fade, and yellow when exposed to sunlight.

System H. Epoxy Primer

Paint layers:
One coat epoxy primer: 4–6 mils dft (100–150 μm)
Properties:
Normally applied to carbon steel or concrete in protected areas such as the interiors of structures, behind walls and ceilings, or for temporary protection during construction.
Limitations:
This is a two component material requiring mixing just prior to application.

System I. Epoxy Novalac/Epoxy Novalac

Paint layers:
One coat epoxy novalac: 6–8 mils dft (150–200 μm)
One coat epoxy novalac: 6–8 mils dft (150–200 μm)
Properties:
An exceptional industrial coating for a wide range of chemical resistance and physical abuse resistance. Has a higher temperature resistance than standard epoxy. May be used to protect insulated piping or for secondary containment. Normally used on carbon steel and concrete surfaces.

If a new facility is being constructed, it is possible that during erection many areas may become enclosed or covered, or so positioned that access is difficult or impossible. These structures must be painted prior to installation.

When all of this information has been collected, the appropriate paint systems may be selected. In most instances it will not be practical or possible to select one single coating system for the entire plant. There will be areas requiring systems to provide protection from aggressive chemicals, while other areas may require coating systems simply for esthetics. If an area is a combination of mild and aggressive conditions, a coating system should be selected that will be resistant to the most aggressive condition.

Several typical industrial environmental areas have been illustrated to which coating systems may be exposed with recommendations for paint systems to be used in these areas. The paint systems are shown in Tables 14.1 through 14.6 with the appropriate surface preparation. The tables have been arranged based on surface preparation. Each coating system shown in a particular table requires the same surface preparation.

AREA 1. MILD EXPOSURE

This is an area where structural steel is embedded in concrete, encased in masonry, or protected by noncorrosive type fireproofing. In many instances no coating will be applied to the steel. However, it is a good idea to coat the steel substrate with a protective coating to protect them during construction and in case they end up being exposed either intentionally or unintentionally.

A good practice would be to apply a general use epoxy primer 3–5 mils dry film thickness (dft) (75–125 μm). If the surface cannot be abrasive blasted, a surface tolerant epoxy mastic may be used.

Recommended systems are found in Table 14.4, systems A and C.

AREA 2. TEMPORARY PROTECTION; NORMALLY DRY INTERIORS

This area consists of office space or dry storage areas (warehouses) or other locations exposed to generally mild conditions, or areas where oil based paints presently last for 10 or more years. If located in an industrial environment there is the possibility of exposure to occasional fumes, splashing, or spillage of corrosive materials. Because of this it is suggested that an industrial grade acrylic coating system or a single coat of epoxy be applied.

This recommendation is not suitable for interior surfaces that are frequently cleaned or exposed to steam cleaning. Refer to area 4. Recommended for this area are systems A and C in Table 14.4.

TABLE 14.2 Multilayer Paint Systems Requiring Surface to Be Abrasive Blasted in Accordance with SSPC-SP-10 Near White Blast

---

System A. Aluminum–Epoxy Mastic/Aluminum–Epoxy Mastic

Paint layers:
 One coat aluminum-epoxy mastic: 4–6 mils dft (100–150 μm)
 One coat aluminum-epoxy mastic: 4–6 mils dft (100–150 μm)

Properties:
 Tolerates poorly prepared surfaces and provides excellent barrier protection. A third coat may be added for additional protection. Can be used on carbon steel and concrete. Concrete must be clean, rough, and cured at least 28 days. If necessary this system can be applied to surfaces that are pitted or cannot be blasted. However, the service life will be reduced.

System B. Epoxy Phenolic Primer/Epoxy Phenolic Finish/Epoxy Phenolic Finish

Paint layers:
 One coat epoxy phenolic primer: 8 mils dft (200 μm)
 One coat epoxy phenolic finish: 8 mils dft (200 μm)
 One coat epoxy phenolic finish: 8 mils dft (200 μm)

Properties:
 Because of this system's outstanding chemical resistance it is often used in areas subject to frequent chemical spills. The finish coats are available in a limited number of colors. Normally used on carbon steel and concrete. Concrete must be clean, rough, and cured at least 28 days.

System C. Epoxy Phenolic Primer/Epoxy Phenolic Lining/Epoxy Phenolic Lining

Paint layers:
 One coat epoxy phenolic primer: 8 mils dft (200 μm)
 One coat epoxy phenolic lining: 8 mils dft (200 μm)
 One coat epoxy phenolic lining : 8 mils dft (200 μm)

Properties:

Because of the system's outstanding overall chemical resistance it is suitable for lining areas subject to flowing or constant immersion in a variety of chemicals. Normally used on carbon steel and concrete. When used on concrete the surface must be clean, rough, and cured at least 28 days.

System D. Epoxy/Epoxy

Paint layers:

One coat epoxy: 4–6 mils dft (100–150 μm)
One coat epoxy: 4–6 mils dft (100–150 μm)

Properties:

Two coats of the same product are applied providing a high build protection. Can be used in immersion service without the addition of corrosion inhibitors. When used in potable water systems the product must meet Federal Standard 61. A third coat may be added for additional protection. Normally used on carbon steel and concrete.

System E. Coal Tar Epoxy/Coal Tar Epoxy

Paint layers:

One coat coal tar epoxy: 8 mils dft (200 μm)
One coat coal tar epoxy: 8 mils dft (200 μm)

Properties:

Provides excellent barrier protection and is the most economical of the water lining systems or for water immersion. Normally used on carbon steel and concrete.

System F. Solventless Elastomeric Polyurethane

Paint layers:

One coat elastomeric polyurethane: 20–250 mils dft (500–6250 μm)

Properties:

Excellent barrier protection. Normally used on carbon steel and concrete.

Limitations:

Must be applied by a knowledgeable contractor.

TABLE 14.3 Multilayer Paint Systems Requiring Surface to Be Clean, Dry, and Free of Loose Dirt, Oil, and Chemical Contamination

System A. Aluminum Epoxy Mastic/Polyurethane Finish

Paint layers:
   One coat aluminum epoxy mastic: 4–6 mils dft (100–150 μm)
   One coat polyurethane finish: 2–4 mils dft (50–100 μm)
Properties:
   Excellent over tight rust. Tolerant of minimally prepared steel. May be applied to a wide range of surfaces, but normally used on carbon steel and concrete. This is a premium system to use when cleaning must be minimal.
Limitations:
   In order to cure properly, temperature must be above 50°F/10°C. For lower temperature requirements other aluminum epoxy/urethane mastics may be substituted.

AREA 3. NORMALLY DRY EXTERIORS

This includes such locations as parking lots, water storage tanks, exterior storage sheds, and lighting or power line poles, which are exposed to sunlight in a relatively dry location. Under these conditions oil based paints should last 6 or more years. These materials that are resistant to ultraviolet rays and are normally rated for exterior use include acrylics, alkyds, silicones, and polyurethanes.

Epoxies will lose gloss, normally chalk, and fade rapidly when exposed to UV rays. Recommended systems include A in Table 14.3, C in Table 14.4, A in Table 14.5, and A in Table 14.6.

AREA 4. FRESHWATER EXPOSURE

Under this category the surface to be protected is frequently wetted by freshwater from condensation, splash, or spray. Included are interior or exterior areas that are frequently exposed to cleaning or washing, including steam cleaning.

The systems used for these surfaces make use of inorganic zinc as a primer. Inorganic zinc is the best coating that can be applied to steel since it provides the longest term of protection. In some situations it may be necessary to substitute an organic zinc (an organic binder such as epoxy or polyurethane with zinc added) for the inorganic zinc.

Recommended systems are B, C, and E of Table 14.1 and A of Table 14.3.

AREA 5. SALTWATER EXPOSURE

This area includes interior or exterior locations on or near a seacoast or industrial environments handling brine or other salts. Under these conditions the surfaces are frequently wet by salt water and include condensation, splash, or spray.

TABLE 14.4 Multilayer Paint Systems for New Clean Surfaces, Free of Chemical Contamination

---

System A. Epoxy Mastic

Paint layers:
  One coat epoxy mastic: 3–5 mils dft (75–125 µm)
Properties:
  Good color selection, excellent chemical resistance, good physical characteristics, ease of maintenance. Used on carbon steel, concrete masonry units, masonry block (a filler is recommended), sheet rock (a sealer is required), wood, polyvinyl chloride, galvanized steel, and other surfaces.
Limitations:
  This is a two-component material that is mixed just prior to application. Additional equipment is required and more expertise to apply than a single packaged product. Epoxy solvents may be objectionable to some people.

System B. Acrylic Primer/Acrylic Intermediate/Acrylic Finish

Paint layers:
  One coat acrylic primer: 2–3 mils dft (50–75 µm)
  One cost acrylic intermediate: 2–3 mils dft (50–75 µm)
  One coat acrylic finish: 2–3 mils dft (50–75 µm)
Properties:
  This is a single package, water base, low odor, semigloss paint. It possesses excellent weathering and acidic acid resistance. Can be used on most surfaces including carbon steel, concrete, concrete masonry units, masonry block (a block filler is recommended), sheet rock (a sealer is required), wood, polyvinyl chloride, galvanized steel, stainless steel, copper, and fabric. May be applied over existing coatings of any type including inorganic zinc.
Limitations:
  Must be protected from freezing during shipping and storage. For application, temperature must be above 60°F/16°C and will remain so for 2–3 hr after application.

System C. Acrylic Primer/Acrylic Finish

Paint layers:
  One coat acrylic primer: 2–3 mils dft (50–75 µm)
  One coat acrylic finish: 2–3 mils dft (50–75 µm)
Properties:
  Excellent weathering and acidic chemical resistance, with good color selection.
Limitations:
  For best performance metallic surfaces should be abrasive blasted. For mild conditions hand or power cleaning may be sufficient. Paint must be applied when temperature exceeds 60°F/16°C.

TABLE 14.5 Multilayer Paint Systems Requiring an Abrasive Blast to the Substrate Surface

System A. Epoxy Primer/Polyurethane Finish

Paint layers:
  One coat epoxy primer: 3–5 mils dft (75–125 μm)
  One coat polyurethane finish: 2–3 mils dft (50–75 μm)
Properties:
  Two coat protection is provided with excellent high gloss finish and long-term color gloss retention. Normally applied to carbon steel and concrete.
Limitations:
  Since these are two component materials they must be mixed just prior to application and require additional equipment and more expertise to apply.

Conditions in this area are essentially the same as for freshwater and the comments for area 4 apply here. Because of the more severe conditions it is recommended that two coats of the primer be applied for system E of Table 14.1 and system A of Table 14.3.

Recommended systems are B, C, and E of Table 14.1 and A of Table 14.3.

AREA 6. FRESHWATER IMMERSION

Wastewaters are also a part of this area. Included are all areas that remain underwater for periods longer than a few hours at a time. Potable and nonpotable water, sanitary sewage, and industrial waste liquids are all included.

TABLE 14.6 Multilayer Paint Systems for Previously Painted Surfaces That Have Had Loose Paint and Rust Removed by Hand Cleaning

System A. Oleoresin

Paint layers:
  One coat oleoresin: 2–4 mils dft (50–100 μm)
Properties:
  This very slow drying material penetrates and protects existing surfaces that cannot be cleaned properly with a single coat. Provides long-term protection without peeling, cracking, and other such problems. Easy to apply by spray, brush, roller, or glove. Normally used on carbon steel and weathering galvanized steel.
Limitations:
  This material is designed to protect steel that will not see physical abuse. It also stays soft for an extended period of time.

If the systems recommended are to be used as a tank lining material, it is important that the application be done by experienced workers. In addition, if the coating to be applied is to be in contact with potable water, it is important that the material selected meets the necessary standards and is approved for use by the local health department. Two coats of epoxy (system D in Table 14.2) is frequently used in this service.

Recommended systems are F of Table 14.1 and A, D, E, and F of Table 14.2.

AREA 7. SALTWATER IMMERSION

Areas that remain under water in a coastal environment or industrial area, or that are constantly subjected to flowing salt or brine laden water are included in this category. Because of the increased rate of corrosion a third coat may be added to system F of Table 14.2 and systems A and E of Table 14.1 as additional protection against this more severe corrosion.

System D of Table 14.1 and systems A, E, and F of Table 14.2 are recommended for this service.

AREA 8. ACIDIC CHEMICAL EXPOSURE (pH 2.0–5.0)

In the chemical process industries this is one of the most severe environments to be encountered. When repainting, it is important that all surfaces are clean of any chemical residue. Inorganic zinc and zinc filled coatings are not recommended for application in this area.

The system selected will be dependent upon the quality of surface preparation, length of chemical exposure, and housekeeping procedures. Decreased cleanup and longer exposure times require a more chemical resistant coating system such as system I in Table 14.1.

Other recommendations for this area include systems D, G, and I in Table 14.1 and system B in Table 14.4.

AREA 9. NEUTRAL CHEMICAL EXPOSURE (pH 5.0–10.0)

This is an area that is not subject to direct chemical attack but may be subject to fumes, spillage, or splash. Under these conditions more protection is required than will be provided by a standard painting system. This would include such locations as clean rooms, packaging areas, hallways, enclosed process areas, instrument rooms, electrical load centers, and other similar locations.

A list of potential chemicals that may contact the coating aids in the coating selection. Knowledge of cleanup procedures will also prove helpful. It may be possible to use systems requiring less surface preparation such as system D in

Table 14.1, system A in Table 14.3, system A in Table 14.4, and system A in Table 14.5.

Recommendations for area 9 are systems A and D in Table 14.1, system A in Table 14.3, systems A and C in Table 14.4, and system A in Table 14.5.

AREA 10. EXPOSURE TO MILD SOLVENTS

This is intended for locations subject to intermittent contact with aliphatic hydrocarbons such as mineral spirits, lower alcohols, glycols, etc. Such contact can be the result of splash, spillage, or fumes.

Cross-linked materials, such as epoxies, are best for this service since solvents will dissolve single package coatings. A single coat of inorganic zinc is an excellent choice for immersion service in solvents or for severe splashes and spills.

The gloss of a coating system is often reduced as a result of solvent splashes or spills. However, this is a surface effect which usually does not affect the overall protective properties of the coating.

Recommended systems for use in this location are A, D, and G of Table 14.1.

AREA 11. EXTREME pH EXPOSURE

This covers locations that are exposed to strong solvents, extreme pHs, oxidizing chemicals, or combinations thereof with high temperatures. The usual choice for coating these areas are epoxy novalacs, epoxy phenolics, and high build polyurethanes. Other special coatings such as the polyesters and vinyl esters may also be considered. However, these systems require special application considerations.

Regardless of which coating system is selected, surface preparation is important. An abrasive blast, even on concrete, is required. In addition, all surface contaminents must be removed. When coating concrete a thicker film is required. System F in Table 14.2 is recommended for optimum protection.

Recommended for this location is system I in Table 14.1 and systems B, C, and F in Table 14.2.

The foregoing have been generalizations as to what environmental conditions may be encountered, along with suggested coating systems to protect the substrate. Data presented will act as a guide in helping the reader to select the proper coating system. Keep in mind that the surface preparation is critical and should not be skimped on.

15

Conversion Coatings

The term *conversion coating* is used to describe coatings in which the substrate metal provides ions which become part of the protective coating. The coating layers are composed of inorganic compounds that are chemically inert. These inert compounds on the surface reduce both anodic and cathodic areas and delay the transit of reactive species to the base metal. This results in increases in the slopes of anodic and cathodic polarization curves, thereby decreasing the rate of corrosion of the substrate.

Conversion layers are used for various reasons:

1. To improve the adherence of the organic layers
2. To obtain electrically insulating barrier layers
3. To provide a uniform grease free surface
4. To provide active corrosion inhibition by reducing the rate of the oxygen reduction reaction, or by passivating the metallic substrate

Conversion coatings belonging in this group are phosphate, chromate, oxide, and anodized coatings. These coatings are composed of corrosion products that have been formed artificially by chemical or electrochemical reactions in selected solutions. The corrosion products thus formed build a barrier protection for the substrate metal. This barrier reduces the active surface area on the base metal, thereby delaying the transport of oxidizers and aggressive species. By so doing, the coating inhibits the formation of corrosion cells. The degree of secondary barrier action depends on the compactness, continuity, and stability of the corrosion product layer.

Each conversion coating protects the base metal against corrosion with two or three of the following protective abilities:

1. Secondary barrier action of corrosion products
2. Inhibiting action of soluble compounds contained in the corrosion products
3. Improvement in paint adhesion by the formation of a uniform corrosion product layer

PHOSPHATE COATING

When a metal surface is treated with a weak phosphoric acid solution of iron, zinc, or manganese phosphate, phosphate layers are formed. These phosphate coatings are applied to iron and steel, zinc, aluminum, and their alloys.

Phosphate films are formed by the dissolution of base metal and the precipitation of phosphate films. The metal surface must be free of greases, oils, and other carbonaceous material before immersion in the phosphating solution or before spray application. Baths operating at 120°F/50°C have pH values of approximately 2, while those operating below 120°F/50°C have pH values of approximately 3.

The zinc phosphate coating is basically the result of a corrosion process. Reactions of iron and steel in a zinc phosphate solution are as follows:

1. The dissolution of base metal at the anodic sites:

$$Fe + 2H_3PO_4 \rightarrow Fe(H_2PO_4)_2 + H_2 \tag{15.1}$$

Promotion by the activator:

$$2Fe + 2H_2PO_4^- + 2H^+ + 3NO_2 \rightarrow 2FePO_4 + 3H_2O + 3NO \tag{15.2}$$

$$4Fe + 3Zn^{2+} + 6H_2PO_4^- + 6NO_2 \rightarrow \\ 4FePO_4 + Zn_3(PO_4)_2 + 6H_2O + 6NO \tag{15.3}$$

2. Precipitation of phosphate films at the cathodic sites:

$$2Zn(H_2PO_4)_2 + Fe(H_2PO_4)_2 + 4H_2O \rightarrow \underbrace{Zn_2Fe(PO_4)_2 \cdot 4H_2O}_{Phosphophyllite}$$

$$+ 4H_3PO_4 \tag{15.4}$$

$$3Zn(H_2PO_4)_2 + 4H_2O \rightarrow \underbrace{Zn_3(PO_4)_2 \cdot 4H_2O}_{Hopeite} + 4H_3PO_4 \tag{15.5}$$

With these reactions the phosphate film consists of phosphophyllite and hopeite. Phosphate solubilities are lowest in the pH range of 6–8. They are stable in neutral environments and are nonelectric conductive compounds. Phosphate film deposits on cathodic areas and anodic sites remain in the form of pinholes. Consequently the continuity of phosphate films is not as good as those of anodic oxide and chromate films.

Since the barrier action of a conversion film is dependent on its solubility and continuity, it is evident that the phosphate films provide only limited protection. However, they do provide an excellent base for paint, plastic, and rubber coatings.

The chemical effects of phosphating on the surface is to convert the surface to a nonalkaline condition, protecting the surface from reactions with oils in paint and to protect against the spread of corrosion from defects. Alkaline residues on the surface of the base metal lead to underfilm corrosion.

Phosphating increases the uniformity in the surface texture and surface area, which improves paint adhesion, which in turn increases the service life of a paint film.

CHROMATE COATING

Chromate conversion coatings are formed on aluminum and its alloys, magnesium, zinc, and cadmium. These coatings provide good corrosion protection and improve the adhesion of organic layers. A chromate coating is composed of a continuous layer consisting of insoluble trivalent chromium compounds and soluble hexavalent chromium compounds. The coating structure provides a secondary barrier, inhibiting action, and also good adhesion for lacquer films.

Chromate coatings provide their corrosion resistance based on the following three properties:

1. Cr(III) oxide, which is formed by the reduction of Cr(IV) oxide, has poor solubility in aqueous media and thereby provides a barrier layer.
2. Cr(VI) will be included in the conversion coating and will be reduced to Cr(III) to passivate the surface when it is damaged, preventing hydrogen gas from developing.
3. The rate of cathodic oxygen reactions is strongly reduced.

Most chromate conversion coatings are amorphous gel-like precipitates, so they are excellent in continuity. The service life is dependent on thickness, characteristics of the base metal, coating conditions—particularly dry heat—and the environmental conditions under which the chromated products are used.

When a chromated product is exposed to the atmosphere, hexavalent chromium slowly leaches from the film, with the result that the surface appearance changes from irridescent yellow to either a green color or to clear. The structure of the film consists of more of the insoluble trivalent chromium compounds. Passivation is provided for any damaged areas by the leached hexavalent chromium.

The longer the time of wetness, the shorter the service life of the coating since chromate coatings absorb moisture and moisture results in the leaching of hexavalent chromium. The leaching behavior of a chromate film is also affected by its aging process, drying process, and long-term storage. Aging of a chromate coating reduces its protective ability.

Chrome baths always contain a source of hexavalent chromium ion (e.g., chromate, dichromate, or chromic acid) and an acid to produce a low pH which usually is in the range of 0–3. A source of fluoride ions is also usually present. These fluoride ions will attack the original (natural) aluminum oxide film, exposing the base metal substrate to the bath solution. Fluoride also prevents the aluminum ions (which are released by the dissolution of the oxide layer) from precipitating by forming complex ions. The fluoride concentration is critical. If the concentration is too low, a conversion layer will not form because of the failure of the fluoride to attack the natural oxide layer, while too high a concentration results in poor adherence of the coating due to reaction of the fluoride with the aluminum metal substrate.

During the reaction, hexavalent chromium is partially reduced to trivalent chromium, forming a complex mixture consisting largely of hydrated hydroxides of both chromium and aluminum:

$$6H^+ + H_2Cr_2O_7 + 6e^- \rightarrow 2Cr(OH)_3 + H_2O \tag{15.6}$$

There are two types of processes by which conversion coatings can be produced: chromic acid processes and chromic acid–phosphoric acid processes. In the formation of the chromic acid based conversion coating the following overall equation governs:

$$6H_2Cr_2O_7 + 3OHF + 12Al + 18HNO_3 \rightarrow$$
$$3Cr_2O_3 + Al_2O_3 + 10AlF_3 + 6Cr(NO_3)_3 + 30H_2O \tag{15.7}$$

The oxide Cr_2O_3 is better described as an amorphous chromium hydroxide, $Cr(OH)_3$. The conversion coating is yellow to brown in color.

In the chromic acid–phosphoric acid process the following reaction governs:

$$2H_2Cr_2O_7 + 10H_3PO_4 + 12HF + 4Al \rightarrow$$
$$CrPO_4 + 4AlF + 3Cr(H_2PO_4)_3 + 14H_2O$$

This conversion coating is greenish in color and consists primarily of hydrated chromium phosphate with hydrated chromium oxide concentrated toward the metal.

The barrier action of a chromate coating increases with its thickness. Chromate conversion coatings can be used as a base for paint or alone for corrosion protection. Previously it was described how the leached hexavalent chromium acts as an anodic inhibitor, by forming passive films over defects in the coating. Since the films formed on aluminum by the chromic acid–phosphoric acid process contain no hexavalent chromium, they do not provide self-healing from defects.

The service life of a chromate coating depends on the coating thickness. Chromate coatings absorb moisture, and moisture results in the leaching of hexavalent chromium. Therefore, the longer the time of wetness, the shorter the life of the coating. However, as long as the leaching of the hexavalent chromium continues, the base metal is protected.

Environmental conditions, particularly time of wetness and temperature, determine the leaching rate. In natural environments the leaching rate is commonly low. Pollutants in the atmosphere, particularly chloride ions, also increase the rate of deterioration of the film. Chromate conversion coatings provide good corrosion resistance in a mild atmosphere, such as indoor atmospheres, and surface appearance. They also provide a good base for organic films.

Chromate conversion coatings are usually applied to zinc and its alloy coated sheets, to protect against staining during storage, and to products of zinc-die castings, aluminum and its alloys, and magnesium and its alloys.

OXIDE COATINGS

Iron or steel articles to be coated are heated in a closed retort to a temperature of 1600°F/871°C after which superheated steam is admitted. This results in the formation of red oxide (Fe_2O_3) and magnetic oxide (Fe_3O_4). Carbon monoxide is then admitted to the retort, reducing the red oxide to magnetic oxide, which is resistant to corrosion. Each operation takes approximately 20 min.

Iron and steel may also be oxide coated by electrolytic means. The article to be coated is made the anode in an alkaline solution (anodic oxidation). These coatings are primarily for appearance, such as for cast iron stove parts.

ANODIZED COATINGS

The electrochemical treatment of a metal serving as an anode in an electrolyte is known as anodizing. Since aluminum's electrode potential is negative and its oxide film is stable in natural environments, surface treatments have been developed for the purpose of producing more stable oxide films. The anodic films formed can be either porous or nonporous depending upon which electrolyte is used.

Porous films result when electrolytes such as sulfuric acid, oxalic acid, chromic acid, and phosphoric acid are used. These films have the advantage of being able to be dyed.

Sulfuric acid is the most widely used electrolyte. A large range of operating conditions can be utilized to produce a coating to meet specific requirements. Hard protective coatings are formed which serve as a good base for dyeing. In order to obtain the maximum corrosion resistance, the porous coating must be sealed after dyeing. The anodic coating formed, using sulfuric acid as the electrolyte, is clear and transparent on pure aluminum. Aluminum alloys containing silicon or manganese and the heterogeneous aluminum-magnesium alloy yield coatings that range from gray to brown and may be patchy in some cases. The adsorptive power of these coatings make them excellent bases for dyes, especially if they are sealed in nickel or cobalt acetate solution.

It is not recommended to use sulfuric acid as the electrolyte for anodizing work containing joints which can retain the sulfuric acid after removal from the bath. The retained electrolyte will provide sites for corrosion.

When chromic acid is used as the electrolyte, the coatings produced are generally opaque, gray, and irridescent, with the quality being dependent on the concentration and purity of the electrolyte. These are unattractive as compared to those formed using sulfuric acid as the electrolyte. When 0.03% sulfate is added to the electrolyte, colorless and transparent coatings are formed. These coatings are generally thin, of low porosity, and hence difficult to dye. Black coatings can be obtained in concentrated solutions at elevated temperatures. Attractive opaque surfaces can be obtained by adding titanium, zirconium, and thallium compounds to the electrolyte.

The chromic acid anodizing process is the only one that can be used on structures containing blind holes, crevices, or difficult to rinse areas. Chromic acid anodizing generally increases fatigue strength, while sulfuric acid anodizing may produce decreases in fatigue strength.

Boric acid electrolytes produce a film that is irridescent and oxides in the range of 2500–7500 Å. The coating is essentially nonporous.

Oxalic and other organic acids are electrolytes used to produce both protective and decorative films. Unsealed coatings are generally yellow in color. These films are harder and more abrasion resistant than the conventional sulfuric acid films. However, the specially hard coatings produced under special conditions in sulfuric acid electrolytes are superior.

The anodized coating consists of two major components: the nonporous barrier layer adjoining the metal and a porous layer extending from the barrier layer to the outer portion of the film. Sulfuric, chromic, and oxalic acid electrolytes form both barrier and porous layers while boric acid electrolytes produce only barrier films.

Anodizing of aluminum provides long-term corrosion resistance and decorative appearance. Corrosion of the anodized film is induced by SO_x gas and depositions of grime, sulfates, and chlorides. These depositions promote corrosion since they tend to absorb aggressive gases and moisture, thereby increasing the time of wetness and decreasing the pH of the electrolyte at the interface between the depositions and the surface.

Although rain increases the time of wetness, it has the effect of cleaning the surface rather than making the surface corrosive. Some of the depositions are removed by the rain. Cleaning with water is one method which helps to protect the anodized aluminum from corrosion. In marine atmospheres the depositions can be removed with water since the depositions are primarily soluble chlorides. However, in industrial atmospheres detergents are needed because the deposits are greasy.

SO_x gas is the most aggressive pollutant for anodic films. The corrosive effect is dependent on the concentration, with the corrosion area increasing linearly with concentration.

16

Metallic Coatings

Metallic coatings are applied to metal substrates for several purposes. Typical purposes include improved corrosion resistance, wear resistance, and appearance. Of primary concern is corrosion resistance.

By providing a barrier between the substrate and the environment, or by cathodically protecting the substrate, metallic coatings protect the substrate from corrosion. Coatings of chromium, copper, and nickel provide increased wear resistance and good corrosion resistance. However, these noble metals make the combination of the substrate (mostly steel or an aluminum alloy) with the protective layer sensitive to galvanically induced local corrosion. Nonnoble metallic layers such as zinc or cadmium provide good cathodic protection but show poor wear resistance.

A coating of a corrosion resistant metal on a corrosion prone substrate can be formed by various methods. The choice of coating material and selection of an application method are determined by the end use.

METHODS OF PRODUCING COATINGS

Electroplating

This is one of the most versatile methods. The metal to be coated is made the cathode in an electrolytic cell. A potential is applied between the cathode on which the plating occurs and the anode, which may be the same metal or an inert material such as graphite. This method can be used for all metals which can be electrolytically reduced from the ionic state to the metallic state when present in an electrolyte. Aluminum, titanium, sodium, magnesium, and calcium cannot be electrodeposited from aqueous solution because the competing cathodic reaction, $2H^+ + 2e^- = H_2$, is strongly thermodynamically favored and takes place in preference to the reduction of the metal ion. These metals can be electrodeposited from conducting organic solutions or molten salt solutions in which the H^+ ion concentration is negligible.

Many alloys may be electrodeposited, including copper-zinc, copper-tin, lead-tin, cobalt-tin, nickel-cobalt, nickel-iron, and nickel-tin. The copper-zinc alloys are used to coat steel wire used in tire-cord. Lead-tin alloys are known as terneplate and have many corrosion resistant applications.

The thickness of coating can be accurately controlled since the amount deposited is a function of the number of coulombs passed.

Electroless Plating

This method is also known as chemical plating or immersion plating. It is based on the formation of metal coatings resulting from chemical reduction of metal ions from solution. The surface which is to be coated must be catalytically active and remain catalytically active as the deposition proceeds in a solution, which must contain a reducing agent. If the catalyst is a reduction product (metal) itself, autocatalysis is ensured, and in this case, it is possible to deposit a coating, in principle, of unlimited thickness. The advantages of electroless plating are:

1. Deposits have fewer pinholes.
2. Electric power supply is not required.
3. Nonconductive materials are metallized.
4. A functional layer is deposited.
5. A uniform layer is deposited, even on complex parts.
6. The equipment for electroless plating is simple.

Electroless plating is limited by the fact that:

1. It is more expensive than electroplating since the reducing agents cost more than an equivalent amount of electricity.
2. It is less intensive since the metal deposition rate is limited by metal ion reduction in the bulk of the solution.

Copper, silver, cobalt, and palladium are the most commonly plated metals using this process. The silvering of mirrors falls into this category.

Hypophosphite, amine boranes, formaldehyde, borohydride, and hydrozine are typical reducing agents. Deposits of nickel formed with hypophosphite as a

TABLE 16.1   Coatings Obtained by Electroless Plating

Metal	Reducing agent						
	$H_2PO_4^-$	N_2H_2	CH_2O	BH_4^-	RBH_3	Me ions	Others
Ni	Ni-P	Ni		Ni-B	Ni-B		
Co	Co-P	Co	Co	Co-B	Co-B		
Fe				Fe-B			
Cu	Cu	Cu	Cu	Cu	Cu	Cu	
Ag		Ag	Ag	Ag	Ag	Ag	Ag
Au		Au	Au	Au	Au		Au
Pd	Pd-P	Pd	Pd	Pd-B	Pd-B		
Rh		Rh					Rh
Ru				Ru			
Pt		Pt		Pt			Pt
Sn						Sn	
Pb			Pb				

reducing agent contain phosphorus. This alloying constituent determines many of the properties. Table 16.1 shows coatings obtained by electroless plating.

Electrophoretic Deposition

Finely divided materials suspended in an electrolyte develop a charge as a result of asymmetry in the charge distribution caused by the selective adsorption of one of the constituent ions. When the substrate metal is immersed in the electrolyte, and a potential is applied, a coating will be formed. If the particles have a negative charge they will be deposited on the anode and if they have a positive charge they will be deposited on the cathode. Commercial applications of this method in the case of metals is limited.

Cathodic Sputtering

This method is carried out under a partial vacuum. The substrate to be coated is attached to the anode. Argon, or a similar inert gas is admitted at low pressure. A discharge is initiated and the positively charged gas ions are attracted to the cathode. Atoms are dislodged from the cathode as the gas ions collide with the cathode. These atoms are attracted to the anode and coat the substrate. This method can be used for nonconducting as well as conducting materials. The major disadvantages are the heating of the substrate and low deposition rates.

Some of the most commonly used metals deposited by sputtering are aluminum, copper, chromium, gold, molybdenum, nickel, platinum, silver, tantalum, titanium, tungsten, vanadium, and zirconium.

Sputtered coatings are used for a wide variety of applications, among which are:

1. Metals and alloys are used as conductors, contacts, and resistors and in other components such as capacitors.
2. Some high performance magnetic data storage media are deposited via sputtering.
3. Thin metal and dielectric coatings are used to construct mirrors, antireflection coatings, light valves, laser optics, and lens coatings.
4. Hard coatings such as titanium carbide, nitride, and carbon produce wear resistant coatings for cutting tools.
5. Thin film coatings can be used to provide high temperature environmental corrosion resistance for aerospace and engine parts, gas barrier layers, and lightweight battery components.
6. Titanium nitride is deposited on watch bands and jewelry as a hard gold colored coating.

Diffusion Coating

This method requires a preliminary step followed by thermal treatment and diffusion of the coating metal into the substrate. A commercial material known as galvannealed steel is made by coating steel with zinc followed by heat treatment and the formation of an iron-zinc intermetallic coating by diffusion.

Flame Spraying

The coating metal is melted and kept in the molten condition until it strikes the substrate to be coated. Aluminum and zinc are applied in this manner. Flame sprayed aluminum has a lower density than pure aluminum because of voids in the coating.

Plasma Spraying

This method is similar to flame spraying except that forms of heating other than a flame are used.

Hot Dipping

Zinc coatings are applied to steel sheets by immersion of the steel in a molten bath of zinc to form galvanized steel. A small amount of aluminum is added to the bath to establish a good bond at the zinc-steel interface. The thickness of the zinc coating is controlled by rigid control of the galvanizing bath temperature, the speed of passage through the bath, the temperature of the steel sheet before it enters the bath, and the use of air jets which exert a wiping action as the steel sheet exits the bath.

Vacuum and Vapor Deposition

This method is used primarily for the formation of metallic coatings on nonconductive substrates. Common deposited coatings using this method include aluminum coatings on plastics and rhodium coatings on mirrors.

Gas Plating

Some metal compounds can be decomposed by heat to form the metal. Typical examples are metal carbonyls, metal halides, and metal methyl compounds.

Fusion Bonding

Coatings of low melting metals such as tin, lead, zinc, and aluminum may be applied by cementing the metal as a powder on the substrate then heating the substrate to a temperature above the melting point of the coating metal.

Explosion Bonding

This method produces a bond between two metals by the exertion of a strong force that compresses the two metals sufficiently to develop a strong interfacial interaction.

Metal Cladding

The most common method is roll bonding which produces full sized sheets of clad (coated) material. The bond formed is partly mechanical and partly metallurgical; consequently, metallurgically incompatible materials cannot be produced.

NOBLE COATINGS

Because of the high corrosion resistance of the noble metals, these materials are used where a high degree of corrosion resistance and decorative appearance are requirements. They find application in domestic appliances, window frames, bicycles, motorbikes, parts for car bodies, furniture, tools, flanges, hydraulic cylinders, shock absorbers in cars, and parts of equipment for the chemical and food processing industries.

Noble metal coatings protect substrates from corrosion by means of anodic control or EMF control. Coating metals that provide protection by means of anodic control include nickel, chromium, tin, lead, and their alloys.

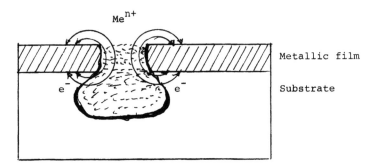

FIGURE 16.1   Dissolution of substrate metal in coating defect.

They can protect the substrate metal as a result of their resistance to corrosion insofar as they form a well adhering and nonporous barrier layer. However, when the coating is damaged, galvanically induced corrosion will lead to severe attack. This corrosion process is extremely fast for coated systems due to the high current density in the defect as a result of the large ratio between the surface areas of the cathodic outer surface and the anodic defect, as shown in Figure 16.1. In order to compensate for these defects in the coating, multilayer coating systems have been developed. The corrosion resistance of a single layer noble metal coating results from the original barrier action of the noble metal, the surface of the noble metal being passivated. With the exception of lead, a secondary barrier of corrosion products is not formed. Noble metals do not provide cathodic protection for the steel substrate since their corrosion potential is more noble than those of iron and steel in a natural environment. Refer to Table 16.2. In multilayer coating systems a small difference in potential between coating layers results in galvanic action in coating layers.

TABLE 16.2   Corrosion Potential of
Noble Metals

pH	Corrosion potential, V, SCE	
	2.9	6.5
Chromium	−0.119	−0.186
Nickel	−0.412	−0.430
Tin	−0.486	−0.554
Lead	−0.435	−0.637
Steel	−0.636	−0.505

TABLE 16.3 Standard Single Potentials, $E°$ (V, SHE, 25°C)

Inert Electrode	$E°$
H_2/H^+	±0
Cu/Cu^{2+}	+0.34
Cu/Cu^+	+0.52
Ag/Ag^+	+0.799
Pt/Pt^{2+}	+1.2
Au/Au^{3+}	+1.42
Au/Au^+	+1.7

Noble coating metals that provide corrosion protection by means of EMF control include copper, silver, platinum, gold, and their alloys. The standard single potentials of these metals are more noble than those of hydrogen (refer to Table 16.3). Therefore, the oxidizer in corrosion cells formed on these metals in a natural environment, containing no other particular oxidizers, is dissolved oxygen. Consequently, the electromotive force that causes corrosion is so small that coating with noble metals is an effective means of providing corrosion protection. With the exception of copper the other members of this group are precious metals, and are used primarily for electrical conduction and decorative appearance.

NICKEL COATINGS

There are three types of nickel coatings: bright, semibright, and dull bright. The difference between the coatings is in the quantity of sulfur contained in them as shown below:

Bright nickel deposits	>0.04% sulfur
Semibright nickel deposits	<0.005% sulfur
Dull bright nickel deposits	<0.001% sulfur

The corrosion potentials of the nickel deposits are dependent on the sulfur content. Figure 16.2 shows the effect of sulfur content on the corrosion potential of a nickel deposit. A single layer nickel coating must be greater than 30 μm to ensure absence of defects.

As the sulfur content increases, the corrosion potential of a nickel deposit becomes more negative. A bright nickel coating is less protective than a semibright or dull nickel coating. The difference in the potential of bright nickel and semibright nickel deposits is more than 50 mV.

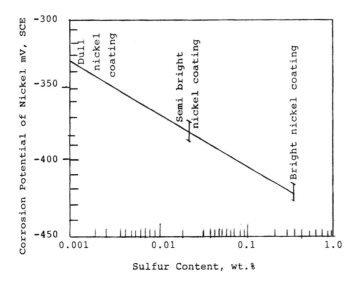

FIGURE 16.2 Effect of sulfur content on corrosion protection of nickel.

Use is made of the differences in the potential in the application of multilayer coatings. The more negative bright nickel deposits are used as sacrificial intermediate layers. When bright nickel is used as an intermediate layer, the corrosion behavior is characterized by a sideways diversion. Pitting corrosion is diverted laterally when it reaches the more noble semibright nickel deposit. Thus the corrosion behavior of bright nickel prolongs the time for pitting penetration to reach the base metal.

The most negative of all nickel deposits is trinickel. In the triplex layer coating system a coating of trinickel approximately 1 μm thick, containing 0.1–0.25% sulfur, is applied between bright nickel and semibright nickel deposits. The high sulfur nickel layer dissolves preferentially, even when pitting corrosion reaches the surface of the semibright nickel deposit. Since the high sulfur layer reacts with the bright nickel layer, pitting corrosion does not penetrate the high sulfur nickel layer in the tunneling form. The application of a high sulfur nickel strike definitely improves the protective ability of a multilayer nickel coating.

In the duplex nickel coating system, the thickness ratio of semibright nickel deposit to bright nickel deposit is nominally 3:1, and a thickness of 20–25 μm is required to provide high corrosion resistance. The properties required for a semibright nickel deposit are as follows:

1. The deposit contains little sulfur.
2. Internal stress must be slight.
3. Surface appearance is semibright and extremely level.

For a trinickel (high sulfur) strike the following properties are required:

1. The deposit contains a stable 0.1–0.25% sulfur.
2. The deposit provides good adhesion for semibright nickel deposits.

Nickel coatings can be applied by electrodeposition or electrolessly from an aqueous solution without the use of an externally applied current.

Depending on the production facilities and the electrolyte composition, electrodeposited nickel can be relatively hard (120–400 HV). Despite competition from hard chromium and electroless nickel, electrodeposited nickel is still being used as an engineering coating because of its relatively low price. Some of its properties are:

1. Good general corrosion resistance
2. Good protection from fretting corrosion
3. Good machineability
4. The ability of layers of 50–75 μm to prevent scaling at high temperatures
5. Mechanical properties, including the internal stress and hardness, that are variable and that can be fixed by selecting the manufacturing parameters
6. Excellent combination with chromium layers
7. A certain porosity
8. A tendency for layer thicknesses below 10–20 μm on steel to give corrosion spots due to porosity

The electrodeposition can be either directly on steel or over an intermediate coating of copper. Copper is used as an underlayment to facilitate buffing, because it is softer than steel, and to increase the required coating thickness with a material less expensive than nickel.

The most popular electroless nickel plating process is the one in which hypophosphite is used as the reducer. Autocatalytic nickel ion reduction by hypophosphite takes place in both acid and alkaline solutions. In a stable solution with a high coating quality, the deposition rate may be as high as 20–25 μm/h. However, a relatively high temperature of 194°F/90°C is required. Since hydrogen ions are formed in the reduction reaction,

$$Ni^{2+} + 2H_2PO_2^- + 2H_2O \rightarrow Ni + 2H_2 + 2H^-$$

A high buffering capacity of the solution is necessary to ensure a steady state process. For this reason acetate, citrate, propionate, glycolate, lactate, or amino-acetate is added to the solutions. These substances along with buffering may form complexes with nickel ions. Binding Ni^{2+} ions into a complex is required in alkaline solutions (here ammonia and pyrophosphate may be added in addition to citrate and amino-acetate). In addition, such binding is desirable in acid solutions because free nickel ions form a compound with the reaction product (phosphate) which precipitates and prevents further use of the solution.

When hypophosphite is used as the reducing agent, phosphorus will be present in the coating. Its amount (in the range of 2–15 mass percent) depends on pH, buffering capacity, ligands, and other parameters of electroless solutions.

Borohydride and its derivatives may also be used as reducing agents. When borohydride is used in the reduction, temperatures of 140°F/60°C to 194°F/90°C are required. The use of dimethylaminoborane (DMAB) enables the deposition of Ni-B coatings with a small amount of boron (0.5–1.0 mass percent at temperatures in the range of 86°F/30°C to 104°F/60°C. Neutral and alkaline solutions may be used.

Depending upon exposure conditions certain minimum coating thicknesses to control porosity are recommended for the coating to maintain its appearance and have a satisfactory life:

Indoor exposures	0.3–0.5 mil (0.008–0.013 mm)
Outdoor exposures	0.5–1.5 mil (0.013–0.04 mm)
Chemical industry	1–10 mil (0.025–0.25 mm)

For applications near the seacoast, thicknesses in the range of 1.5 mil (0.04 mm) should be considered. This also applies to automobile bumpers and applications in general industrial atmospheres.

Nickel is sensitive to attack by industrial atmospheres and forms a film of basic nickel sulfate that causes the surface to "fog" or lose its brightness. To overcome this fogging, a thin coating of chromium (0.01–0.03 mil/0.003–0.0007 mm) is electrodeposited over the nickel. This finish is applied to all materials for which continued brightness is desired.

Single layer coatings of nickel exhibit less corrosion resistance than multilayer coatings due to their discontinuities. The electroless plating process produces a coating with fewer discontinuous deposits. Therefore, the single layer deposited by electroless plating provides more corrosion resistance than does an electroplated single layer.

Most electroless plated nickel deposits contain phosphorus which enhances corrosion resistance. In the same manner an electroplated nickel deposit containing phosphorus will also be more protective.

SATIN FINISH NICKEL COATING

A satin finish nickel coating consists of nonconductive materials such as aluminum oxide, kaolin, and quartz which are codeposited with chromium on the nickel deposit. Some particles are exposed on the surface of the chromium deposit, so the deposit has a rough surface. Since the reflectance of the deposit is decreased to less than half of that of a level surface, the surface appearance looks like satin.

A satin finish nickel coating provides good corrosion resistance due to the discontinuity of the top coat of chromium.

NICKEL-IRON ALLOY COATING

In order to reduce production costs of bright nickel, the nickel-iron alloy coating was developed. The nickel-iron alloy deposits full brightness, high leveling, and excellent ductility and good reception for chromium.

This coating has the disadvantage of forming red rust when immersed in water consequently nickel-iron alloy coating is suitable for use in mild atmospheres only. Typical applications include kitchenware and tubular furniture.

CHROMIUM COATINGS

In the northern parts of the United States, immediately after World War II it was not unusual for the chromium plated bumpers on the most expensive cars to show severe signs of rust within a few months of winter exposure. This was partially the result of trying to extend the short supply of strategic metals by economizing on the amount used. However, the more basic reason was the lack of sufficient knowledge of the corrosion process in order to control the attack by the atmosphere. Consequently, an aggressive industrial program was undertaken to obtain a better understanding of the corrosion process and ways to control it.

Chromium plated parts on automobiles consist of steel substrates with an intermediate layer of nickel, or in some cases, layered deposits of copper and nickel. The thin chromium deposit provides bright appearance and stain free surface while the nickel layer provides the corrosion protection to the steel substrate. With this system it is essential that the nickel cover the steel substrate completely because the iron will be the anode and nickel the cathode. Any breaks or pores in the coating will result in the condition shown in Figure 16.3. This figure illustrates the reason for the corrosion of chrome trim on automobiles after World War II.

The corrosion problem was made worse by the fact that addition agents used in the plating bath resulted in a bright deposit. Bright deposits contain sulfur, which makes the nickel more active from a corrosion standpoint. From a corrosion standpoint this is discouraging. However, it occurred to the investigators that this apparent disadvantage of bright nickel could be put to good use.

To solve this problem, a duplex nickel coating was developed as shown in Figure 16.4. An initial layer of sulfur-free nickel is applied to the steel substrate, followed by an inner layer of a bright nickel containing sulfur, with an outer layer of microcracked chromium.

Any corrosion that takes place is limited to the bright nickel layer containing sulfur. The corrosion spreads laterally between the chromium and sulfur

Cathodic reaction takes place on chromium or nickel

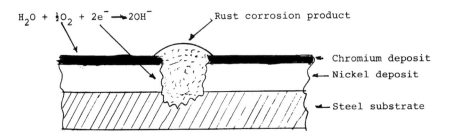

FIGURE 16.3 Corrosion of steel at breaks in a nickel/chromium coating when exposed to the atmosphere.

free nickel deposits because the outer members of this sandwich (chromium and sulfur free nickel) are cathodic to the sulfur containing nickel.

A potential problem that could result from this system of corrosion control would be the undermining of the chromium and the possibility that the brittle chromium deposits could flake off the surface. This potential problem was prevented by the development of a microcracked or microporous chromium coating. These coatings contain microcracks or micropores that do not detract from the bright appearance of the chromium. They are formed very uniformly over the

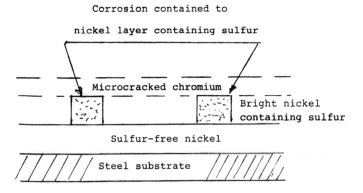

FIGURE 16.4 Duplex nickel electrode deposit to prevent corrosion of steel substrate.

exterior of the plated material and serve to distribute the corrosion process over the entire surface. The result has been to extend the life of chromium plated steel exposed to outdoor atmospheric conditions.

Microcracked chromium coatings are produced by first depositing a high stress nickel strike on a sulfur free nickel layer and then a decorative chromium deposit. The uniform crack network results from the interaction of the thin chromium layer and the high stress nickel deposit. The result is a mirror surface as well as a decorative chromium coating.

Microporous chromium coatings are produced by first electroplating a bright nickel layer containing suspended nonconductive fine particles. Over this a chromium layer is deposited which results in a mirror finish. As the chromium thickness increases the number of pores decrease. For a chromium deposit of 0.25 μm thickness a porosity of more than 10,000 pores/cm^2 are required. A porosity of 40,000 pores/cm^2 provides the best corrosion resistance.

Hard (engineering) chromium layers are also deposited directly on a variety of metals. The purpose in applying these layers is to obtain wear resistant surfaces with a high hardness or to restore original dimensions to a workpiece. In addition, the excellent corrosion resistance resulting from these layers make them suitable for outdoor applications.

Thick chromium deposits have high residual internal stress and may be brittle due to the electrodeposition process, in which hydrogen can be incorporated in the deposited layer. Cracks result during plating when the stress exceeds the tensile stress of the chromium. As plating continues some cracks are filled. This led to the development of controlled cracking patterns, which produce wettable porous surfaces that can spread oil, which is important for engine cylinders, liners, etc.

Some of the properties of engineering chromium layers are:

1. Excellent corrosion resistance.
2. Wear resistance.
3. Hardness up to 950 HV.
4. Controlled porosity is possible.

TIN COATINGS (TINPLATE)

Tinplate is produced mainly by the electroplating process. Alkaline and acid baths are used in the production line. The acid baths are classified as either ferrostan or halogen baths.

A thermal treatment above the melting point of tin follows the electrolytic deposition. The intermetallic compound $FeSn_2$ forms at the interface between the iron and tin during this thermal processing. The corrosion behavior of the tinplate is determined by the quality of the $FeSn_2$ formed, particularly when the amount of the free tin is small. The best performing tinplate is that in which the $FeSn_2$ uniformly covers the steel so that the area of iron exposed is very small in case the tin should dissolve. Good coverage requires good and uniform nucleation of $FeSn_2$. Many nuclei form when electrodeposition of tin is carried out from the alkaline stannate bath.

Compared to either iron or tin, $FeSn_2$ is chemically inert in all but the strongest oxidizing environments.

Most of the tinplate (tin coating on steel) produced is used for the manufacture of food containers (tin cans). The nontoxic nature of tin salts makes tin an ideal material for the handling of foods and beverages.

An inspection of the galvanic series will indicate that tin is more noble than steel and, consequently, the steel would corrode at the base of the pores. On the outside of a tinned container this is what happens—the tin is cathodic to the steel. However, on the inside of the container there is a reversal of polarity because of the complexing of the stannous ions by many food products. This greatly reduces the activity of the stannous ions, resulting in a change in the potential of tin in the active direction.

This change in polarity is absolutely necessary because most tin coatings are thin and therefore porous. To avoid perforation of the can, the tin must act as a sacrificial coating. Figure 16.5 illustrates this reversal of activity between the outside and inside of the can.

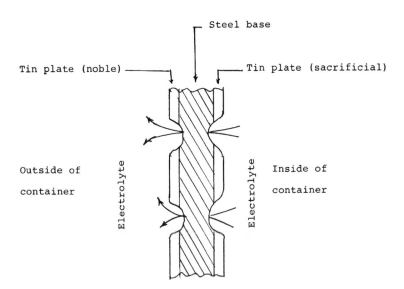

FIGURE 16.5 Tin acting as both a noble and sacrificial coating.

The environment inside a hermetically sealed can varies depending upon the contents which include general foods, beverages, oils, aerosol products, liquid gases, etc. For example, pH values vary for different contents as shown below:

Acidic beverage	2.4–4.5
Beer and wine	3.5–4.5
Meat, fish, marine products, and vegetables	4.1–7.4
Fruit juices, fruit products	3.1–4.3
Nonfood products	1.2–1.5

The interior of cans is subject to general corrosion, localized corrosion, and discoloring. The coating system for tinplate consists of tin oxide, metallic tin, and alloy. The dissolution of the tin layer in acid fruit products is caused by acids such as citric acid. In acid fruit products the potential reversal occurs between the tin layer and the steel substrate, as shown in Figure 16.6. The potential reversal of a tin layer for steel substrate occurs at a pH range <3.8 in a citric acid solution. This phenomenon results from the potential shift of the tin layer to a more negative direction. Namely, the activity of the stannous ion, Sn^{2+}, is reduced by the formation of soluble tin complexes, and thereby the corrosion potential of the tin layer becomes more negative than that of steel. Thus the tin layer acts as a sacrificial anode for steel so that the thickness and density of the pores in the tin layer are important factors affecting the service life of the coating. A thicker tin layer prolongs the service life of a tin can. The function of the alloy layer (FeSn) is to reduce the active area of steel by covering it since it is inert in acid fruit products. When some parts of the steel substrate are exposed, the corrosion of the tin layer is accelerated by galvanic coupling with the steel. The corrosion

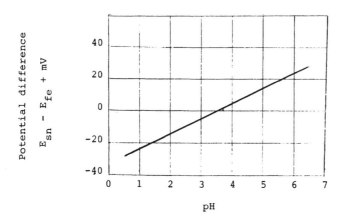

FIGURE 16.6 Potential reversal in tinplate.

potential of the alloy layer is between that of the tin layer and that of the steel. A less defective layer exhibits potential closer to that of the tin layer. Therefore, the covering with alloy layer is important to decrease the dissolution of the tin layer.

In carbonated beverages the potential reversal does not take place; therefore, the steel dissolves preferentially at the defects in the tin layer. Under such conditions pitting corrosion sometimes results in perforation. Consequently, except for fruit cans almost all tinplate cans are lacquered.

When tinplate is to be used for structural purposes such as roofs, an alloy of 12–25 parts of tin to 88–75 parts of lead is frequently used. This is called terneplate. It is less expensive and more resistant to the weather than a pure tin coating. Terneplate is used for fuel tanks of automobiles and is also used in the manufacture of fuel lines, brake lines, and radiators in automobiles.

LEAD COATINGS

Coatings of lead and its alloy (5–10% Sn) protect steel substrate especially in industrial areas having an SO_x atmosphere. At the time of initial exposure, pitting occurs on the lead surface, however, the pits are self healed and then the lead surface is protected by the formation of insoluble lead sulfate. Little protection is provided by these coatings when in contact with the soil.

Lead coatings are usually applied by either hot dipping or by electrodeposition. When the coating is to be applied by hot dipping, a small percentage of tin is added to improve the adhesion to the steel plate. If 25% or more of tin is added, the resulting coating is termed terneplate.

Caution: Do not use lead coatings where they will come into contact with drinking water or food products. Lead salts can be formed that are poisonous.

TERNEPLATE

Terneplate is a tin-lead alloy coated sheet steel, and is produced either by hot dipping or electrodeposition. The hot dipping process with a chloride flux is used to produce most terneplates. The coating layer, whose electrode potential is more noble than that of the steel substrate, contains 8–16% Sn. Since the electrode potential of the coating layer is more noble than the steel substrate, it is necessary to build a uniform and dense alloy layer ($FeSn_2$) in order to form a pinhole free deposit.

Terneplate exhibits excellent corrosion resistance, especially under wet conditions, excellent weldability and formability, with only small amounts of corrosion products forming on the surface. A thin nickel deposit can be applied as an undercoat for the terne layer. Nickel reacts rapidly with the tin-lead alloy to form a nickel-tin alloy layer. This alloy layer provides good corrosion resistance and inhibits localized corrosion.

The main application for terneplate is in the production of fuel tanks for automobiles.

GOLD COATINGS

Gold electrodeposits are primarily used to coat copper in electronic applications to protect the copper connectors and other copper components from corrosion. It is desirable to obtain the corrosion protection with the minimum thickness of gold because of the cost of the gold. As the thickness of the electrodeposit is decreased, there is a tendency for the deposit to provide inadequate coverage of the copper. For this reason it is necessary that there be a means whereby the coverage of the copper can be determined. Such a test, using corrosion principles as a guide, has been developed. In a 0.1 M NH_4Cl solution, gold serves as the cathode and copper serves as the anode. At a high cathode/anode surface area fraction, the corrosion potential is linearly related to the area fraction of copper exposed, as shown in Figure 16.7. By measuring the corrosion potential of the gold plated copper in a 0.1 M NH_4Cl solution, the area fraction of copper exposed is determined.

Gold coatings can also be deposited by means of electroless plating. Borohydride or DMAB are used as reducers with a stable gold cyanide complex. Thin gold coatings may be deposited on plastics by an aerosol spray method using gold complexes with amines and hydrazine as a reducer. A relatively thick coat may be obtained.

COPPER COATINGS

Even though copper is soft, it has many engineering applications in addition to its decorative function. One such application is the corrosion protection of steel. It can be used as an alternative to nickel to prevent fretting and scaling corrosion. Copper can be deposited electrochemically from various aqueous solutions. The properties of the deposit will depend on the chosen bath and the applied procedures. The hardness of the layers varies from 40 to 160 HV.

Since copper is very noble it causes extreme galvanically induced local corrosion of steel and aluminum substrates. Because of this, extreme care must be taken to produce well-adhering nonporous layers.

The corrosion protection provided by a copper coating is twofold, consisting of an original barrier action of the coating layer and a secondary barrier action of corrosion products. The low EMF of copper is responsible for the forma-

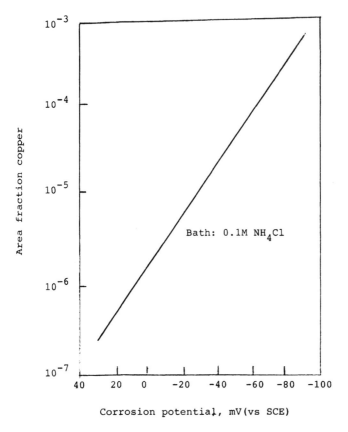

FIGURE 16.7 Data showing that the fractional exposed area of copper in copper/gold system is linearly related to the corrosion potential at low exposed copper areas.

tion of the original barrier action. The electrochemical reactions in the corrosion cells on copper are as follows:

Anodic reaction: $Cu \rightarrow Cu^+ + e$
$$Cu \rightarrow Cu^{2+} + 2e$$

Cathodic reaction: $O_2 + H_2O \rightarrow 4e + 4OH^-$

Chloride ions in a natural environment stabilize cuprous ions. Cupric ions are more stable. Since the EMF of corrosion on copper is less than that on iron, the reactivity of a steel surface is decreased by coating it with copper.

Over a period of time, corrosion products gradually build up a secondary layer against corrosion. Initially, a cuprous oxide layer is formed, followed by the copper surface covered with basic copper salts. Pollutants in the atmosphere determine the formation of basic copper salts as follows:

Mild atmosphere	Malachite $CuCO_3:Cu(OH)_2$
SO_x atmosphere	Brochanite $CuSO_4:3Cu(OH)_2$
Chloride atmosphere	Atacamite $CuCl_2:3Cu(OH)_2$

In most coastal areas the amount of sulfates in the atmosphere exceeds the amount of chlorides. As a layer of copper salt grows on the surface of the corrosion product layer, the protective ability of the corrosion layer is increased. As the exposure time increases, the average corrosion rate of copper gradually decreases. After 20 years the corrosion rate of copper is reduced to half the value of the first year as a result of the secondary barrier of corrosion products.

The initial corrosion rate of a copper coating is dependent on atmospheric conditions such as time of wetness and type and amount of pollutants. Time of wetness is the most important factor affecting the corrosion rate of copper. The corrosion rate of copper usually obeys parabolic law:

$$M^2 = kt$$

where M = mass increase
k = a constant
t = exposure time

Accordingly, the average corrosion rate decreases with increased exposure time, which means the surface of the copper is covered with basic salts by degrees and thereafter the corrosion rate approaches a constant value.

Copson [6] conducted 20 year exposure tests and found the average corrosion rate of copper to be as follows:

0.0034 mil/year in dry rural atmospheres
0.143 mil/year in rural atmospheres
0.0476–0.515 mil/year in industrial atmospheres
0.0198–0.0562 mil/year in marine atmospheres

Until the base metal is exposed, the corrosion process of a copper coated layer is similar to that of copper plate. Galvanic corrosion of copper coated steel is induced when the steel substrate is exposed. However, in the case of copper coated stainless steel the occurrence of galvanic action is dependent on the composition of the stainless steel.

In chloride atmospheres galvanic pitting takes place at the pores in copper layers and galvanic tunneling at cut edges on types 409 and 430 stainless steels, whereas in SO_x atmospheres uniform corrosion takes place on the copper coating.

Copper coatings are used both for decorative and for corrosion protection from the atmosphere. Copper coated steels are used as roofs, flashings, leaders, gutters, and architectural trim. Copper undercoats also improve the corrosion resistance of multilayered coatings, specifically in the plating of nickel and chromium.

NONNOBLE COATINGS

Nonnoble metals protect the substrate by means of cathodic control. Cathodic overpotential of the surface is increased by coating which makes the corrosion potential more negative than that of the substrate. The coating metals used for cathodic control protection are zinc, aluminum, manganese, and cadmium—and their alloys, of which the electrode potentials are more negative than those of iron and steel. Consequently, the coating layers of these metals act as sacrificial anodes for iron and steel substrates, when the substrates are exposed to the atmospheres. The coating layer provides cathodic protection for the substrate by galvanic action. These metals are called sacrificial metals.

The electrical conductivity of the electrolyte, the temperature, and the surface condition determines the galvanic action of the sacrificial metal coating. An increase in the cathodic overpotential is responsible for the corrosion resistance of the coating layer. Figure 16.8 shows the principle of cathodic control protection by a sacrificial metal coating.

The corrosion rate of zinc coated iron $i_{\text{corr. of zinc coating}}$ becomes lower than that of uncoated iron $i_{\text{corr. of uncoated iron}}$ since the cathodic overpotential of the surface is increased by zinc coating and the exchange current density of dissolved oxygen on zinc $i_{\text{oc on zinc}}$ is lower than that on iron $i_{\text{oc on iron}}$. If a small part of iron is exposed to the atmosphere, the electrode potential of the exposed iron is equal to the corrosion potential of zinc $E_{\text{corr. of zinc coating}}$ since the exposed iron is polarized cathodically by the surrounding zinc, so that little corrosion occurs on the exposed iron $i_{\text{corr. of exposed iron}}$. Zinc ions dissolved predominately from the zinc coating form the surrounding barrier of corrosion products at the defect and thereby protecting the exposed iron.

Sacrificial metal coatings protect iron and steel by two or three protective abilities such as:

1. Original barrier action of coating layer
2. Secondary barrier action of corrosion product layer
3. Galvanic action of coating layer

The surface oxide film and the electrochemical properties based on the metallography of the coating metal provide the original barrier action.

An air formed film of Al_2O_3 approximately 25 Å thick forms on aluminum. This film is chemically inert, and its rapid formation of oxide film by a self-healing ability leads to satisfactory performance in natural environments.

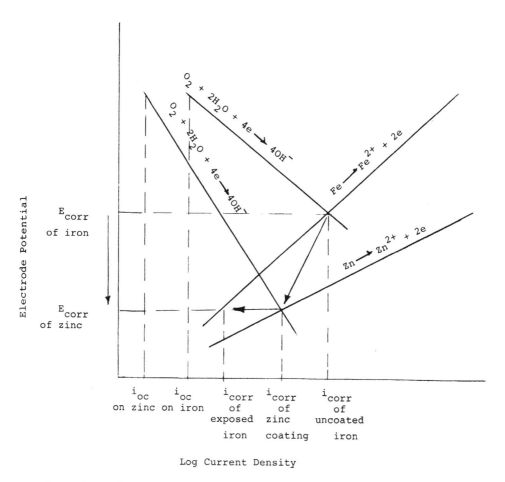

FIGURE 16.8 Cathodic control protection.

Zinc however, does not produce a surface oxide film that is as effective a barrier as the oxide film on aluminum. The original barriers of zinc and zinc alloy coatings result from electrochemical properties based on the structure of the coating layer.

Nonuniformity of the surface condition generally induces the formation of a corrosion cell. Such nonuniformity results from defects in the surface oxide film, localized distribution of elements, and the difference in crystal face or phase. These nonuniformities of surface cause the potential difference between portions of the surface, thereby promoting the formation of a corrosion cell.

Many corrosion cells are formed on the surface, accelerating the corrosion rate, as a sacrificial metal and its alloy-coated materials are exposed in the natural

TABLE 16.4   Corrosion Products Formed on Various Sacrificial Metal Coatings

Metal	Corrosion product
Al	Al_2O_3, $\beta Al_2O_3 \cdot 3H_2O$, $\alpha AlOOH$, $Al(OH)_3$, amorphous Al_2O_3
Zn	ZnO, $Zn(OH)_2$, $2ZnCO_3 \cdot 3Zn(OH)_2$, $ZnSO_4 \cdot 4Zn(OH)_2$, $ZnCl_2 \cdot 4Zn(OH)_2$, $ZnCl_2 \cdot 6Zn(OH)_2$
Mn	γMn_2O_3, $MnCO_3$, $\gamma\text{-}MnOOH$
Cd	CdO, $CdOH_2$, $2CdCO_3 \cdot 3Cd(OH)_2$

atmosphere. During this time corrosion products are gradually formed and converted to a stable layer after a few months of exposure. Typical corrosion products formed are shown in Table 16.4. Once the stable layer has been formed, the corrosion rate becomes constant. This secondary barrier of corrosion protection regenerates continuously over a long period of time. In most cases the service life of a sacrificial metal coating depends on the secondary barrier action of the corrosion product layer.

Sacrificial metal coatings are characterized by their galvanic action. Exposure of the base metal, as a result of mechanical damage, polarizes the base metal cathodically to the corrosion potential of the coating layer, as shown in Figure 16.8, so that little corrosion takes place on the exposed base metal. A galvanic couple is formed between the exposed part of the base metal and the surrounding coating metal. Since sacrificial metals are more negative in electrochemical potential than iron or steel, a sacrificial metal acts as an anode and the exposed base metal behaves as a cathode. Table 16.5 shows the corrosion potentials of sacrificial metals and steel in a 3% NaCl solution. Consequently, the dissolution of the coating layer around the defect is accelerated and the exposed part of base

TABLE 16.5   Corrosion Potentials of Sacrificial Metals in a 3% NaCl Solution

Metal	Corrosion potential, V, SCE
Mn	-1.50
Zn	-1.03
Al	-0.79
Cd	-0.70
. . . .	
Steel	-0.61

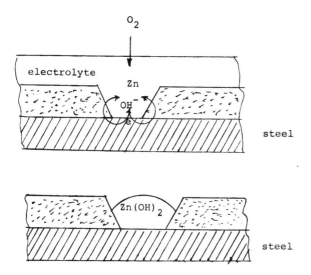

FIGURE 16.9 Schematic illustration of galvanic action of sacrificial metal coating.

metal is protected against corrosion. Figure 16.9 shows a schematic illustration of the galvanic action of a sacrificial metal coating.

The loss of metal coating resulting from corrosion determines the service life of the coating. The degree of loss is dependent on the time of wetness on the metal surface and the type of concentration of pollutants in the atmosphere. Table 16.6 shows the average corrosion losses of zinc, aluminum, and 55% Al-Zn coatings in various locations and atmospheres. The losses were calculated from the mean values of time of wetness and the average corrosion rate during wet duration. The time of wetness of walls is 40% of that of roofs. Coating metals and coating thicknesses can be decided from Table 16.6 since the corrosion losses of zinc, aluminum, and Al-Zn alloy are proportional to exposure time.

As can be seen from the table, a G90 sheet, which has a 1 mil zinc coating, cannot be used for a roof having a durability of 10 years in any atmosphere except in a rural area. Were this sheet to be used in an urban, marine, or industrial atmosphere, it would have to be painted for protection.

Aluminum and 55% Al-Zn alloy provide galvanic protection for the steel substrate. In rural and industrial atmospheres an aluminum coating does not act as a sacrificial anode. However, in a chloride atmosphere, such as a marine area, it does act as a sacrificial anode.

The choice as to which sacrificial metal coating to use will be based on the environment to which it will be exposed and the service life required. The service life required will also determine the coating thickness to be applied, which

TABLE 16.6 Average Corrosion Losses of Sacrificial Metal Coatings for 10 Years

Location	Atmosphere	Zinc Roof	Zinc Wall	55% AlZn Roof	55% AlZn Wall	Aluminum Roof	Aluminum Wall
Inland	Rural	0.42	0.17	0.15	0.06	0.06	0.02
	Urban	1.48	0.59				
	Industrial	1.40	0.56	0.25	0.06	0.06	0.02
	Severe industrial	1.59	0.64				
Inland shore of lake	Rural	0.59	0.24	0.20	0.08	0.07	0.03
or marsh	Urban	1.97	0.79				
	Industrial	1.40	0.56	0.20	0.08	0.08	0.03
	Severe industrial	2.12	0.85				
Coast	Rural	0.74	0.23	0.25	0.10	0.08	0.04
	Urban	2.47	0.99				
	Industrial	1.75	0.70	0.25	0.10	0.10	0.04
	Severe industrial	2.65	1.06				
Seashore	Severe industrial	2.06	0.82	0.46	0.18	0.19	0.07

Average corrosion loss (mil[a]/10 yr)

[a] 1 mil = 25.4 μm
Source: Ref. 2.

in turn will influence the coating process to be used. Sacrificial metal coatings have been used successfully for roofs, walls, ducts, shutters, doors, and window frames in the housing industry; and on structural materials such as transmission towers, structural members of a bridge, antennae, chimney structures, grandstands, steel frames, high-strength steel bolts, guard rails, corrugated steel pipe, stadium seats, bridge I beams, footway bridges, road bridges, and fencing.

ZINC COATINGS

Approximately half of the world's production of zinc is used to protect steel from rust. Zinc coatings are probably the most important type of metallic coating for corrosion protection of steel. The reasons for the wide application are:

1. Prices are relatively low.
2. Due to large reserves, an ample supply of zinc is available.
3. There is great flexibility in application procedures resulting in many different qualities with well-controlled layer thicknesses.
4. Steel provides good cathodic protection.

5. Many special alloy systems have been developed with improved corrosion protection properties.

The ability to select a particular alloy or to specify a particular thickness of coating depends on the type of coating process used. Zinc coatings can be applied in many ways. The six most commonly used procedures follow.

Hot Dipping

This is a process in which cleaned steel is immersed in molten zinc or zinc alloy, and a reaction takes place to form a metallurgically bonded coating.

The coating is integral with the steel because the formation process produces zinc-iron alloy layers overcoated with zinc. Continuity and uniformity is good since any discontinuities are readily visible as "black spots."

Coating thicknesses can be varied from approximately 50–125 μm on tube and products. Thicker coatings up to 250 μm can be obtained by grit-blasting before galvanizing. Sheet and wire normally receive thicknesses of 10–30 μm.

Conventional coatings that are applied to finished articles are not formable. The alloy layer is abrasion resistant but brittle on bending. Special coatings with little or no alloy layer are readily formed (e.g., on sheet) and resistance welded.

A chromate conversion coating over the zinc coating prevents wet storage stain while phosphate conversion coatings provide a good base (on a new sheet) for paints. Weathered coatings are often painted after 10–30 years for longer service.

Hot dip galvanizing is the most important zinc coating process. All mild steels and cast iron can be coated by this process. The thickness and structure of the coating will depend on the alloying elements. Approximately half of the steel that is coated is in the form of sheet, approximately one quarter is fabricated work while the remainer is tube or wire. Metallurgically the processes used for tubes and fabricated work is similar, while the process used for sheet has small additions to the zinc which reduces the quantity of iron-zinc alloy in the coating which provides flexibility.

Zinc Electroplating

This process is sometimes mistakenly referred to as electrogalvanizing. In this process zinc salt solutions are used in the electrolytic deposition of a layer of zinc on a cleaned steel surface.

This process provides good adhesion, comparable with other electroplated coatings. The coating is uniform within the limitations of "throwing power" of the bath. Pores are not a problem as exposed steel is protected by the adjacent zinc.

Coating thickness can be varied at will but is usually 2.5–15 µm. Thicker layers are possible but are not generally economical.

Electroplated steel has excellent formability and can be spot welded. Small components are usually finished before being plated.

Chromate conversion coatings are used to prevent wet storage stain while phosphate conversion coatings are used as a base for paint.

The process is normally used for simple, fairly small components. It is suitable for barrel plating or for continuous sheet and wire. No heating is used in this process except for hydrogen embrittlement relief on high strength steels.

Electroplated zinc is very ductile, and consequently this process is widely used for the continuous plating of strip and wire, where severe deformation may be required.

The coating on steel from this process gives a bright and smooth finish. It is used for decorative effect to protect delicate objects where rough or uneven finishes cannot be tolerated (e.g., instrument parts). It is also used for articles that cannot withstand the pretreatment or temperatures required in other coating processes.

It was previously mentioned that the term "electrogalvanizing" is sometimes used to describe this process. This is misleading since the chief characteristic of galvanizing is the formation of a metallurgical bond at the zinc-iron interface. This does not occur in electroplating.

Mechanical Coating

This process involves the agitating of suitably prepared parts to be coated with a mixture of nonmetallic impactors (e.g., glass beads), zinc powder, a chemical promoter, and water. All types of steel can be coated. However, this process is less suitable for parts heavier than ½ lb (250 g) because the tumbling process reduces coating thickness at the edges.

The adhesion is good compared with electroplated coatings. Thickness can be varied as desired from 5 µm to more than 70 µm. However, the coating is not alloyed with the steel, nor does it have the hard, abrasion resistance iron-zinc alloy layers of galvanized or sherardized coatings. Conversion coatings can be applied.

Sherardizing

The articles to be coated are tumbled in a barrel containing zinc dust at a temperature just below the melting point of zinc, usually around 716°F/380°C. In the case of spring steels the temperature used is somewhat lower. By means of a diffusion process the zinc bonds to the steel forming a hard, even coating of zinc-iron compounds. The coating is dull gray in color and can readily be painted if necessary.

The finish is continuous and very uniform, even on threaded and irregular parts. This is a very useful finish for nuts and bolts which, with proper allowance for thickness of coats, can be sherardized after manufacture and used without retapping the threads.

The thickness of coating can be controlled. Usually a thickness of 30 μm is used for outdoor applications while 15 μm is used indoors.

Thermally Sprayed Coatings

In the process droplets of semimolten zinc are sprayed from a special gun that is fed with either wire or powder onto a grit-blasted surface. The semimolten droplets coalesce with some zinc oxide present at each interface between droplets. Electrical continuity is maintained both throughout the coating and with the iron substrate so that full cathodic protection can be obtained since the zinc oxide forms only a small percentage of the coating.

The sprayed coating contains voids (typically 10–20% by volume) between coalesced particles. These voids have little effect on the corrosion protection since they soon fill up with zinc corrosion products and are thereafter impermeable. However, the use of a sealer to fill the voids improves appearance in service and adds to life expectancy, but more important it provides a better surface for subsequent application of paints.

There are no size or shape limitations regarding the use of this process.

Zinc Dust Painting

Zinc dust paints may be used alone for protection or as a primer followed by conventional top coats. More details will be found in Chapter 13.

Corrosion of Zinc Coatings

In general zinc coatings corrode in a similar manner as solid zinc. However, there are some differences. For example, the iron-zinc alloy present in most galvanized coatings has a higher corrosion resistance than solid zinc in neutral and acid solutions. At points where the zinc coating is defective, the bare steel is cathodically protected under most conditions.

The corrosion of zinc coatings in air is an approximate straight line relationship between weight loss and time. Since the protective film on zinc increases with time in rural and marine atmospheres of some types, under these conditions the life of the zinc may increase more than in proportion to thickness. However, this does not always happen.

Zinc coatings are used primarily to protect ferrous parts against atmospheric corrosion. These coatings have good resistance to abrasion by solid pollutants in the atmosphere. General points to consider are:

1. Corrosion increases with time of wetness.
2. The corrosion rate increases with an increase in the amount of sulfur compounds in the atmosphere. Chlorides and nitrogen oxides usually have a lesser effect but are often very significant in combination with sulfates.

Zinc coatings resist atmospheric corrosion by forming protective films consisting of basic salts, notably carbonate. The most widely accepted formula is $3Zn(OH)_2 \cdot 2ZnCO_3$. Environmental conditions that prevent the formation of such films, or conditions that lead to the formation of soluble films, may cause rapid attack on the zinc.

The duration and frequency of moisture contact is one such factor. Another factor is the rate of drying because a thin film of moisture with high oxygen concentration promotes reaction. For normal exposure conditions the films dry quite rapidly. It is only in sheltered areas that drying times are slow, so that the attack on zinc is accelerated significantly.

The effect of atmospheric humidity on the corrosion of a zinc coating is related to the conditions that may cause condensation of moisture on the metal surface and to the frequency and duration of the moisture contact. If the air temperature drops below the dew point, moisture will be deposited. The thickness of the piece, its surface roughness, and its cleanliness also influence the amount of dew deposited. Lowering the temperature of a metal surface below the air temperature in a humid atmosphere will cause moisture to condense on the metal. If the water evaporates quickly, corrosion is usually not severe and a protective film is formed on the surface. If water from rain or snow remains in contact with zinc when access to air is restricted and the humidity is high, the resulting corrosion can appear to be severe (wet storage stain, known as "white rust") since the formation of a protective basic zinc carbonate is prevented.

In areas having atmospheric pollutants, particularly sulfur oxides and other acid forming pollutants, time of wetness becomes of secondary importance. These pollutants can also make rain more acid. However, in less corrosive areas time of wetness assumes a greater proportional significance.

In the atmospheric corrosion of zinc, the most important atmospheric contaminent to be considered is sulfur dioxide. At relative humidities of about 70% or above, it usually controls the corrosion rate.

Sulfur oxides and other corrosive species react with the zinc coating in two ways: dry deposition and wet deposition. Sulfur dioxide can deposit on a dry surface of galvanized steel panels until a monolayer of SO_2 is formed. In either case the sulfur dioxide that deposits on the surface of the zinc forms a sulfurous or other strong acid, which reacts with the film of zinc oxide, hydroxide, or basic carbonate to form zinc sulfate. The conversion of sulfur dioxide to sulfur based acids may be catalyzed by nitrogen compounds in the air (NO_X compounds). This factor may affect corrosion rates in practice. The acids partially destroy the film of corrosion products, which will then reform from the underlying metal, thereby causing continuous corrosion by an amount equivalent to the film dissolved, hence the amount of SO_2 absorbed.

Chloride compounds have less effect than sulfur compounds in determining the corrosion rate of zinc. Chloride is most harmful when combined with acidity due to sulfur gases. This is prevalent on the coast in highly industrial areas.

Atmospheric chlorides will lead to the corrosion of zinc, but to a lesser degree than the corrosion of steel, except in brackish water and flowing seawater. Any salt deposit should be removed by washing. The salt content of the atmosphere will usually decrease rapidly inland further away from the coast. Corrosion also decreases with distance from the coast, but the change is more gradual and erratic because chloride is not the primary pollutant affecting zinc corrosion. Chloride is most harmful when combined with acidity resulting from sulfur gases.

Other pollutants also have an effect on the corrosion of galvanized surfaces. Deposits of soot or dust can be detrimental because they have the potential to increase the risk of condensation onto the surface and hold more water in position. This is prevalent on upward-facing surfaces. Soot (carbon) absorbs large quantities of sulfur, which are released by rainwater.

In rural areas overmanuring of agricultural land tends to increase the ammonia content of the air. The presence of normal atmospheric quantities of ammonia does not accelerate zinc corrosion, and petrochemical plants where ammonium salts are present show no accelerated attack on galvanized steel. However, ammonia will react with atmospheric sulfur oxides, producing ammonium sulfate, which accelerates paint film corrosion as well as zinc corrosion. When ammonium reacts with NO_x^- compounds in the atmosphere, ammonium nitrite and nitrate are produced. Both compounds increase the rate of zinc corrosion, but less than SO_2 or SO_3.

Because of the Mears effect (wire corrodes faster per unit of area than more massive materials) galvanized wire corrodes some 10–80% faster than galvanized sheet. However, the life of rope made from galvanized steel wires is greater than the life of the individual wire. This is explained by the fact that the parts of the wire that lie on the outside are corroded more rapidly and when the zinc film is penetrated in those regions, the uncorroded zinc inside the rope provides cathodic protection for the outer regions.

Table 16.7 lists the compatibility of zinc coatings with various corrodents.

ZINC–5% ALUMINUM HOT DIP COATINGS

This zinc alloy coating is known as Galfan. Galfan coatings have a corrosion resistance up to three times that of galvanized steel. The main difference between

TABLE 16.7 Compatibility of Galvanized Steel with Selected Corrodents

Acetic acid	U	Chromium chloride	U
Acetone	G	Chromium sulfate solution	U
Acetonitrile	G	Copper chloride solution	U
Acrylonitrile	G	Decyl acrylate	G
Acrylic latex	U	Diamylamine	G
Aluminum chloride 26%	U	Dibutylamine	G
Aluminum hydroxide	U	Dibutyl cellosolve	G
Aluminum nitrate	U	Dibutyl phthalate	G
Ammonia, dry vapor	U	Dichloroethyl ether	G
Ammonium acetate solution	U	Diethylene glycol	G
Ammonium bisulfate	U	Dipropylene glycol	G
Ammonium bromide	U	Ethanol	G
Ammonium carbonate	U	Ethyl acetate	G
Ammonium chloride 10%	U	Ethyl acrylate	G
Ammonium dichloride	U	Ethyl amine 69%	G
Ammonium hydroxide		N-Ethyl butylamine	G
Vapor	U	2-Ethyl butyric acid	G
Reagent	U	Ethyl ether	G
Ammonium molybdate	G	Ethyl hexanol	G
Ammonium nitrate	U	Fluorine, dry, pure	G
Argon	G	Formaldehyde	G
Barium hydroxide		Fruit juices	S
Barium nitrate solution	S	Hexanol	G
Barium sulfate solution	S	Hexylamine	G
Beeswax	U	Hexylene glycol	G
Borax	S	Hydrochloric acid	U
Bromine, moist	U	Hydrogen peroxide	S
2-Butanol	G	Iodine, gas	U
Butyl acetate	G	Isohexanol	G
Butyl chloride	G	Isooctanol	G
Butyl ether	G	Isopropyl ether	G
Butylphenol	G	Lead sulfate	U
Cadmium chloride solution	U	Lead sulfite	S
Cadmium nitrate solution	U	Magnesium carbonate	S
Cadmium sulfate solution	U	Magnesium chloride 42.5%	U
Calcium hydroxide,		Magnesium fluoride	G
sat. solution	U	Magnesium hydroxide sat.	S
20% solution	S	Magnesium sulfate	
Calcium sulfate, sat.	U	2% solution	S
solution		10% solution	U
Cellosolve acetate	G	Methyl amyl alcohol	G
Chloric acid 20%	U	Methyl ethyl ketone	G
Chlorine, dry	G	Methyl propyl ketone	G
Chlorine water	U	Methyl isobutyl ketone	G

TABLE **16.7**   Continued

Nickel ammonium sulfate	U	Propyl acetate	G
Nickel chloride	U	Propylene glycol	G
Nickel sulfate	S	Propionaldehyde	G
Nitric acid	U	Propionic acid	U
Nitrogen, dry, pure	G	Silver bromide	U
Nonylphenol	G	Silver chloride	
Oxygen		pure, dry	S
dry, pure	G	moist, wet	U
moist	U	Silver nitrate solution	U
Paraldehyde	G	Sodium acetate	S
Perchloric acid solution	S	Sodium aluminum sulfate	U
Permanganate solution	S	Sodium bicarbonate solution	U
Peroxide		Sodium bisulfate	U
pure, dry	S	Sodium carbonate solution	U
moist	U	Sodium chloride solution	U
Phosphoric acid 0.3–3%	G	Sodium hydroxide solution	U
Polyvinyl acetate latex	U	Sodium nitrate solution	U
Potassium carbonate		Sodium sulfate solution	U
10% solution	U	Sodium sulfide	U
50% solution	U	Sodium sulfite	U
Potassium chloride solution	U	Styrene, monomeric	G
Potassium bichromate		Styrene oxide	G
14.7%	G	Tetraethylene glycol	G
20%	S	1,1,2 Trichloroethane	G
Potassium disulfate	S	1,2,3 Trichloropropane	G
Potassium fluoride 5–20%	G	Vinyl acetate	G
Potassium hydroxide	U	Vinyl ethyl ether	G
Potassium nitrate		Vinyl butyl ether	G
5–10% solution	S	Water	
Potassium peroxide	U	potable, hard	G
Potassium persulfate 10%	U		

G, suitable application; S, borderline application; U, not suitable.

these two coatings lies in the degree of cathodic protection they afford. This increase in corrosion protection is evident in both a relatively mild urban industrial atmosphere as can be seen in Table 16.8. The latter is particularly significant, because unlike galvanizing the corrosion rate appears to be slowing after about 4 years, and conventional galvanized steel would show rust in 5 years. See Figure 16.10. The slower rate of corrosion also means that the zinc–5% aluminum coatings provide full cathodic protection to cut edges over a longer period. Refer to Table 16.9.

TABLE 16.8 Five Year Outdoor Exposure Results of Galfan Coating

Atmosphere	Thickness loss (μm) Galvanized	Galfan	Ratio of improvement
Industrial	15.0	5.2	2.9
Severe marine	>20.0	9.5	>2.1
Marine	12.5	7.5	1.7
Rural	10.5	3.0	3.5

Source: Ref. 4.

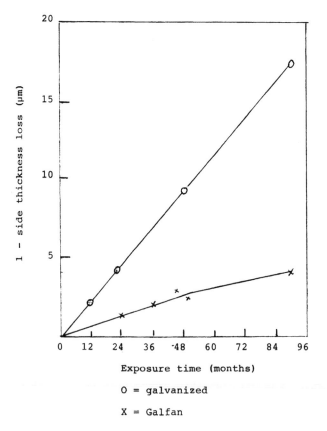

Exposure time (months)

O = galvanized

X = Galfan

FIGURE 16.10 Seven year exposure of Galfan and galvanized steel in a severe marine atmosphere.

TABLE 16.9 Comparison of Cathodic
Protection for Galfan and Galvanized Coatings

| Environment | Amount (mm) of bare edges exposed after 3 years (coating recession from edge) | |
	Galvanized	Galfan
Severe marine	1.6	0.1
Marine	0.5	0.06
Industrial	0.5	0.05
Rural	0.1	0

Source: Ref. 4.

Because Galfan can be formed with much smaller cracks than can be obtained in conventional galvanized coatings, it provides excellent protection at panel bulges. This reduced cracking means that less zinc is exposed to the environment which increases the relative performance factor compared with galvanized steel.

ZINC–55% ALUMINUM HOT DIP COATINGS

These coatings are known as Galvalume and consist of zinc–55% aluminum–1.5% silicon. This alloy is sold under such tradenames as Zaluite, Aluzene, Alugalva, Algafort, Aluzink, and Zincalume. Galvalume exhibits superior corrosion resistance over galvanized coatings in rural, industrial, marine, and severe marine environments. However, this alloy has limited cathodic protection and less resistance to some alkaline conditions and is subject to weathering discoloration and wet storage staining. The latter two disadvantages can be overcome by chromate passivation which also improves its atmospheric corrosion resistance.

Initially, a relatively high corrosion loss is observed for Galvalume sheet as the zinc-rich portion of the coating corrodes and provides sacrificial protection at cut edges. This takes place in all environments, whereas aluminum provides adequate galvanic protection only in marine chloride environments. After approximately 3 years, the corrosion-time curves take on a more gradual slope reflecting a change from active, zinclike behavior to passive, aluminumlike behavior as the interdentric regions fill with corrosion products. It has been predicted that Galvalume sheets should outlast galvanized sheets of equivalent thickness by at least two to four times over a wide range of environments. Figures comparing the corrosion performance of galvanized sheet and Galvalume sheet are shown in Figure 16.11.

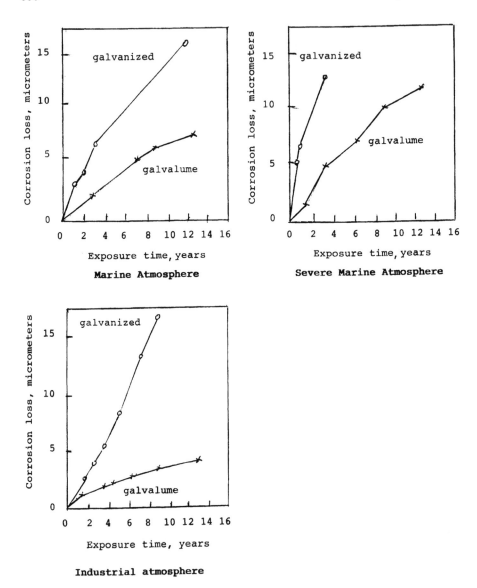

FIGURE 16.11 Thirteen year exposure of Galvalume in marine and industrial atmospheres.

Galvalume sheets provide excellent cut-edge protection in very aggressive conditions, where the surface does not remain too passive. However, it does not offer as good a protection on the thicker sheets in mild rural conditions, where zinc–5% aluminum coatings provide good general corrosion resistance and when sheared edges are exposed or localized damage to the coating occurs during fabrication or service, the galvanic protection is retained for a longer period.

ZINC–15% ALUMINUM THERMAL SPRAY

Zinc–15% aluminum coatings are available as thermally sprayed coatings. These coatings have a two-phase structure consisting of a zinc-rich and an aluminum-rich phase. The oxidation products formed are encapsulated in the porous layer formed by the latter and do not build up a continuous surface layer as with pure zinc coatings. As a result no thickness or weight loss is observed even after several years of exposure in atmospheric field testing.

It is normally recommended that thermally sprayed coatings be sealed to avoid initial rust stains, to improve appearance, and to facilitate maintenance painting. Sealing is designed to fill pores and give only a thin overall coating, too thin to be directly measurable. Epoxy or acrylic system resin, having a low viscosity, is used as a sealer.

ZINC-IRON ALLOY COATINGS

As compared with pure zinc, the zinc-iron alloy coatings provide increased corrosion resistance in acid atmospheres but slightly reduced corrosion resistance in alkaline atmospheres.

Electroplated zinc-iron alloy layers containing more than 20% iron provide a corrosion resistance 30% higher than zinc in industrial atmospheres. In other atmospheres the zinc-iron galvanized coatings provide as good a coating as coatings with an outer zinc layer. Sherardized coatings are superior to electroplated coatings and equal to galvanized coatings of the same thickness. However, the structure of the alloy layer, and its composition, affects the corrosion resistance.

If the zinc layer of a galvanized coating has weathered, or the zinc-iron alloy layer forms the top layer after galvanizing, brown areas may form. Brown staining can occur on sherardized or hot dip galvanized coatings in atmospheric corrosion through the oxidation of iron from the zinc-iron alloy layers or from the substrate. Such staining is usually a dull brown, rather than the bright red-brown of uncontrolled rust. Usually there is a substantial intact galvanized layer underneath, leaving the life of the coating unchanged. Unless the aesthetic appearance is undesirable, no action need be taken.

ALUMINUM COATINGS

Aluminum coatings protect steel substrates by means of cathodic control, with an original barrier action of an air formed film which is chemically inert, and the rapid formation of oxide film by a self-healing ability. Aluminum coatings are excellent in general corrosion resistance. However, they do not act as a sacrificial anode in rural and industrial atmospheres but do in a chloride area, such as a marine environment. In a nonchloride environment, the formation of red rust occurs at sheared edges and in other defects of an aluminum coating layer. However, the growth of red rust is slow.

Aluminum coatings, sealed with organic or composite layers such as etch primer, zinc chromate, etc. will provide long service in seawater environments. Recommended coating thickness plus sealing for the splash zone and submerged zone is 150 μm.

CADMIUM COATINGS

Cadmium coatings are produced almost exclusively by electrodeposition. A cadmium coating on steel does not provide as much cathodic protection to the steel as does a zinc coating because the potential between cadmium and iron is not as great as between zinc and iron. Therefore, it becomes important to minimize defects in the cadmium coating.

Unlike zinc, a cadmium coating will retain a bright metallic appearance. It is more resistant to attack by salt spray and atmospheric condensate than zinc. In aqueous solutions, cadmium will resist attack by strong alkalies, but will be corroded by dilute acids and aqueous ammonia.

Cadmium coatings should not be allowed to come into contact with food products because cadmium salts are toxic. This coating is commonly used on nuts and bolts, but because of its toxicity, usage is declining.

MANGANESE COATINGS

Manganese is very active, having an electrode potential more negative than zinc (Mn: -1.5 m, Zn -1.03 V, SCE). In a natural atmosphere a dense corrosion layer builds on the surface of manganese during a very short period. However, defects in the coating accelerate the anodic dissolution of manganese, thereby shortening the life of the coating. Therefore, manganese is combined with zinc to form a duplex Mn/Zn alloy coating. The types of corrosion products found on these coatings were shown in Table 16.4. The compound γ-Mn_2O_3 is effective for the formation of a barrier. The more γ-Mn_2O_3 in the corrosion products, the denser the layer on a Mn/Zn coating.

Manganese is so negative in electrochemical potential, and active, that its alloy and duplex coatings provide galvanic protection. Mn/Zn alloy coating exhibits high corrosion resistance, and the corrosion potential of manganese is more negative than that of zinc; therefore, this alloy coating provides cathodic protection for a steel substrate. The structure of Mn/Zn alloy coating is composed of the single phase of ε in the manganese content range <20%, and ε and γ phases in the range above 20%. As the manganese content in the deposit is increased, the percentage of γ-Mn phase is increased.

REFERENCES

1. Satas, D. *Coatings Technology Handbook*, Marcel Dekker, New York, 1991.
2. Suzuki, Ichiro. *Corrosion Resistant Coatings Technology*, Marcel Dekker, New York, 1989.
3. Mansfield, Florian. *Corrosion Mechanisms*, Marcel Dekker, New York, 1987.
4. Porter, Frank C. *Corrosion Resistance of Zinc and Zinc Alloys*, Marcel Dekker, New York, 1994.
5. Schweitzer, Philip A. *Corrosion and Corrosion Protection Handbook*, 2nd ed., Marcel Dekker, New York, 1988.
6. Copson, H. R. Report of Subcommittee VI of Committee B-3 on Atmospheric Corrosion of Nonferrous Metals, ASTM Annual Meeting, Atlantic City, NJ, June 25, 1955.

17

Cementitious Coatings

Cementitious coatings provide corrosion resistance to substrates such as steel by maintaining the pH at the metal/coating interface above 4, a pH range where steel corrodes at a low rate. The proper selection of materials and their application is necessary if the coating is to be effective. To select the proper material, it is necessary to define the problem:

1. Identify all chemicals that will be present and their concentrations. Knowing pH alone is not sufficient. All pH tells you is whether the environment is acid, neutral, or alkaline. It does not identify whether the environment is oxidizing, organic or inorganic, alternately acid or alkaline.
2. Is the application fumes, splash, or total immersion?
3. What are the minimum and maximum temperatures to which the coating will be subjected?
4. Is the installation indoors or outdoors? Thermal shock and ultraviolet exposure can be deleterious to many resins.
5. How long is the coating expected to last? This can have an effect on the cost.

Surface preparation prior to application of the coating is essential. The surface must be free of mill scale, oil, grease, and other chemical contaminents.

The surface must also be roughened by sandblasting and the coating applied immediately after surface preparation. An intermediate bonding coating is used when adhesion between the substrate and the coating is poor, or where thermal expansion characteristics are incompatible. Coatings are installed in thicknesses of $1/16$–$1/2$ in. (1.5–13 mm). They may be applied by casting, troweling, or spraying. The spraying process, known as Gunite or Shotcrete, is particularly useful on systems with unusual geometry or with many sharp bends or corners. It has the advantage that there are no seams, which are often weak points as far as corrosion resistance is concerned.

SILICATES

These materials are noted for their resistance to concentrated acids, except hydrofluoric acid and similar fluorinated chemicals, at elevated temperatures. They are also resistant to many aliphatic and aromatic solvents. They are not intended for use in alkaline or alternately acid and alkaline environments. This category of coatings includes:

1. Sodium silicate
2. Potassium silicate
3. Silica (silica sol)

The alkali silicates form a hard coating by a polymerization reaction involving repeating units of the structure.

$$-\overset{\displaystyle OH}{\underset{\displaystyle OH}{\overset{|}{\underset{|}{Si}}}}-O-$$

The sodium and potassium silicates are available as two-component systems: filler and binder with the setting agent in the filler. Sodium and potassium silicates are referred to as soluble silicates because of their solubility in water. This prevents their use in many dilute acid services while they are not affected by strong concentrated acids. This disadvantage becomes an advantage for formulating single component powder systems. All that is required is the addition of water at the time of use. The fillers of these materials are pure silica.

The original sodium silicate acid resistant coating uses an inorganic silicate base consisting of two components, a powder and a liquid. The powder is basically quartzite of selected gradation and a setting agent. The liquid is a special sodium silicate solution. When the coating is used, the two components are mixed together and hardening occurs by a chemical reaction.

This coating may be cast, poured, or may be applied by guniting. It has excellent acid resistance and is suitable for use over a pH range of 0.0–7.0.

The sodium silicates can be produced over a wide range of compositions of the liquid binder. These properties and new hardening systems have significantly improved the water resistance of some sodium silicate coatings. These formulations are capable of resisting dilute as well as concentrated acids without compromising physical properties.

The potassium silicate materials are less versatile in terms of formulation flexibility than the sodium silicate materials. However, they are less susceptible to crystallization in high concentrations of sulfuric acid so long as metal ion contamination is minimal. Potassium silicate materials are available with halogen free hardening systems, thereby removing the remote possibility for catalyst poisoning in certain chemical processes.

Chemical setting potassium silicate materials are supplied as two component systems that comprise the silicate solution and the filler powder and setting agent. Setting agents may be inorganic, organic, or a combination of both. The properties of the coating are determined by the setting agent and the alkali-silica ratio of the silicate used. Properties such as absorption, porosity, strength, and water resistance are affected by the choice of setting agent. Organic setting agents will burn out at low temperatures, thereby increasing porosity and absorption. Organic setting agents are water soluble and can be leached out if the coating is exposed to steam or moisture. Coatings that use inorganic setting agents are water and moisture resistant.

Silicate formulations will fail when exposed to mild alkaline mediums, such as bicarbonate of soda. Dilute acid solutions, such as nitric acid, will have a deleterious effect on sodium silicates unless the water resistant type is used. Table 17.1 points out the differences between the various silicate coatings.

Silica, or silica sol, types of coatings are the newest of this class of material. They consist of a colloidal silica binding instead of the sodium or potassium silicates, with a quartz filler. These materials are two-component systems that comprise a powder composed of high quality crushed quartz and a hardening agent, which are mixed with colloidal silica solution to form the coating. These coatings are recommended for use in the presence of hot concentrated sulfuric acid. They are also used for weak acid conditions up to a pH of 7.

CALCIUM ALUMINATE

Coatings of this type consist of a calcium aluminate based cement and various inert ingredients and are supplied in powder form to be mixed with water when used. They may be applied by casting, pouring, or guniting. Calcium aluminate based coatings are hydraulic and consume water in their reaction mechanism to form hydrated phases. This is similar to portland cement compositions; however,

TABLE 17.1 Comparative Chemical Resistance: Silicate Coatings

	Sodium		Potassium	
Medium, R.T.	Normal	Water resistant	Normal	Halogen free
Acetic acid, glacial	G	G	R	R
Chlorine dioxide, water sol.	N	N	R	R
Hydrogen peroxide	N	R	N	N
Nitric acid, 5%	C	R	R	R
Nitric acid, 20%	C	R	R	R
Nitric acid, over 20%	R	R	R	R
Sodium bicarbonate	N	N	N	N
Sodium sulfite	R	R	N	N
Sulfates, aluminum	R	R	R	R
Sulfates, copper	G	G	R	R
Sulfates, iron	G	G	R	R
Sulfates, magnesium	G	G	R	R
Sulfates, nickel	G	G	R	R
Sulfates, zinc	G	G	R	R
Sulfuric acid, to 93%	G	G	R	R
Sulfuric acid, over 93%	G	G	R	R

R.T., room temperature; R, recommended; N, not recommended; G, potential failure, crystalline growth; C, conditional.

their rates of hardening are very rapid. Essentially full strength is reached within 24 hr at 73°F (23°C).

They have better mild acid resistance than portland cement, but they are not useful in acids below pH 4.5–5.0. They are not recommended for alkali service above a pH of 10, nor are they recommended for halogen service. Refer to Table 17.2 for the chemical resistance of calcium aluminate and portland cement.

PORTLAND CEMENT

Portland cement is made from limestone or other natural sources of calcium carbonate, clay (a source of silica), alumina, ferric oxide, and minor impurities. After grinding, the mixture is fired in a kiln at approximately 2500°F/1137°C. The final product is ground to a fineness of about 10 μm, and mixed with gypsum to control setting. When mixed with water, the portland cement forms a hydrated phase and hardens. As the cement hardens, chemical reactions take place. The two most important reactions are the generation of calcium hydroxide and of tricalcium silicate hydrate. The calcium hydroxide generated could theoretically

TABLE 17.2 Chemical Resistance of Calcium Aluminate and Portland Cement

Cement type	Calcium aluminate	Portland cement
pH range	4.5–10	7–12
Water resistance	E	E
Sulfuric acid	X	X
Hydrochloric acid	X	X
Phosphoric acid	P	X
Nitric acid	X	X
Organic acids	F	F
Solvents	G	G
Ammonium hydroxide	F	G
Sodium hydroxide	F	F
Calcium hydroxide	F	G
Amines	F	G

E, excellent; G, good; F, fair; P, poor; X, not recommended.

be as high as 20% of the weight of the cement, producing an alkalinity that at the solubility of lime results in an equilibrium pH of 12.5. Steel that has been coated with the cement is passivated as a result of the hardened materials 12.5 pH. The alkalinity of the coating is provided by the presence of calcium oxide (lime). Any material that will cause the calcium oxide or hydroxide to be removed, lowering the pH, will prove detrimental and cause solution of the cement hydrates. Contact with inorganic or organic acids can cause this to happen. Organic acids can be generated when organic materials ferment.

When carbon dioxide dissolves in water that may be present on the cement, weak carbonic acid is produced. The weak carbonic acid lowers the pH of the cement solution allowing the coated steel to corrode. This is sometimes referred to as the carbonation of cement.

Sulfates will also cause portland cement to deteriorate. In addition to being able to produce sulfuric acid, which is highly corrosive to portland cement, sulfates are also reactive with some additives used in the formulations. Refer to Table 17.2.

18

Coatings for Concrete

The life of concrete can be prolonged by providing a coating which will be resistant to the pollutants present. These coatings are referred to as monolithic surfacings. Before selecting an appropriate coating, consideration must be given to the condition of the concrete and the environment to which the concrete will be exposed. Proper surface preparation is essential. Surface preparations can be different for freshly placed concrete than for old concrete.

When concrete is poured, it is usually held in place by means of steel or wood forms that are removed when the concrete is still in its tender state. To facilitate their removal, release agents are applied to the forms prior to pouring. Oils, greases, and proprietary release agents are left on the surface of the concrete. These must be removed if they will interfere with the adhesion of subsequent coatings.

Quite often curing compounds are applied to fresh concrete as soon as practical after the forms have been removed. These are liquid membranes based on waxes, resins, chlorinated rubber, or other film formers, usually in a solvent. Pretesting is necessary to determine whether or not they will interfere with the coating to be applied.

Generally, admixtures that are added to concrete mixes in order to speed up or slow down the cure, add air to the mix, or obtain special effects will not interfere with surface treatments to improve durability. The concrete supplier can furnish specific data regarding the admixtures. If in doubt, try a test patch of the coating material to be used.

It is essential that water from subslab ground sources be eliminated or minimized since migration through the concrete may create pressures at the bond line of water resistant barriers.

As discussed in Chapter 17 under portland cement coating, when water is added to portland cement a chemical reaction takes place during the hardening. This reaction produces calcium hydroxide and tricalcium silicate hydrate. The alkalinity of concrete is provided by the presence of calcium oxide from the cement. Consequently, concrete attack can be due to chemicals that react with the portland cement binder and form conditions that physically deteriorate the material. Any material that will cause the calcium oxide or hydroxide to be removed, lowering the pH of the cement mix, will cause instability and solution of the cement hydrates.

The most common cause for the deterioration of concrete results from contact with inorganic and organic acids. Those that form soluble salts with calcium oxide or calcium hydroxide are the most aggressive. Typical compounds that can cause problems include sour milk, industrial wastes, fruit juices, some ultrapure waters, and organic materials that ferment and produce organic acids. Typical chemical families found in various types of chemical processing industry plants and their effect on concrete are shown in Table 18.1.

SURFACE PREPARATION

It is essential that the concrete surface be properly prepared prior to application of the coating. The surfaces of cement containing materials may contain defects that require repair before the application of the coating. In general, the surface must be thoroughly cleaned and all cracks repaired.

Unlike specifications for the preparation of steel prior to coating, there are no detailed standard specifications for the preparation of concrete surfaces. In

TABLE 18.1 Effects of Various Chemicals on Concrete

Chemical	Effect on concrete
	Chemical plants
Acid waters pH 6.5 or less	Disintegrates slowly
Ammonium nitrate	Disintegrates
Benzene	Liquid loss by penetration
Sodium hypochlorite	Disintegrates slowly
Ethylene	Disintegrates slowly
Phosphoric acid	Disintegrates slowly
Sodium hydroxide 20% and above	Disintegrates slowly

TABLE 18.1   Continued

Chemical	Effect on concrete
	Food and beverage plants
Almond oil	Disintegrates slowly
Beef fat	Solid fat disintegrates slowly, melted fat more readily
Beer	May contain, as fermentation products, acetic, carbonic, lactic, or tannic acids which disintegrate slowly
Buttermilk	Disintegrates slowly
Carbonic acid (soda water)	Disintegrates slowly
Cider	Disintegrates slowly
Coconut oil	Disintegrates slowly
Corn syrup	Disintegrates slowly
Fish oil	Disintegrates slowly
Fruit juices	Disintegrates
Lard or lard oil	Lard disintegrates slowly, lard oil more quickly
Milk	No effect
Molasses	Disintegrates slowly above 120°F/49°C
Peanut oil	Disintegrates slowly
Poppyseed oil	Disintegrates slowly
Soybean oil	Disintegrates slowly
Sugar	Disintegrates slowly
	Electric generating utilities
Ammonium salts	Disintegrates
Coal	Sulfides leaching from damp coal may oxidize to sulfurous or sulfuric acid, disintegrates
Hydrogen sulfide	Dry, no effect; in moist oxidizing environments converts to sulfurous acid and disintegrates slowly
Sulfuric acid (10–80%)	Disintegrates rapidly
Sulfur dioxide	With moisture forms sulfurous acid which disintegrates rapidly
	Pulp and paper mills
Chlorine gas	Slowly disintegrates moist concrete
Sodium hypochlorite	Disintegrates slowly
Sodium hydroxide	Disintegrates concrete
Sodium sulfide	Disintegrates slowly
Sodium sulfate	Disintegrates concrete of inadequate sulfate resistance
Tanning liquor	Disintegrates if acid

most instances it is necessary to follow the instructions supplied by the coating manufacturer. Specifications can range from the simple surface cleaning that provides a clean surface without removing concrete from the substrate, to surface abrading that provides a clean roughened surface, to acid etching that also provides a clean roughened surface.

Surface Cleaning

Surface cleaning is accomplished by one of the following means:

1. Broom sweeping
2. Vacuum cleaning
3. Air blast
4. Water cleaning
5. Detergent cleaning
6. Steam cleaning

Water cleaning and detergent cleaning will not remove deep embedded soils in the pores of the concrete surfaces. In addition, these methods saturate the concrete, which then requires a drying period (not always practical).

Surface Abrading

Surface abrading can be accomplished by

1. Mechanical abrading
2. Water blasting
3. Abrasive blasting

Of the three methods to produce a roughened surface, the most technically effective is abrasive blasting by means of sandblasting or shot blasting followed by broom sweeping, vacuum cleaning, or an air blast to remove the abrasive. However, in some instances this may not be a practical approach.

Acid Etching

Acid etching is a popular procedure used for both new and aged concrete. It must be remembered that during this process acid fumes will be evolved that may be objectionable. Also thorough rinsing is required, which saturates the concrete. This may necessitate a long drying period, depending on the coating to be used.

COATING SELECTION

The coating to be selected will depend on the physical properties and conditions of the concrete, as well as the environmental conditions.

Factors such as alkalinity, porosity, water adsorption and permeability, and weak tensile strength must be considered. The tendency of concrete to crack, particularly on floors, must also enter into the decision. Floor cracks may develop as a result of periodic overloading or from drying shrinkage. Drying shrinkage is not considered a working movement, while periodic overloading is. In the former a rigid flooring system could be used while in the latter an elastomeric caulking of the moving cracks would be considered.

Selection may be influenced by the presence of substrate water during coating. If the concrete cannot be dried, one of the varieties of water based or water tolerant systems should be considered.

When aggressive environments are present, the surface profile and surface porosity of the concrete must be taken into account. If complete coverage of the substrate is required, specification of film thickness or number of coats may need to be modified. Block fillers may be required. In a nonaggressive atmosphere an acrylic latex coating may suffice.

Specific environmental conditions will dictate the type of coating required. Not only do normal atmospheric pollutants have to be considered but any specific local pollutants will also have to be taken into account. Also to be considered are the local weather conditions, which will result in minimum and maximum temperatures as seasons change. Also to be considered are the possibilities of spillage of chemicals on the surface. Coatings can be applied in various thicknesses depending upon the environment and contaminants.

Thin film coatings are applied at less than 20 mils dry film thickness. Commonly used are epoxies that may be formulations of polyamides, polyamines, polyesters, or phenolics. These coatings will protect against spills of hydrocarbon fuels, some weak solutions of acids and alkalies, and many agricultural chemicals. Epoxies can also be formulated to resist spills of aromatic solvents such as xylol or toluol.

Most epoxies will lose some of their gloss and develop a "chalk face" when exposed to weather. However, this does not affect their chemical resistance.

Medium film coatings are applied at approximately 20–40 mils dry film thickness. Epoxies used in this category are often flake filled to give them rigidity, impact strength, and increased chemical resistance. The flakes can be mica, glass, or other inorganic platelets.

Vinyl esters are also used in this medium film category. These coatings exhibit excellent resistance to many acids, alkalies, hypochlorites, and solvents. Vinyl esters may also be flake filled to improve their resistance.

Some vinyl esters require the application of a low viscosity penetrating primer to properly cleaned and profiled concrete before application, while others may be applied directly.

Thick film coatings are installed by two means. The specialty epoxy types are mixed with inorganic aggregates and trowel applied. The polyesters and vinyl esters are applied with a reinforcing fiberglass mat.

Table 18.2 provides a guideline for specifying film thickness.

The most popular monolithic surfacings are formulated from the following resins:

1. Epoxy, including epoxy novolac
2. Polyester
3. Vinyl ester, including vinyl ester novolac
4. Acrylic
5. Urethane
6. Phenolic novolacs

TABLE 18.2 Guidelines for Specifying Film Thickness

Contaminant	Film thickness		
	Thin	Medium	Thick
Aliphatic hydrocarbons	X	X	X
Aromatic hydrocarbons	X	X	
Organic acids:			
Weak	X		
Moderate	X	X	
Strong			X
Inorganic acids:			
Weak	X		
Moderate		X	
Strong			X
Alkalies:			
Weak	X		
Moderate		X	X
Strong			X
Bleach liquors	X		
Oxygenated fuels	X	X	
Fuel additives	X	X	
Deionized water			X
Methyl ethyl ketone	X	X	
Fermented beverages	X	X	
Seawater	X		
Hydraulic/brake fluids	X	X	X

INSTALLATION OF COATINGS

As with the application of any coating system, proper surface preparation and installation techniques are essential. Basic preparations for the application of a coating are as follows:

1. Substrate must be free of cracks and be properly sloped to drains.
2. New as well as existing slabs should have a coarse surface profile, be clean and dry, and be free of contaminants.
3. In general, slab and materials to be installed should have a temperature of 65–85°F/18–29°C. If necessary, special catalysts and hardening systems are available to accommodate higher or lower temperatures.
4. Manufacturers' directions should be adhered to regarding the priming of the substrate prior to applying the monolithic surfacing.
5. Individual and combined components should be mixed thoroughly at a maximum speed of 500 rpm to minimize air entrapment.
6. Prior to curing, the surface must be protected from moisture and contamination.

Monolithic surfacings may be installed by a variety of methods, many of which are the same methods used in the portland cement concrete industry. The primary methods are:

1. Hand troweled
2. Power troweled
3. Spray
4. Pour in place/self-level
5. Broadcast

Small areas, or areas with multiple obstructions, should be hand troweled. Topcoat sealers should be applied to provide imperviousness and increased density, with a smooth, easy-to-clean finish.

Large areas with a minimum of obstructions are best handled with power troweling. The minimum thickness would be $1/4$ in. (6.5 mm). The density of the finish can be improved by the use of appropriate sealers.

For areas subjected to aggressive corrosion, spray application is recommended. The consistency of the material can be formulated to control slump and type of finish. This permits the material to be sprayed on vertical and overhead surfaces including structural components, with thicknesses of $1/16$–$3/32$ in. (1.5–2.4 mm) as is suitable for light duty areas.

Economical and aesthetically attractive floors can be applied by the broadcast system, in which resins are "squeegee" applied to the concrete slab. Filler or colored quartz aggregates of varying color and size are sprinkled or broadcast into the resin. Excess filler and quartz are vacuumed or swept from the floor after the resin has set. This results in a floor thickness of $3/32$–$1/8$ in. (2–3 mm). This type of floor is outstanding for light industrial and interior floors.

EPOXY AND EPOXY NOVOLAC COATINGS

The three most often used epoxy resins for monolithic surfacings are the bisphenol A, bisphenol F (epoxy novolac), and epoxy phenol novolac. These base components are reacted with epichlorhydrin to form resins of varying viscosity and molecular weight. The hardening system employed to effect the cure or solidification will determine the following properties of the cured system:

1. Chemical and thermal resistance
2. Physical properties
3. Moisture tolerance
4. Workability
5. Safety during use

Bisphenol A epoxy is the most popular, followed by the bisphenol F, which is sometimes referred to as an epoxy novolac. The epoxy phenol novolac is a higher viscosity resin that requires various types of diluents or resin blends for formulating coatings.

The bisphenol A resin uses the following types of hardeners:

1. Aliphatic amines
2. Modified aliphatic amines
3. Aromatic amines
4. Others

Table 18.3 shows effects of the hardener on the chemical resistance of the finished coating of bisphenol A systems for typical compounds. Table 18.4 pro-

TABLE 18.3 Types of Epoxy Hardeners and Their Effect on Chemical Resistance

Medium	Hardeners		
	Aliphatic amines	Modified aliphatic amines	Aromatic amines
Acetic acid, 5–10%	C	N	R
Benzene	N	N	R
Chromic acid, <5%	C	N	R
Sulfuric acid, 25%	R	C	R
Sulfuric acid, 50%	C	N	R
Sulfuric acid, 75%	N	N	R

R, recommended; N, not recommended; C, conditional.
Source: Ref. 2.

TABLE 18.4 Comparative Chemical and Thermal
Resistance of Bisphenol A, Aromatic Amine–Cured vs.
Bisphenol F (Epoxy Novolac)

Medium, R.T.	Bisphenol A	Bisphenol F
Acetone	N	N
Butyl acetate	C	E
Butyl alcohol	C	E
Chromic acid, 10%	C	E
Formaldehyde, 35%	E	G
Gasoline	E	E
Hydrochloric acid, to 36%	E	E
Nitric acid, 30%	N	C
Phosphoric acid, 50%	E	E
Sulfuric acid, to 50%	E	E
Trichloroethylene	N	G
Max. temp., °F/°C	160/71	160/71

R.T., room temperature; C, conditional; N, not recommended; E, excellent; G, good.
Source: Ref. 2.

vides a comparison of the general chemical resistance of optimum chemical resistant bisphenol A, aromatic amine cured, with bisphenol F resin systems.

Amine hardening systems are the most popular for ambient temperature curing epoxy coatings. These systems are hygroscopic and can cause allergenic responses to sensitive skin. These responses can be minimized or virtually eliminated by attention to personal hygiene and the use of protective creams on exposed areas of skin, i.e., face, neck, arms, and hands. Protective garments, including gloves, are recommended when using epoxy materials.

Epoxies are economical, available in a wide range of formulations and properties, and offered from many manufacturers. Formulations are available for interior as well as exterior applications. However, when installed outside, moisture from beneath the slab can affect adhesion and cause blistering.

The typical epoxy system is installed in layers of primer, base, and finish coats. Overall, the installation can take several days. Epoxies should not be applied to new concrete before it has reached full strength (approximately 28 days).

Table 18.5 provides the atmospheric corrosion resistance of monolithic floor surfacings.

The bisphenol Fs (epoxy novolacs) are essentially premium grade epoxy resins providing an increased chemical resistance. They are also available in a wide range of formulations and from many manufacturers. The primary advantages in the use of epoxy novolacs is in their improved resistance to higher con-

TABLE 18.5 Atmospheric Corrosion Resistance of Monolithic Concrete Surfacings

Surfacing	NO$_x$	H$_2$S	SO$_2$	CO$_2$	UV	Chloride salt	Weather	Ozone
Epoxy–bisphenol A Aromatic amine hardener	R	X	X	R	R	R	R	R
Epoxy novolac	R	X	X	R	R	R	R	R
Polyesters:								
Isophthalic	R	R	X	R	RS	R	R	R
Chlorendic	R	R	X	R	RS	R	R	R
Bisphenol A fumarate	R	R	X	R	RS	R	R	R
Vinyl esters	R	R	R	R	R	R	R	R
Acrylics	R		R	R	R	R	R	R
Urethanes	R	X			R	R		

R, resistant; X, not resistant; RS, resistant when stabilized.
Source: Ref. 1.

TABLE 18.6 Corrosion Resistance of Bisphenol A and Bisphenol F Epoxies

	Hardeners			
			Aromatic amines	
			Bisphenol	
Corrodent at room temp.	Aliphatic amines	Modified aliphatic amines	A	F
Acetic acid 5–10%	C	U	R	
Acetone	U	U	U	U
Benzene	U	U	R	R
Butyl acetate	U	U	U	R
Butyl alcohol	R	R	R	R
Chromic acid 5%	U	U	R	R
Chromic acid 10%	U	U	U	R
Formaldehyde 35%	R	R	R	R
Gasoline	R	R	R	R
Hydrochloric acid to 36%	U	U	R	R
Nitric acid 30%	U	U	U	U
Phosphoric acid 50%	U	U	R	R
Sulfuric acid 25%	R	U	R	R
Sulfuric acid 50%	U	U	R	R
Sulfuric acid 75%	U	U	U	U
Trichloroethylene	U	U	U	R

R, recommended; U, unsatisfactory.
Source: Ref. 3.

centrations of oxidizing and nonoxidizing acids, and aliphatic and aromatic solvents. Refer to Table 18.6.

The novolacs are more expensive than the bisphenol A epoxies and can discolor from contact with sulfuric and nitric acids. For exposure to normal atmospheric pollutants the bisphenol A epoxies are satisfactory. However, if the surface is to be exposed to other more aggressive contaminents, then the novolacs should be considered.

Table 18.7 provides a comparison of the chemical resistance of various monolithic surfacings.

TABLE 18.7 Comparative Chemical Resistance

1-A = bisphenol A epoxy—aliphatic amine hardener
1-B = bisphenol A epoxy—aromatic amine hardener
1-C = bisphenol F epoxy (epoxy novolac)
2-D = polyester resin—chlorendic acid type
2-E = polyester resin—bisphenol A fumarate type
3-F = vinyl ester resin
3-G = vinyl ester novolac resin

| | 1 | | | 2 | | 3 | |
Medium, R.T.	A	B	C	D	E	F	G
Acetic acid, to 10%	R	R	R	R	R	R	R
Acetic acid, 10–15%	C	R	C	R	R	C	R
Benzene	C	R	R	R	N	R	R
Butyl alcohol	R	C	R	R	R	N	R
Chlorine, wet, dry	C	C	C	R	R	R	R
Ethyl alcohol	R	C	R	R	R	R	R
Fatty acids	C	R	C	R	R	R	R
Formaldehyde, to 37%	R	R	R	R	R	R	R
Hydrochloric acid, to 36%	C	R	R	R	R	R	R
Kerosene	R	R	R	R	R	R	R
Methyl ethyl ketone, 100%	N	N	N	N	N	N	N
Nitric acid, to 20%	N	N	R	R	R	R	R
Nitric acid, 20–40%	N	N	R	R	N	N	C
Phosphoric acid	R	R	R	R	R	R	R
Sodium hydroxide, to 25%	R	R	R	N	R	R	R
Sodium hydroxide, 25–50%	R	C	R	N	R	C	R
Sodium hypochlorite, to 6%	C	R	R	R	R	R	R
Sulfuric acid, to 50%	R	R	R	R	R	R	R
Sulfuric acid, 50–75%	C	R	R	R	C	R	R
Xylene	N	R	R	R	R	N	R

R.T., room temperature; R, recommended; N, not recommended; C, conditional.
Source: Ref. 2.

POLYESTER COATINGS

Polyester coatings were originally developed to resist chlorine dioxide. The three types of unsaturated polyesters most commonly used are

1. Isophthalic
2. Chlorendic acid
3. Bisphenol A fumarate

The chlorendic and bisphenol A resins offer improved chemical resistance, higher thermal capabilities, and improved ductility with less shrinkage. The bisphenol A resins provide improved resistance to alkalies and essentially equivalent resistance to oxidizing mediums. Refer to Table 18.5 for the resistance of the polyester resins to atmospheric contaminents and to Table 18.8 for the comparative chemical resistance of the various polyester resins.

Polyester resins are easily pigmented for aesthetic considerations. The essentially neutral curing systems provide compatibility for application to many substrates including concrete. Properly formulated polyester resins provide installation flexibility to a wide range of temperatures, humidities, and contaminents encountered on most construction sites.

Polyester formulations have limitations such as

1. Strong aromatic odor that can be offensive for certain indoor and confined space applications
2. Shelf life limitations that can be controlled by low temperature storage (below 60°F/15°C) of the resin component

Table 18.9 provides the chemical resistance of chlorendic and bisphenol A fumarate resins in the presence of selected corrodents.

TABLE 18.8 Comparative Chemical Resistance of Various Polyester Resins

Medium, R.T.	Isophthalic	Chlorendic	Bisphenol A fumarate
Acids, oxidizing	R	R	R
Acids, nonoxidizing	R	R	R
Alkalies	N	N	R
Salts	R	R	R
Bleaches	R	R	R
Max. temp., °F/°C	225/107	260/127	250/121

R.T., room temperature; R, recommended; N, not recommended.
Source: Ref. 2.

TABLE 18.9 Chemical Resistance of Chlorendic
and Bisphenol A Fumarate Resins

Corrodent at room temp.	Polyester	
	Chlorendic	Bisphenol A fumarate
Acetic acid, glacial	U	U
Benzene	U	U
Chlorine dioxide	R	R
Ethyl alcohol	R	R
Hydrochloric acid 36%	R	R
Hydrogen peroxide	R	U
Methanol	R	R
Methyl ethyl ketone	U	U
Motor oil and gasoline	R	R
Nitric acid 40%	R	U
Phenol 5%	R	R
Sodium hydroxide 50%	U	R
Sulfuric acid 75%	R	U
Toluene	U	U
Triethanolamine	U	R
Vinyl toluene	U	U

R, recommended; U, unsatisfactory.
Source: Ref. 3.

VINYL ESTER/VINYL ESTER NOVOLAC COATINGS

Vinyl ester resins are addition reactions of methacrylic acid and epoxy resin. These resins have many of the same properties as the epoxy, acrylic, and bisphenol A fumarate resins.

The vinyl ester resins are the most corrosion resistant of any of the monolithic surfacing systems, and they are also the most expensive and difficult to install. They are used when extremely corrosive conditions are present. The finished flooring is vulnerable to hydrostatic pressure and vapor moisture transmission. Refer to Table 18.5 for their resistance to atmospheric corrosion and Table 18.10 for their resistance to selected corrodents.

The major advantage of these resins are their resistance to most oxidizing mediums and high concentrations of sulfuric acid, sodium hydroxide, and many solvents.

Vinyl ester resins also have the disadvantages of having

1. Strong aromatic odor for indoor or confined space applications
2. Shelf life limitation of the resins require refrigerated storage below 60°F/15°C to extend its useful life

TABLE 18.10 Resistance of Vinyl Ester and Vinyl
Ester Novolac to Selected Corrodents

| | Vinyl ester | |
Corrodent	Vinyl ester	Novolac
Acetic acid, glacial	U	R
Benzene	R	R
Chlorine dioxide	R	R
Ethyl alcohol	R	R
Hydrochloric acid 36%	R	R
Hydrogen peroxide	R	R
Methanol	U	R
Methyl ethyl ketone	U	U
Motor oil and gasoline	R	R
Nitric acid 40%	U	R
Phenol 5%	R	R
Sodium hydroxide 50%	R	R
Sulfuric acid 75%	R	R
Toluene	U	R
Triethanolamine	R	R
Vinyl toluene	U	R
Max. temp., °F/°C	220/104	230/110

R, recommended; U, unsatisfactory.
Source: Ref. 3.

ACRYLIC COATINGS

Acrylic monolithic coatings are suitable for interior or exterior applications with
relatively benign atmospheric exposures. Refer to Table 18.5. They excel at water
and weather resistance, and are best at "breathing" in the presence of a moisture
transmissive problem in the slab. They are intended for protection against moderate corrosion environments. The advantages for their use are as follows:

1. They are the easiest of the resin systems to mix and apply by using
    pour-in-place and self-leveling techniques.
2. Because of their outstanding weather resistance, they are equally ap-
    propriate for indoor or outdoor applications.
3. They are the only system that can be installed at below freezing temper-
    atures (25°F/−4°C) without having to use special hardening or catalyst
    systems.
4. They are the fastest set and cure of all the resins.

5. They are the easiest to pigment and with the addition of various types of aggregate can be aesthetically attractive.
6. They are equally appropriate for maintenance and new construction, and bond well to concrete.

The disadvantage of the acrylic coating system is the aromatic odor in indoor or confined spaces.

URETHANE COATINGS

Monolithic urethane flooring systems offer the following advantages:

1. They are easy to mix and apply using the pour-in-place, self-level technique.
2. Systems are available for indoor or outdoor installations.
3. The elastomeric quality of the systems provides underfoot comfort.
4. They have excellent sound-deadening properties.
5. They have excellent resistance to impact and abrasion.
6. They are excellent waterproof flooring systems for above grade light and heavy duty floors.
7. They are capable of bridging cracks in concrete $1/16$ in. (1.5 mm) wide.

As with the acrylics, the urethanes are intended for protection against moderate to light corrosion environments. Standard systems are effective at temperatures of $10-140°F/-24-60°C$. High temperature systems are available with a range of $10-180°F/-24-82°C$.

Refer to Table 18.11 for the comparative chemical resistance of urethane and acrylic systems.

The acrylic and urethane systems have substantially different physical properties. The acrylic flooring systems are extremely hard and are too brittle for applications subjected to excessive physical abuse such as impact, while the inherent flexibility and impact resistance of the urethanes offer potential for this type of application. Table 18.12 provides physical and thermal properties for the acrylic and urethane flooring systems.

PHENOLIC/EPOXY NOVOLAC COATINGS

In order to satisfy the need to provide a coating system that has the ability to bridge cracks and provide improved corrosion resistance, medium build coating systems have been developed. The phenolic/epoxy novolac system is capable of bridging cracks and providing outstanding corrosion resistance.

One such system uses a low viscosity penetrating epoxy primer. Low viscosity allows the primer to be used for areas that need a fast turnaround by quickly

TABLE 18.11 Comparative Chemical Resistance: Urethane vs. Acrylic Systems

Medium, R.T.	Acrylic	Urethane Standard	Urethane High temperature
Acetic acid 10%	G	G	C
Animal oils	G	G	N
Boric acid	E	E	E
Butter	G	F	N
Chromic acid 5–10%	C	C	C
Ethyl alcohol	N	N	N
Fatty acids	F	F	N
Gasoline	E	N	N
Hydrochloric acid 20–36%	F	C	C
Lactic acid, above 10%	F	C	C
Methyl ethyl ketone 100%	N	N	N
Nitric acid 5–10%	G	C	F
Sulfuric acid 20–50%	G	C	C
Water, fresh	E	E	E
Wine	G	G	F

R.T., room temperature; E, excellent; G, good; F, fair; C, conditional; N, not recommended.
Source: Ref. 2.

TABLE 18.12 Minimum Physical and Thermal Properties of Acrylic Monolithic Surfacing and Urethane Monolithic Surfacings

Property	Acrylics Monolithic	Urethanes Standard	Urethanes High temperature
Tensile, psi (MPa) ASTM Test Method C-307	1000 (7)	650 (5)	550 (5)
Flexural, psi (MPa) ASTM Test Method C-580	2500 (17)	1100 (8)	860 (6)
Compressive, psi (MPa) ASTM Test Method C-579	8000 (55)	2500 (17)	1500 (10)
Bond to Concrete	Concrete fails	Concrete fails	Concrete fails
Max. temp., °F/°C	150/66	140/60	180/82

Source: Ref. 2.

TABLE **18.13** Compatibility of Phenolic with Selected Corrodents[a]

Chemical	Max. temp.		Chemical	Max. temp.	
	°F	°C		°F	°C
Acetic acid 10%	212	100	Hydrobromic acid to 50%	200	93
Acetic acid, glacial	70	21	Hydrochloric acid to 38%	300	149
Acetic anhydride	70	21	Hydrofluoric acid	x	x
Acetone	x	x	Lactic acid 25%	160	71
Aluminum sulfate	300	149	Methyl isobutyl ketone	160	71
Ammonium carbonate	90	32	Muriatic acid	300	149
Ammonium chloride to sat.	80	27	Nitric acid	x	x
Ammonium hydroxide 25%	x	x	Phenol	x	x
Ammonium nitrate	160	71	Phosphoric acid 50–80%	212	100
Ammonium sulfate	300	149	Sodium chloride	300	149
Aniline	x	x	Sodium hydroxide	x	x
Benzene	160	71	Sodium hypochlorite	x	x
Butyl acetate	x	x	Sulfuric acid 10%	250	121
Calcium chloride	300	149	Sulfuric acid 50%	250	121
Calcium hypochlorite	x	x	Sulfuric acid 70%	200	93
Carbonic acid	200	93	Sulfuric acid 90%	70	21
Chromic acid	x	x	Sulfuric acid 98%	x	x
Citric acid, conc.	160	71	Sulfurous acid	80	27
Copper sulfate	300	149			

[a] The chemicals listed are in the pure state or in a saturated solution unless otherwise indicated. Compatibility is shown to the maximum allowable temperature for which data are available. Incompatibility is shown by an x.

"wetting out" the substrate. If surface deterioration or preparation presents an unacceptably rough surface, the surface can be smoothed further by using a pigmented, high solid/high build epoxy polyamide filler/sealer.

For deep pits, the crack filler used is a two component epoxy paste developed specifically for sealing and smoothing out applications on concrete.

The crack filler can be used to fill and smooth hairline cracks, bug holes, gouges, or divots when minimal movement of the substrate is expected.

To provide maximum corrosion resistance a two component phenolic/ epoxy novolac coating can be used. The phenolic coating provides resistance to high concentrations of acids, particularly to sulfuric acid at elevated temperatures. Refer to Table 18.13 for the compatibility of phenolic with selected corrodents.

REFERENCES

1. Schweitzer, Philip A. *Atmospheric Degradation and Corrosion Control*, Marcel Dekker, New York, 1999.
2. Boova, Augustus A. Chemical Resistant Mortars, Grouts and Monolithic Surfacings, in *Corrosion Engineering Handbook* (P. A. Schweitzer, ed.), Marcel Dekker, New York, 1996, pp. 459–487.
3. Schweitzer, Philip A. *Encyclopedia of Corrosion Technology*, Marcel Dekker, New York, 1998.

Index